T0174480

Cybersexualities

CYBERSEXUALITIES

A READER ON FEMINIST THEORY, CYBORGS AND CYBERSPACE

Edited and with an Introduction by
Jenny Wolmark

EDINBURGH UNIVERSITY PRESS

© Jenny Wolmark, introduction,
selection and editorial materials, 1999

Transferred to Digital Print 2011

Edinburgh University Press
22 George Square, Edinburgh

Typeset in Sabon and Gill Sans
by Bibliocraft Ltd, Dundee, and
printed and bound in Great Britain
by CPI Antony Rowe, Chippenham and Eastbourne

A CIP record for this book is available
from the British Library

ISBN 0 7486 1118 5 (hardback)
ISBN 0 7486 1117 7 (paperback)

The right of the contributors to be
identified as authors of this work has
been asserted in accordance with the
Copyright, Designs and Patents Act
1988.

CONTENTS

COPYRIGHT ACKNOWLEDGEMENTS

Grateful acknowledgement is made to the following sources for permission to reproduce material in this book previously published elsewhere. Every effort has been made to trace copyright holders, but if any have been inadvertently overlooked the publisher will be pleased to make the necessary arrangement at the first opportunity.

1. Mary Ann Doane, 'Technophilia: Technology, Representation and the Feminine', from M. Jacobus et al (eds), *Body/Politics*, New York and London: Routledge, 1990.
2. Claudia Springer, 'The Pleasure of the Interface', from *Screen*, 32, No. 3, Oxford: Oxford University Press, 1991.
3. Zoë Sofia, 'Virtual Corporeality: A Feminist View', from *Australian Feminist Studies*, 15, Adelaide: University of Adelaide, 1992.
4. Allucquere Rosanne Stone, 'Will the Real Body Please Stand Up? Boundary Stories about Virtual Cultures', from M. Benedikt (ed.), *Cyberspace: First Steps*, Cambridge, Mass: MIT Press, 1991.
5. Sadie Plant, 'The Future Looms: Weaving Women and Cybernetics', from M. Featherstone and R. Burrows (eds), *Cyberspace, Cyberbodies, Cyberpunks*, London, Thousand Oaks, New Delhi: Sage, 1995.
6. Elisabeth Grosz, 'Space, Time, and Bodies', from *Space, Time and Perversion* by E. Grosz, New York and London: Routledge, 1995.
7. Anne Balsamo, 'Reading Cyborgs Writing Feminism', from *Communication*, 10, Lausanne: Gordon and Breach Publishing Group, 1988.
8. N. Katherine Hayles, 'The Life Cycle of Cyborgs: Writing the Posthuman', from M. Benjamin (ed.), *A Question of Identity*, New Jersey: Rutgers University Press, 1993, © N. Katherine Hayles.
9. Veronica Hollinger, 'Cybernetic Deconstructions: Cyberpunk and Postmodernism', from L. McCaffery (ed.), *Storming the Reality Studio*, Durham, NC: Duke University Press, 1991.
10. Nicola Nixon, 'Cyberpunk: Preparing the Ground for Revolution or Keeping the Boys Satisfied?', from *Science Fiction Studies*, 19, Greencastle: De Pauw University, 1992.
11. Thomas Foster, 'Meat Puppets or Robopaths? Cyberpunk and the Question of Emobodiment', from *Genders*, 18, Austin, Texas: University of Texas Press, 1993.

12. Jenny Wolmark, 'The Postmodern Romances of Feminist Science Fiction', from L. Pearce and J. Stacey (eds), *Romance Revisited*, London: Lawrence and Wishart, 1995.
13. Chela Sandoval, 'New Sciences: Cyborg Feminism and the Methodology of the Oppressed', from C. Gray (ed.), *The Cyborg Handbook*, London and New York: Routledge, 1995.
14. Jennifer González, 'Envisioning Cyborg Bodies: Notes from Current Research', from C. Gray (ed.), *The Cyborg Handbook*, London and New York: Routledge, 1995.
15. Kathleen Woodward, 'From Virtual Cyborgs to Biological Time Bombs: Technocriticism and the Material Body', from G. Bender and T. Druckrey (eds), *Culture on the Brink*, Seattle: Bay Press, 1994.
16. Donald Morton, 'Birth of the Cyberqueer', from *PMLA*, 110, New York: Modern Language Association of America, 1995.
17. Donna Haraway, 'The Promises of Monsters: A Regenerative Politics for Inappropriate/d Others', from L. Grossberg, C. Nelson and P. Treichler (eds), *Cultural Studies*, New York and London: Routledge, 1992.

ILLUSTRATION ACKNOWLEDGEMENTS

2.1: *Metropolis*, courtesy of the Museum of Modern Art/Film Stills Archive, New York; 2.2 and 2.3: *Cyberpunk*, courtesy of the Innovative Corporation, Scott Rockwell and Daryl Banks; 2.4: *Interface*, courtesy of James D. Hudnall and Paul Johnson; 14.1: *L'Horlogère*, courtesy of Bibliotheque Nationale de Paris; 14.2: Hannah Höch, *Das schöne Mädchen* (The Beautiful Girl), 1919–20, 35 × 29 cm, private collection; 14.3: Raoul Hausmann, *Tête Méchanique. L'Esprit de notre temps*, courtesy of Musee National d'Art Moderne, Centre Georges Pompidou, Paris; 14.4: Robert Longo: *All Your Zombies: Truth Before God*, courtesy of Robert Longo; 14.5 and 14.6: *Silent Mobius* (Part One, Issue 6), courtesy of Kadokawa Shoten Publishing Co, Tokyo 102-0071, Japan; 17.1: A Few Words About Reproduction from a Leader in the Field, courtesy of Logic General; 17.2: Ortho-mune, courtesy of Logic General; 17.3: There are no missing links in MacroGene Workstation, courtesy of Pharmacia LKB Biotechnology Electrophoresis; 17.4: Realize the potential of your cell line, courtesy of BioResponse; 17.5: The link between science and tomorrow: Guaranteed. Pure., courtesy of Vega Technologies; 17.7: Understanding is Everything, courtesy of Gulf Oil Corporation; 17.8: HAM, courtesy of World Wide Pictures, © Henry Burroughs, 1961; 17.10: Nilsson: *The Body Victorious* 1987, courtesy of Bonnier Fakta Bokforlag AB, Sweden; 17.11: Evolution of Recognition Systems, from *Immunology at a Glance*, Blackwell, Oxford; 17.12: Cyborg, courtesy of Lynn Randolph.

INTRODUCTION AND OVERVIEW

Jenny Wolmark

Over the past ten years or so, there has been a great deal of general interest within feminist and cultural theory in two related but distinct metaphors, that of cyberspace, and the cyborg. In the context of the contemporary post-industrial, postmodern environment, it is perhaps not altogether surprising that these particular metaphors should have acquired such cultural resonance. They are an especially apt expression of the simultaneous fascination with, and anxiety about, the rapid changes brought about by the new information and biotechnologies, the development of which seems to occur both haphazardly and without restraint in the global environment of transnational corporate capitalism. In the face of such rapid change, the once stable boundaries between time and space, human and machine, self and other have become increasingly uncertain, raising interesting and challenging questions about accepted definitions of space, place and identity in the post-Enlightenment present. Despite, or perhaps because of, their science fictional overtones, these metaphors have become increasingly visible across a range of cultural texts; they operate on a number of different levels, and are capable of carrying complex and often contradictory meanings. As such, they have become highly charged signifiers which are not only fully implicated in the gendered politics of science and technology, but are also eminently suited to deliver a critique of those same politics. Since the mid-1980s there have been a number of occasionally unexpected but always highly engaging and creative encounters between cultural theory, feminist theory, cyborgs and cyberspace, and it is the purpose of this collection of essays to chart some of the key issues that have emerged from those encounters.

For general readers who have recently encountered these influential metaphors, it is, I think, important to have some idea of the theoretical matrix within which they exist. This introductory chapter therefore provides a general overview of the critical and theoretical environment within which the metaphors of the cyborg and cyberspace emerged and in which they have continued to thrive. The Reader has also been structured in such a way that it can convey some idea of the range and complexity of critical responses that have emerged in recent years. I have chosen to group the essays according to the particular concerns and emphases that the writers put on the interface between the body and technology, rather than arrange them in chronological order. These concerns are signalled in the title of each of the three parts into which the book is divided: thus, the key issue in Part One, 'Technology, Embodiment and Cyberspace', is the impact of both technology and shifting definitions of spatiality and temporality on notions of representation and embodiment. In Part Two, 'Cybersubjects: Cyborgs and Cyberpunks', the essays are concerned with the emergence of different subjectivities in the new spaces and places of information technology, particularly as they are articulated within science fiction texts. In Part Three, the essays explore the usefulness of the metaphors of cyberspace and cyborgs to those who seek to challenge hegemonic notions of otherness in contemporary western culture. Some of the theoretical issues explored in the essays, arising from the interaction of feminist theory and cultural theory with the metaphors of the cyborg and cyberspace, may not be immediately accessible to the general reader, since they often assume a familiarity with what has been called 'technotheory' or 'technocriticism'. I have, therefore, summarised the main arguments of each of the essays in the three separate parts, in order to provide an introduction to the main concerns and issues contained therein. These summaries are intended to provide basic guidelines only, and are, of course, no substitute for reading the essays themselves. While many of the essays are undoubtedly complex, they are consistently challenging, and as I have already indicated, they present some of the most exciting and provoking critical writing of recent years.

There are two seminal texts in the ongoing history of cyborgs and cyberspace that are not included in this Reader, but which nevertheless require some comment, not least because of their continuing impact on both feminist and cultural theory. The provocative metaphor of the cyborg was first used by Donna Haraway in her essay, 'A Manifesto for Cyborgs: Science, Technology, and Socialist Feminism in the 1980s', published in 1985. In the essay, she employs the metaphor in order to argue, firstly, for a reconsideration of Marxist and feminist analyses of the social relations of science and technology which rely on a received model of domination and subordination and, secondly, for the development of an innovative socialist–feminist political strategy that is not dependent on totalising theories and in which the formation of new and unexpected alliances and coalitions are prioritised. William

Gibson's science fiction cyberpunk novel *Neuromancer*, published in 1984, contains the first use of the term 'cyberspace'. Gibson coined the term to describe the 'consensual hallucination' that there is a space/place behind the computer screen, commenting in an interview that 'Everyone I know who works with computers seems to develop a belief that there's some kind of *actual space* behind the screen, someplace that you can't see but you know is there.'[1] Both works had a considerable impact when they were first published, though admittedly for different reasons, and they have continued to inform the critical and imaginative vocabularies of those who are working at the various points of intersection between feminist theory, cultural theory and, of course, science fiction. This suggests that one of the reasons for the enduring interest in these works is that both Haraway and Gibson were able to deploy precisely those metaphors that most accurately and imaginatively embodied the actual experience of living within an information saturated environment. The technology of computer generated imagery, or virtual reality, had been developing rapidly up to the point at which Gibson's novel was published, as was the existence of electronic user groups or communities; interest in and awareness of the possibilities of both virtual reality and the electronic community of the net was increasing exponentially, like the technology itself. However, there was an absence of any kind of shared or even appropriate conceptual vocabulary that could address a range of different cultural and political constituencies, or, to put it another way, that could allow the new spaces and places of information technology to be redefined to encompass difference. In my view, this is the real significance of the metaphors of the cyborg and cyberspace – not only did they embody the lived experience of information technology, but they also offered a means of reconceptualising that experience in potentially non-hierarchical and non-binary terms.

Although I have chosen not to include Haraway's 'A Manifesto for Cyborgs' in this collection because it has already been extensively anthologised, it is nevertheless appropriate at this point to discuss it further, particularly since many of the essays in this collection are a response, both direct and indirect, to the ideas contained in that work. I will return to William Gibson's novel at a later stage. It would be useful to try to establish why the 'Manifesto' has had such a long-standing significance for feminist theory and the politics of difference. A comment from Haraway describes the initial context for the essay: 'The essay originated as a response to a call for political thinking about the 1980s from socialist–feminist points of view, in hopes of deepening our political and cultural debates in order to renew commitments to fundamental social change in the face of the Reagan years' (173).[2] This context partly explains the elaborately rhetorical style of writing used in the essay, in which Haraway employs irony chiefly as 'a rhetorical strategy and a political method' (173). The metaphor of the cyborg is itself used ironically in order to challenge, and ultimately to subvert, the binarisms inherent in contemporary culture, such as those between human and machine, self and other, inside and outside, nature and culture. For Haraway, the transgressive nature of the cyborg means that it can provide a crucial means of contesting meanings, and thus of developing a more radical set of inclusive politics that cuts across traditional categories of difference such as race, gender and class. As Haraway puts it 'by the late twentieth century, our time, a mythic time, we are all chimeras, theorized and fabricated hybrids of machine and organism; in short, we are cyborgs. The cyborg is our ontology; it gives us our politics' (174). Cyborg politics are described as being inclusive rather than exclusive, sympathetic to difference rather than hostile to it, and necessarily based on the possibility that new and strategic alliances can be forged between unexpected groups, no matter how partial or contradictory those alliances may be.

By its very nature, the cyborg is a contradictory boundary creature: on the one hand, it is the product of the masculinist technologies that, in the 1980s, sought to produce a so-called defensive Star Wars weapon that had every possibility of leading to some kind of final apocalypse. On the other hand, because it is a hybrid creature, the cyborg marks a refusal to sustain the very dualisms that structure existing relations of power and control within science and technology. What Haraway refers to as the 'informatics of domination', or the new and adaptable networks of power that derive from cybernetic systems, are doubly-coded, like everything else in the postmodern environment, and 'the political struggle is to see from both perspectives at once' (179). For Haraway, the contradictions of the cyborg are the key to its usefulness as a metaphor for exploring alternative ways of thinking and acting politically. She argues that, with the collapse of boundaries, binary oppositions, and universalising theories, socialist feminism is free to explore new kinds of local, often temporary and shifting alliances in a coalition politics which challenges existing definitions of gender, race and class.

Haraway incorporates references to science fiction into her discussion, as she does in most of her work, because the efficacy of the cyborg metaphor depends on an understanding of the 'myth and meanings structuring our imaginations' (187). In other words, change is not just about the alteration of material circumstances alone, since possibilities for change have to be imagined before they can be enacted. The usefulness of the cyborg metaphor is that it is capable of operating on two levels at the same time: it can provide an account of the lived experience of the inequalities inherent in the ways in which science and technology structure social relations, as well as providing a means of imaginatively exploring the possibilities for fundamental change within those structures. Since Haraway is particularly concerned with the social relations of dominance within science and technology, she turns specifically to feminist science fiction for examples of what she refers to as 'cyborg monsters' who 'make very problematic the statuses of man or woman, human, artefact, member of a race, individual identity, or body' (201). The cyborg monsters of feminist science fiction are described by Haraway as 'promising and dangerous monsters who help to redefine the pleasures and politics of embodiment and feminist writing' (202) and as such they are 'a myth system waiting to become a political language to ground one way of looking at science and technology and challenging the informatics of domination – in order to act potently' (204). They demonstrate the way in which the cyborg metaphor can encompass the dispersed and fragmented experience of being female in the postmodern, post-industrial environment of information and biotechnology, and in embodying that experience, provide a 'political language' by means of which existing relations of power can be challenged. Haraway's essay has been rightly criticised for its tendency towards technological determinism and the level of generality in its political analysis, but the persuasiveness and force of its utopian vision is, nonetheless, undeniable.

The period in the 1980s during which Ronald Reagan was the ultra-conservative President of the United States certainly provides one highly specific context for Haraway's 'Manifesto' and for the impact of the cyborg metaphor, but there is another, more diffuse set of parameters that informs her writing, that of postmodernism. At the time that Haraway's ambitious and wide-ranging attempt to mark the contours of a new feminist epistemology was published, many other theorists and critics were also recognising the urgency of the need to re-examine orthodoxies. The enormous changes brought about by the new communications technologies in particular, were causing anxiety and alarm, and in many ways they became the focus for a rethinking of established conceptual frameworks. The instantaneous transmission of information made possible by the new technologies destabilised spatial and temporal relations, and undercut the linear logic of time. As signifiers increasingly came adrift from that which they signified, the boundaries between self and other, human and machine, nature and culture, élite art and popular culture, became increasingly hard to sustain. The fragmented subject was seen as having to operate within a temporal discontinuity leading to a subjective and cultural schizophrenia; depth of meaning is replaced by a fascination with surface, and as all sense of critical and historical distance is lost, pastiche and parody become the dominant manifestations of a culture immersed in simulation.

These characteristics of postmodernism are by now reasonably familiar and there is no need to rehearse again arguments that have been made by writers such as Fredric Jameson and Jean Baudrillard, amongst others.[3] In summary, the 'grand narratives' that sustained enlightenment rationality and the notion of the patriarchal unified subject in western culture and society, characterised by Lyotard as 'the dialectics of Spirit, the hermeneutics of meaning, the emancipation of the rational or working subject, or the creation of wealth'[4] were no longer perceived as having the same legitimacy in postmodern, post-industrial societies. Crucially, it is the universalising and totalising tendencies of those grand narratives, in which both Marxism and feminism can be included, that have been most criticised, particularly for their exclusionary effect, which has been especially marked where the lives and histories of women are concerned. Certainly, the destabilisation of the subject as unified and coherent has enabled the complexity of the cultural construction of gender, and the discontinuities inherent in gender identity, to become more apparent. With the collapse of the legitimacy of all such overarching narratives, other partial and fragmented narratives which are concerned with identity and difference are able to claim their own legitimacy in a way that had not been possible before. In the post-modern context, then, Haraway's cyborg metaphor is a playful but deeply political response to a perceived need to construct other, inclusive narratives in which diversity and difference are significant rather than peripheral. The cyborg's propensity to disrupt boundaries and explore differently embodied subjectivities could, therefore, be regarded as its most valuable characteristic, and it is undoubtedly one of the reasons for its continued usefulness in feminist and cultural theory.

William Gibson's novel *Neuromancer*, with its imaginative delineation of the electronic spaces and places of information technology, has also to be set within the context of postmodernism; at the same time, it is part of a newly emerging sub-genre within science fiction known as cyberpunk. Although cyberpunk originated within science fiction, it rapidly extended across a range of other cultural forms such as video, music, comics, and film, and in so doing it acquired a much wider set of cultural meanings, many of which are recognis-ably postmodern. In particular, the thoroughgoing penetration of all aspects of contemporary society and culture by communication technology is taken for granted in cyberpunk fictions, so much so that the future depicted in cyberpunk narratives looks, ironically, very much like our own present. These technolo-gically dominated near futures may be expressive of the collapse of the tem-poral into the spatial, and the erosion of a sense of history often remarked on by theorists writing about postmodernism[5]; on the other hand, the interface with technology is our cultural dominant and there is certain urgency behind the need to explore the possible conditions of existence at that interface. Following Haraway's formulation, we are all, in a sense, cyberpunks, just as we are all cyborgs, and the metaphor of cyberspace, like that of the cyborg, is a device which makes it possible for the experience of living in an information-saturated environment to be articulated from within.

Cyberpunk writers have been described as 'the first generation of artists for whom the technologies of satellite dishes, video and audio players and recorders, computers and video games (both of particular importance), digital watches, and MTV were not exoticisms, but part of a daily "reality Matrix".'[6] The writers themselves readily acknowledged these influences, and Bruce Sterling, the editor of the cyberpunk anthology *Mirrorshades*, suggests that, for cyberpunks 'technology is visceral. It is not the bottled genie of remote Big Science boffins; it is pervasive, utterly intimate. Not outside us, but next to us. Under our skin; often, inside our minds.'[7] In as much as cyberpunk embraces technology as the defining characteristic of contemporary life, then it shares the same world view as Haraway's cyborgs, though not necessarily the same politics, a point that will be made in more detail later in the discussion.

Cyberpunk writers were also influenced by mainstream, non-genre writers such as William Burroughs and Thomas Pynchon, but this was a two-way process, as many non-science fiction writers were also borrowing images and metaphors from science fiction in order to find new ways of describing the multiple disruptions and dislocations occurring within postmodernism. There was, in effect, an interesting exchange taking place between popular and élite culture, and cyberpunk could therefore be thought of as engaging in the kind of 'border crossings' that are made possible by the collapse of accepted categories and boundaries within postmodernism. A more cautious and certainly less generous account of cyberpunk would see it as simply engaged in the process of recycling images and ideas as a form of pastiche and parody, incapable of displaying more than an ambiguous irony at the collapse of the real into the hyperreal, and the future into the present. It was not only the eclecticism of cyberpunk, its fascination with technology, and its predominant tone of detached irony that placed it firmly within the context of postmodernism, it was also the way in which it contributed to the destabilisation of the increasingly tenuous and shifting borders between previously distinct critical categories such as genre, and popular and élite culture.

Although William Gibson has resisted the description of his work as cyberpunk, on the grounds that 'Tying my stuff to *any* label is unfair because it gives people preconceptions about what I'm doing',[8] he was claimed as a cyberpunk writer whether he liked it or not, and this is the context in which the metaphor of cyberspace initially appeared. In the novel *Neuromancer*, Gibson evocatively describes the imaginary space behind our computer screens: 'Cyberspace. A consensual hallucination experienced daily by billions of legitimate operators, in every nation, by children being taught mathematical concepts ... A graphic representation of data abstracted from the banks of every computer in the human system. Lines of light ranged in the nonspace of the mind, clusters and constellations of data. Like city lights, receding ...'[9]

In the following extract, physical space is rendered into data to become 'cyberspace':

> Home.
> Home was BAMA, the Sprawl, the Boston–Atlanta Metropolitan Axis. Program a map to display frequency of data exchange, every thousand megabytes a single pixel on a very large screen. Manhattan and Atlanta burn solid white. Then they start to pulse, the rate of traffic threatening to overload your simulation. Your map is about to go nova. Cool it down. Up your scale. Each pixel a million megabytes. At a hundred million megabytes per second, you begin to make out certain blocks in midtown Manhattan, outlines of hundred-year-old industrial parks ringing the old core of Atlanta ... [10]

The contemporary urban style of Gibson's writing here is one of its most distinguishing features, and it has been likened to the cut-up, fragmented visual style made familiar by MTV, in which the narratives present 'a matrix of images that is more a glitterspace'.[11] It is particularly suited to the kind of near future that Gibson's narratives construct and which his street-wise, super-cool hacker heroes inhabit. In Gibson's cyberpunk trilogy, *Neuromancer*, *Count Zero*, and *Mona Lisa Overdrive*, the console cowboys, or hackers, spend much of their time doing 'biz' in the electronic environment of cyberspace, the geography of which is just as real as that of the other commodified urban spaces described in the novels. Equally, the future time in which the narratives are set is disturbingly familiar, based as it is on transnational corporate control and commodity exchange. The impact of the novels derives from their capacity to destabilise existing notions of time and space, so that the future is now, and electronic space acquires a physicality that is different but equal to geographical space.

Gibson's cyberspace metaphor provided an imaginative depiction of the same electronic environment that programmers working with the technology of virtual reality were in the process of developing, but which as yet only had form in science film and fiction. This, however, is only part of its significance. As with the cyborg, cyberspace is a metaphor with the potential to explore more radical notions of identity and embodiment. The spatial and temporal dislocations that inform the postmodern fictions of cyberpunk are also those which are increasingly instrumental in shaping the social relations of technology, and the implications of this for embodiment are, as yet, unclear. The cyborg metaphor opens up possibilities for imagining embodiment differently, since the cyborg is both a product of and an instrument for the erosion of the borders between self and other, human and machine. The metaphor of cyberspace also emerges from a sense of spatial and temporal dislocation, but additionally it is focused on bodily dislocation, and serves to remind us that embodiment becomes both more complex and more diffuse, as disembodied on-line existence becomes increasingly common. Despite the innovatory depiction of the virtual world of information, however, cyberpunk was forlornly timid in its approach towards other possible consequences of such destabilisation, such as the impact on representations of gender identity. Certainly, the street-wise console cowboys who are the heroes of Gibson's novels are determinedly masculine, and the transition from real to virtual embodiment has no discernable impact on gender identity. It becomes apparent in these texts that leaving the 'meat' behind does not serve to obliterate all the contradictions inherent in culturally constructed masculinity, nor does it enable a less compromised virtual masculinity to be enacted in cyberspace. Instead, cyberspace is inevitably dominated by anxieties about masculinity that can never be resolved, and as the interface between human and technology continues to be structured around masculinity, any attempt to incorporate difference is repressed.[12] This has inevitably placed some restrictions on the way in which the metaphor of cyberspace can be used by feminist writers and critics, but nevertheless it remains an extraordinarily compelling metaphor.

The essays in this book have several things in common: the authors are all concerned with the exploration of the construction of gendered subjectivity, and with the possibilities of formulating alternative ways of thinking about embodiment and the self. These concerns are themselves situated within another shared agenda, that of postmodernism, which has questioned the continuing dominance of the enlightenment orthodoxies of reason, knowledge and truth, as well as the failure of those orthodoxies to recognise difference. The need for absolute categories and universalising theory has become increasingly untenable as both feminist theory and cultural theory have developed other ways of thinking about the politics of the subject. The displacement of the unitary subject within postmodern and feminist theory has occurred during a period in which global political and economic change has precipitated a gradual collapse of cultural consensus about the shape of the future; indeed, the reassuring linearity of time itself can no longer be taken for granted. Concepts of time and space are as much open to question as everything else, and considerations of this sort are an increasingly common feature of contemporary speculation, rather than being restricted to the narratives of science fiction. As Baudrillard has suggested, the future has imploded into the present and the present itself has become increasingly like science fiction[13]; under these circumstances, the use of metaphors from science fiction is not surprising. The metaphors of the cyborg and of cyberspace seem particularly relevant to the times we are living in: they signify the temporal, spatial and cultural shifts and dislocations of postmodernism, while also operating as a means of understanding those same shifts and dislocations. They allow us to explore the new critical contours of postmodernity, in which alternative constructions of difference and identity become possible. The writers included in this Reader have used one or both of these metaphors to do exactly this, informed by both feminist theory and cultural theory.

The metaphors of the cyborg and of cyberspace have been widely used since their first appearance, and there is considerable diversity in the way that they have been employed, as might be expected. The concerns of feminist theory with embodiment, for example, focused in the first instance on the female body, but the interaction of feminist theory with both postmodern theory and cultural theory has resulted in more diverse accounts of both identity and sexual difference, particularly in the field of representation. At the same time, the radical instability of critical categories and discipline boundaries within postmodernism has enabled feminist theory to rethink the way in which its own practices may have replicated the assumptions and premises of modernity in ways that have been detrimental to the overall project of feminism. The search for new ways of negotiating with the uncertainties of postmodernism has resulted in the kind of innovative critical narratives that are to be found in this collection of essays, which demonstrate a shared desire to understand and explore the multiplicity of interconnections between feminist theory, postmodern theory and new technologies, chiefly through the metaphors of the cyborg and cyberspace.

REFERENCES

1. Larry McCaffery (ed.) (1990) 'An Interview with William Gibson' in *Across the Wounded Galaxies: Interviews with Contemporary American Science Fiction Writers*, Urbana and Chicago: University of Illinois Press, pp. 130–50.
2. Donna Haraway (1989) 'A Manifesto for Cyborgs: Science, Technology, and Socialist Feminism in the 1980s' in Elizabeth Weed (ed.) *Coming to Terms*, New York and London: Routledge. All further references to Haraway's 'Manifesto' in the Introduction are taken from this anthology, which also contains several usefully critical responses to the 'Manifesto'.
3. See Fredric Jameson (1991) *Postmodern, or, the Cultural Logic of Late Capitalism*, London and New York: Verso, and (1985) 'Postmodernism and Consumer Society' in Hal Foster (ed.) *Postmodern Culture*, London: Pluto Press, pp. 111–25.
4. Quoted in Seyla Benhabib 'Epistemologies of Postmodernism: A Rejoinder to Jean-Francois Lyotard' in L. Nicholson (ed.) (1990) *Feminism/Postmodernism*, New York and London: Routledge, pp. 107–30.
5. See David Harvey (1989) *The Condition of Postmodernity*, Oxford: Basil Blackwell; Paul Virilio (1984) 'The Overexposed City', *Zone*, Part 1/2, pp. 15–31.
6. Larry McCaffery (ed.) (1991) *Storming the Reality Studio: A Casebook of Cyberpunk and Postmodern Fiction*, Durham, NC and London: Duke University Press, p. 12.
7. Bruce Sterling (ed.) (1988) *Mirrorshades*, London: Paladin, p. xi.
8. Larry McCaffery (ed.) 'An Interview with William Gibson', p. 144.
9. William Gibson (1986) *Neuromancer*, London: Grafton, p. 67.
10. Ibid., p. 57.
11. George Slusser, in *Storming the Reality Studio*, p. 334.
12. For further discussion of this issue, see Jenny Wolmark (1993) *Aliens and Others: Science Fiction, Feminism and Postmodernism*, Hemel Hempstead: Harvester Wheatsheaf, Chapter 5, and (1997) 'Rethinking Bodies and Boundaries: Science Fiction, Cyberpunk and Cyberspace' in Mary Maynard (ed.) *Science and the Construction of Women*, London: UCL Press, pp. 162–82.
13. Jean Baudrillard (1988) 'The Year 2000 Has Already Happened' in A. and M. Kroker (eds) *Body Invaders: Sexuality and the Postmodern Condition*, London: Macmillan, and (1991) 'Simulacra and Science Fiction', *Science Fiction Studies*, **18**, pp. 309–13.

PART I
TECHNOLOGY, EMBODIMENT
AND CYBERSPACE

INTRODUCTION

This section is concerned with the contradictions that are inherent in the interface between body and machine, and with the impact of contemporary technological developments, such as virtual reality or cyberspace, on our understanding of the dimensions of spatiality and temporality. The particular consequences of those developments for female embodiment are the central concern of this section, and the essays contain analyses of the inherent masculinity of assumptions about spatial and temporal relations generally, and at the interface in particular.

Mary Ann Doane's chapter explores the relation of technology to the body in the field of representation, in science fiction narratives in particular, which she describes as the genre 'which most apparently privileges technophilia'. She points out that the science fiction genre 'frequently envisages a new, revised body as a direct result of the advance of science', but that, at the same time, anxieties about technology are frequently displaced onto the figure of the woman. This inconsistency is manifest in the tension between science fiction texts, particularly by feminist science fiction writers, in which technology can be the site for the destabilisation of gendered identity, and those texts in which, by contrast, great efforts are made to maintain existing categories of sexual identity. Doane's discussion of the relationship between technology and the body in the field of representation focuses primarily on the fantasy of the creation of an artificial woman, a fantasy which indicates not only the 'machine–woman problematic' but also the 'persistence of the maternal'. She suggests that this particular fantasy can be found in a wide range of cinematic narratives in which the link between technology and reproduction is explored in increasingly complex ways. Even in films such as *Alien* (1979), and *Aliens* (1986), in which there is no artificial woman, the link between technology and reproduction is made clear, because the alien itself evokes the maternal. From this point of view, the motherless reproduction that is represented in *Alien* is particularly graphic: the alien develops within the stomach of a male crew member until it is ready to emerge, at which point it gnaws its way out through his chest in a horrific imitation of the birth process. The apparent confusion of gender identities here does not in any way obviate the presence of those deeply rooted anxieties about technology that are displaced onto the female body, and Doane draws on Kristeva in order to argue that what is really at stake is the confusion of identities that is inherent in the concept of motherhood itself, associated by Kristeva with the abject, when the boundaries between self and other are thrown into question, as is identity. In a discussion of *Blade Runner* (1982), Doane employs the axiom that 'reproduction is a guarantee of a history' to argue that anxieties about technological reproduction are differently located within anxieties about a 'loss of history', and that despite the fact that there are no mothers within the narrative, nevertheless the maternal is present in the way in which both Deckard and the replicants are preoccupied with knowledge of origins and personal history. The contradictions in these particular texts revolve around a simultaneous 'nostalgia for and a terror of the maternal function, both linking it to and divorcing it from the idea of the machine woman.' For Doane, then, there are unresolvable contradictions that result from the intersection of technology and the body, and the way in which they are expressed in film narratives suggests that the question of embodiment is extremely complex. This clearly has implications for the metaphor of the cyborg as proposed by Donna Haraway, in which the contradictions between organic and machine are potentially dissolvable, suggesting a different notion of embodiment. For Doane, the machine–woman problematic, in these films at least, ensures that embodiment remains haunted by unresolvable contradictions about both technology and the maternal which are endlessly re-enacted in the figure of the artificial woman. In Doane's discussion, the figure of the artificial woman, linked to the specificities of film narrative, is a somewhat literal rendering of a cyborg as a machine; although its presence is disruptive, preventing the absolute containment of the technologies of reproduction in the narratives, it remains caught in an Oedipal narrative from which it cannot escape. This cyborg is evidently not capable of transgressing borders and thus it cannot be seen as the kind of oppositional or utopian cyborg for which Haraway argues.

Claudia Springer's discussion of the representation of the 'pleasure of the interface' in popular cyborg imagery also draws on the contention that anxieties about technology are displaced onto female sexuality, and that this results in the technology itself being eroticised. In common with Mary Ann Doane, she draws on psychoanalysis to suggest that representations of technology in popular narratives are structured by the 'simultaneous attraction and dread evoked by the womb', and that reproduction 'is also a site of dispute in cyborg texts'. Springer postulates a significant difference between robots and cyborgs, in that the latter 'incorporate rather than exclude humans, and in so doing erase the distinctions previously assumed to distinguish humanity from technology.' Indeed, within popular culture the technologising of the human body and the body–machine fusion is frequently represented as a transgression of boundaries that is 'a pleasurable experience', one that is 'often represented as a sexual act'. As Springer points out, popularised accounts of developments in the technology of virtual reality have chosen to emphasis the erotic potential of virtual reality, and even the console cowboys in William Gibson's cyberpunk narratives 'jack in' to cyberspace by means of a direct neural link with the computer. Despite the pleasure in the transgression of boundaries evinced in popular narratives, however, representations of gendered cyberbodies continue to reproduce the familiar and stereotypical characteristics of masculinity and femininity, often in an exaggerated form. In a discussion of the hypermasculinity present in films such as *The Terminator* (1984) and *Robocop* (1987), Springer suggests that the transformation of the male body 'into something only minimally human' indicates an 'intense crisis in the construction of masculinity'. Although this cyborg masculinity enables masculine subjectivity itself to be reconstituted, it is at the expense of the coherent male body, and Springer speculates that this contradiction may account in part for the violence in such films. Thus, not only is the cyborg in popular culture 'constituted by paradoxes', but the narratives also articulate a fear of sexuality and, by implication, a fear of the maternal, and Springer refers to Klaus Thewelcit's analysis of male fascist soldiers in support of her contention that fear of the loss of the masculine self may well animate these depictions of the cyborg. The essays by both Doane and Springer are, then, concerned to explore the impact of technology on embodiment, and the contradictions that result from the human–machine interface, although Springer is more concerned with the way in which constructions of masculinity have been thrown into crisis by the engagement with technology, while Doane examines the concurrence of technology with the feminine and the consequent fascination and anxiety which this generates. The popular narratives discussed in Springer's chapter are specifically concerned with the pleasures associated with the possibility of disembodiment, and what this might mean for a further redefinition of the self. In the context of the postwar contemporary environment, the vulnerability of the body has become a major concern, and in response to this anxiety, the narratives discussed by Springer exhibit an overwhelming desire for the separation of mind and body which seems curiously anachronistic, and crucially, also leaves masculinity intact. As such, it is a response which is inevitably at odds with any notion of female embodiment. While it may be the case that, as Springer points out, 'debates over sexuality and gender roles have thus contributed to producing the concept of the cyborg', representations of cyborg embodiment in popular narratives reveal a singular unwillingness to engage in those debates.

The ambiguity of the pleasure of the interface is a central concern in the chapter by Zoë Sofia, who, whilst readily admitting the 'seductive' quality of the possibilities of 'virtual corporeality', also expresses some scepticism towards the postmodern 'fascination with virtual reality and corporeality', and the expectation that information technology *per se* can result in new and different forms of subjectivity. Sofia notes not only the prosthetic qualities of the computer by means of which it can take on the role of a 'second self', but also its 'ambiguous gender', and in order to explore the effect of this doubled framing, she develops a 'mythic iconography' which she calls 'Jupiter Space'. This is a space in which the entirety of knowledge and existence can be claimed as 'the conceptions of a masculine, rational and (increasingly) artificial brain', and in which there is little room for the female body. If the female body and the maternal function appear at all, they are 'displaced onto masculine and corporate technological fertility', as in the example of the caring but psychotic computer HAL in the film *2001: A Space Odyssey* or 'Mother', the computer in *Alien*. In comparison to the analysis put forward by Mary Ann Doane, Sofia argues that the female and maternal figure is indicative of masculine, rather than maternal, excess. From such a perspective, cyberspace itself can, therefore, be thought of as a female, maternal space to be 'penetrated, cut up and manipulated', presumably by predominantly male hackers. She argues that, ultimately, virtual reality and corporeality offer a 'misogynistic fantasy of escape from Earth, gravity, and maternal/material origins'. Her analysis rests on the assumption that the masculinisation of new technologies stems from the framework of corporate and military control within which they are situated, and the ensuing social relations of science and technology; Sofia insists that this perspective provides a necessary corrective to some of the more enthusiastic and less critical responses to communications technology that have emerged from within postmodern theory. She advocates a less celebratory approach and suggests that it is certainly possible for women to appropriate computer technology for creative and life affirming purposes, as long as it is done with an ironic awareness of the ultimate ambivalence of such technologically enhanced corporeality. In common with Donna Haraway, Sofia proposes that irony is a key factor in the subversion of the masculine hegemony of technology.

This ambivalence towards the uses to which technology is put is explored in the chapter by Allucquere Rosanne Stone, who looks at the nature and meaning of embodiment in the context of the new communities evolving in virtual reality. She usefully charts the growth of 'virtual' communities in a reciprocal relationship with technology, from the seventeenth century up to the contemporary period, and presents an informal 'history' of the way in which computer networks and imaging systems, along with cultural texts such as *Star Wars* and *Neuromancer*, provided those who worked in them with both the material and imaginative means for the development of virtual reality. Stone suggests that 'VR touched the same nerve that *Star Wars* had, the englobing specular fantasy made real' and that 'Gibson's novel and the technological and social imaginary that it articulated enabled the researchers in virtual reality – or, under the new dispensation, cyberspace – to recognise and organise themselves as a community.' These developments are the context in which the boundaries between the subject and the world are being redefined, as are the categories within which the world is conceptualised. Stone argues that anxieties about the technologising of 'nature', for example, are unduly alarmist, and can instead be thought of in a more positive way as providing an opportunity for thinking of nature precisely as technology, in a world in which the 'boundaries between subject and environment have collapsed.' In order to examine the consequences for embodiment that are attendant on the development of virtual reality, she finds examples of two groups of people, phone sex workers and virtual reality engineers, who are already operating within a world defined by such collapsing boundaries, and faced with the task of 'representing the human body through limited communication channels', both groups have to draw on existing and gendered paradigms of desire. The effect of this is that, as Claudia Springer has pointed out, the pleasure of the interface is inevitably compromised by the refusal to renegotiate existing gender boundaries. However, Stone suggests that the parameters of the engagement with the virtual body in cyberspace may already be changing, and that the 'refigured and reinscribed embodiments of the cyberspace community' are already in the process of undermining the construction of bodies as stable and unitary. Stone is clear that she is not embracing a romanticised notion of cyberspace which projects the possibility that real bodies can simply be abandoned – as she says, 'life is lived through bodies'. Since virtual embodiment is inextricably linked to real bodies, it is Stone's view that electronic virtual communities have the capacity to make significant strategic interventions into the social, economic and cultural realities of contemporary life.

Some of the same enthusiasm for virtual embodiment can be found in Sadie Plant's chapter, in which she declares herself to be in favour of a 'cybernetic feminism' that explores the 'convergence of woman and machine'. In an extended discussion of the significance of Ada Lovelace's part in the development of Charles Babage's Analytical Engine, she suggests that 'Ada Lovelace may have been the first encounter between woman and computer', and that Ada's software 'encouraged the convergence of nature and intelligence which guides the subsequent development of information technology.' The discussion of this and other such encounters is framed by the overarching metaphor of weaving, and Plant describes the way in which the computer 'emerges out of the history of weaving, the process so often said to be the quintessence of women's work', work that is eventually transformed by Jacquard's loom into a fledgling cybernetic system. Drawing on Freud's suggestion that women's only substantial contribution to civilisation has been weaving, Plant turns this somewhat restricted view into an argument that, in fact, weaving has been integral to the whole of human history and creative and scientific discovery; weaving, she contends, 'has been the art and science of software' and perhaps 'is even the fabric of every other discovery and invention'. Since it seems to be the case that 'weaving is always already entangled with the question of female identity', it could be argued, then, that there is a 'point at which weaving, women and cybernetics converge in a movement fatal to history', a convergence that is crucial to the eventual liberation of women. Like Irigaray, on whose work she draws, Plant is talking about the history that is made as man struggles to escape from what he perceives to be his subordination to nature and to biology; driven by a flight from the material and the maternal, a drive for transcendence becomes a drive for domination. As Plant points out, women have never been the subjects of history, but she suggests that they have, nevertheless, woven themselves into its fabric through mimicry and simulation; like the computer, women can 'mimic any function', thus espousing a particular kind of virtuality. For Plant, this virtual materiality finds its fullest expression in cyberspace, which, remembering the liberating convergence of weaving, women and cybernetics, is the 'place of woman's affirmation'. Unlike other postmodern feminist theorists, however, Plant argues for a version of 'cybernetic feminism' which 'does not ... seek out for woman a subjectivity, an identity or even a sexuality of her own', on the grounds that these are superfluous concerns in the new future of cybernetic systems. Embodiment itself has become something of an irrelevancy in Sadie Plant's version of the matrix, as the self-organising systems and self-arousing machines of cybernetic systems/circuits make the concept of a 'self' outmoded. This is of less concern to women, argues Plant, who have always existed by mimicry, and who therefore have less invested in the necessity of defining or achieving selfhood. Instead, cybernetic systems can lead to the possibility of 'an agency, of sorts, which has no need of a subject position', a view that it is at the other end of the spectrum to that put forward by Zoë Sofia, and one which may appear to depend on precisely the kind of technological fetishism of which Mary Ann Doane was so critical in her essay. On the other hand, through a sustained reading of the way in which women have always, in a sense, been disembodied because they masquerade and simulate, Plant attempts to radicalise the very notion of disembodiment in order to demonstrate that cyberspace itself is emancipatory for women.

Underlying the kinds of postmodern redefinitions of embodiment that are contained in this section are shifting notions of spatiality and temporality, and in her chapter Elizabeth Grosz argues that the way in which we understand space and time has a profound impact on the way in which we understand the world and our place in it. As she puts it, 'The kinds of world that we inhabit, and our understanding of our places in these worlds are to some extent an effect of the ways in which we understand space and time.' I have chosen to include this chapter because it offers a parallel vantage point to that of techno theory from which to consider both the 'space–time of lived bodies' and representations of female corporeality. Although Grosz does not engage in a discussion of embodiment from the specific perspective of cyber-theory, her chapter nevertheless demonstrates a shared perception that women's experiences have been excluded from definitions of space and time, which has had a profoundly negative impact on representations of female corporeality. While cyber-theory concerns itself with the new ontologies of communication systems, it has not necessarily acknowledged the fact that, as Grosz points out, the history of philosophy itself is 'strewn with speculations about the nature of space and time, which form among the most fundamental categories of ontology.' It is also arguable that cyber-theory has, on occasion, underestimated the complex inter-relationship between theoretical notions of space and time and representations of subjectivity. Given the conjectural nature of the cyber-theories included here, it is appropriate, in my view, to include a discussion of some of those founding speculations about the way in which 'we live and move in space as bodies in relation to other bodies.' If these are mapped on to the way in which embodiment is imagined through the metaphor of cyberspace, the effect on both sets of discussions is likely to be intriguing and enhancing. Grosz suggests that psychoanalysis 'has a great deal to say about how the body is lived and positioned as a spatio-temporal being', and particularly in Lacan's work on the mirror stage, 'corporeality and spatiality [have] been related to the construction of personal identity, and the challenge to philosophical dualism undertaken head on.' She points out that Lacan 'never reduce psychological to physiological explanation', which is a useful reminder that, in debates about the nature of cyberspace, elisions of the psychic and the physical take place with a frequency that is both alarming and unhelpful. From Grosz's examination of the historical transitions in scientific, mathematical and philosophical conceptions of space and time, it becomes evident that the contemporary environment of non-Euclidean geometries and post-Newtonian physics has undoubtedly contributed to the development of the imaginative conception of cyberspace. The continued domination of representations of time by those of space, for example, for which there appears to be no substantial argument, is evidenced in the metaphor of cyberspace itself. Grosz makes a general point that, because scientific thought and prevalent social and cultural conceptions are interdependent, there is a likely correlation between conceptions of space and conceptions of subjectivity and social life. Thus, it is perhaps appropriate for the decentred postmodern subject to find its correlative in the virtual environs of cyberspace. As she says, these conceptualisations are 'both ways of negotiating our positions as subjects within a social and cosmic order, and representations that affirm or rupture pe-existing forms of subjectivity.' Grosz argues that, in order to transform representations of female corporeality, it is necessary to question not only gender relations and the dominance of masculinity, but also those notions of space–time within which bodies interact. Located as it is within a multiplicity of different perspectives and cultural positions, this is surely what the metaphor of cyberspace is attempting to do.

I

TECHNOPHILIA: TECHNOLOGY, REPRESENTATION, AND THE FEMININE

Mary Ann Doane

The concept of the 'body' has traditionally denoted the finite, a material limit that is absolute – so much so that the juxtaposition of the terms 'concept' and 'body' seems oxymoronic. For the body is that which is situated as the precise opposite of the conceptual, the abstract. It represents the ultimate constraint on speculation or theorization, the place where the empirical finally and always makes itself felt. This notion of the body as a set of finite limitations is, perhaps, most fully in evidence in the face of technological developments associated with the Industrial Revolution. In 1858, the author of a book entitled *Paris* writes, 'Science; as it were, proposes that we should enter a new world that has not been made for us. We would like to venture into it; but it does not take us long to recognize that it requires a constitution we lack and organs we do not have.'[1] Science fiction, a genre specific to the era of rapid technological development, frequently envisages a new, revised body as a direct outcome of the advance of science. And when technology intersects with the body in the realm of representation, the question of sexual difference is inevitably involved.

Although it is certainly true that in the case of some contemporary science-fiction writers – particularly feminist authors – technology makes possible the destabilization of sexual identity as a category, there has also been a curious but fairly insistent history of representations of technology that work to fortify – sometimes desperately – conventional understandings of the feminine. A certain anxiety concerning the technological is often allayed by a displacement of this anxiety onto the figure of the woman or the idea of the feminine. This has certainly been the case in the cinema, particularly in the genre which most apparently privileges technophilia, science fiction. And despite the emphasis in discourses about technology upon the link between the machine and *production* (the machine as a labor-saving device, the notion of man as a complicated machine which Taylorism, as an early-twentieth-century attempt to regulate the worker's bodily movements, endeavored to exploit), it is striking to note how often it is the woman who becomes the model of the perfect machine. Ultimately, what I hope to demonstrate is that it is not so much *production* that is at stake in these representations as *reproduction*.

The literary text that is cited most frequently as the exemplary forerunner of the cinematic representation of the mechanical woman is *L'Eve future* (*Tomorrow's Eve*), written by Villiers de l'Isle-Adam in 1886. In this novel, Thomas Edison, the master scientist and entrepreneur of mechanical reproduction – associated with both the phonograph and the cinema – is the inventor of the perfect mechanical woman, an android whose difference from the original human model is imperceptible. Far from investing in the type of materialism associated with scientific progress, Villiers is a metaphysician. Edison's creation embodies the Ideal (her name is Hadaly which is, so we are told, Arabic for the Ideal). The very long introductory section of the novel is constituted by Edison's musings about all the voices in history that have been lost and that could have been captured had the phonograph been invented sooner. These include, among others, 'the first vibrations of the good tidings brought to Mary! The resonance of the Archangel saying Hail! a sound that has reverberated through the ages in the Angelus. The Sermon on the Mount! The "Hail, master!" on the Mount of Olives, and the sound of the kiss of Iscariot.'[?] Almost simultaneously, however, Edison realizes that the mechanical recordings of the sounds is not enough: 'To hear the sound is nothing, but the inner essence, which creates these mere vibrations, these veils – that's the crucial thing.'[3] This 'inner essence' is what the human lover of Lord Ewald, Edison's friend, lacks. In Lord Ewald's report, although her body is magnificent, perfect in every detail, the human incarnation of the *Venus Victorious*, she lacks a *soul*. Or, more accurately, between the body and soul of Miss Alicia Clary there is an 'absolute disparity.' Since Lord Ewald is hopelessly in love with the soulless Alicia, Edison takes it upon himself to mold Hadaly to the form of Miss Clary.

A great deal of the novel consists of Edison's scientific explanations of the functioning of Hadaly. As he opens Hadaly up to a dissecting inspection, Lord Ewald's final doubts about the mechanical nature of what seemed to him a living woman are dispelled in a horrible recognition of the compatibility of technology and desire.

> Now he found himself face to face with a marvel the obvious possibilities of which, as they transcend even the imaginary, dazzled his understanding and made him suddenly feel to what lengths a man who wishes can extend the courage of his desires.[4]

Hadaly's interior is a maze of electrical wizardry including coded metal discs that diffuse warmth, motion, and energy throughout the body; wires that imitate nerves, arteries, and veins; a basic electro-magnetic motor, the Cylinder, on which are recorded the 'gestures, the bearing, the facial expressions, and the attitudes of the adored being'; and two golden phonographs that replay Hadaly's only discourse, words 'invented by the greatest poets, the most subtle metaphysicians, the most profound novelists of this century.'[5] Hadaly has no past, no memories except those embodied in the words of 'great men.' As Annette Michelson remarks, in a provocative analysis of the novel,

> Hadaly's scenes, so to speak, are set in place. Hadaly becomes that palimpsest of inscription, that unreasoning and reasonable facsimile, generated by reason, whose interlocutor, Lord Ewald, has only to submit to the range and nuance of mise-en-scene possible in what Edison calls the 'great kaleidoscope' of human speech and gesture in which signifiers will infinitely float.[6]

As Edison points out to Lord Ewald, the number of gestures or expressions in the human repertoire is extremely limited, clearly quantifiable, and hence reproducible. Yet, precisely because Villiers is a metaphysician, something more is needed to animate the machine – a spark, a touch of spirit.

This spark is provided, strangely enough, by an abandoned mother, Mrs Anny Anderson (who, in the hypnotic state Edison maintains her in, takes on the name Miss Anny Sowana). Her husband, Howard, another of Edison's friends, had been seduced and ruined by a beautiful temptress, Miss Evelyn Habal, ultimately committing suicide. Miss Evelyn Habal was in a way the inspiration for the *outer* form of Hadaly, for through his investigations, Edison discovered that her alleged beauty was completely *artificial*. He displays for Lord Ewald's sake a drawer containing her implements: a wig corroded by time, a makeup kit of greasepaint and patches, dentures, lotions, powders, creams, girdles, and falsies, etc. Edison's cinema reveals that, without any of these aids, Evelyn Habal was a macabre figure. The display demonstrates to Ewald that mechanical reproduction suffices in the construction of the forms of femininity. But its spirit, at least, is not scientifically accessible. The abandoned Mrs Anderson, mother of two children, suffers a breakdown after the suicide of her husband. Only Edison is able to communicate with her and eventually her spirit establishes a link with his android Hadaly, animating it, humanizing it. The mother infuses the machine. Perhaps this is why, for Edison, science's most important contribution here is the validation of the dichotomy between woman as mother and woman as mistress:

> Far from being hostile to the love of men for their wives – who are so necessary to perpetuate the race (at least till a new order of things comes in), I propose to reinforce, ensure, and guarantee that love. I will do so with the aid of thousands and thousands of marvelous and completely

innocent facsimiles, who will render wholly superfluous all those beautiful but deceptive mistresses, ineffective henceforth forever.[7]

Reproduction is that which is, at least initially, unthinkable in the face of the woman-machine. Herself the product of a desire to reproduce, she blocks the very possibility of a future through her sterility. Motherhood acts as a limit to the conceptualization of femininity as a scientific construction of mechanical and electrical parts. And yet it is also that which infuses the machine with the breath of a human spirit. The maternal and the mechanical/synthetic coexist in a relation that is a curious imbrication of dependence and antagonism.

L'Eve future is significant as an early signpost of the persistence of the maternal as a sub-theme accompanying these fantasies of artificial femininity. It is also, insofar as Edison (a figure closely associated with the prehistory of cinema) is the mastermind of Hadaly's invention, a text that points to a convergence of the articulation of this obsession and the cinema as a privileged site for its exploration. In Michelson's argument, Hadaly's existence demonstrates the way in which a compulsive movement between analysis and synthesis takes the female body as its support in a process of fetishization fully consistent with that of the cinema:

> We will want once more to note that assiduous, relentless impulse which claims the female body as the site of an analytic, mapping upon its landscape a poetics and an epistemology with all the perverse detail and somber ceremony of fetishism. And may we not then begin to think of that body in its cinematic relations somewhat differently? Not as the mere object of a cinematic *iconography* of repression and desire – as catalogued by now in the extensive literature on dominant narrative in its major genres of melodrama, *film noir*, and so on – but rather as the fantasmatic ground of cinema itself.[8]

Indeed, cinema has frequently been thought of as a prosthetic device, as a technological extension of the human body, particularly the senses of perception. Christian Metz, for instance, refers to the play 'of that *other mirror*, the cinema screen, in this respect a veritable psychical substitute, a prosthesis for our primally dislocated limbs.'[9] From this point of view it is not surprising that the articulation of the three terms – 'woman,' 'machine,' 'cinema' – and the corresponding fantasy of the artificial woman recur as the privileged content of a wide variety of cinematic narratives.

An early instance of this tendency in the science-fiction mode is Fritz Lang's 1926 film, *Metropolis*, in which the patriarch of the future city surveys his workers through a complex audio-visual apparatus resembling television. In *Metropolis*, the bodies of the male workers become mechanized; their movements are rigid, mechanical, and fully in sync with the machines they operate. The slightest divergence between bodily movement and the operation of the machine is disastrous, as evidenced when the patriarch's son, Freder, descends to the realm of the workers and witnesses the explosion of a machine not sufficiently controlled by a worker. Freder's resulting hallucination transforms the machine into a Moloch-figure to whom the unfortunate workers are systematically sacrificed. When Freder relieves an overtired worker, the machine he must operate resembles a giant clock whose hands must be moved periodically – a movement that corresponds to no apparent logic. In a production routine reorganized by the demands of the machine, the human body's relation to temporality becomes inflexible, programmed. The body is tied to a time clock, a schedule, a routine, an assembly line. Time becomes oppression and mechanization – the clock, a machine itself, is used to regulate bodies as machines. *Metropolis* represents a dystopic vision of a city run by underground machines whose instability and apparent capacity for vengeance are marked.

But where the men's bodies are analogous to machines, the woman's body literally becomes a machine. In order to forestall a threatened rebellion on the part of the workers, the patriarch Fredersen has a robot made in the likeness of Maria, the woman who leads and instigates them. Rotwang, who is a curious mixture of modern scientist and alchemist, has already fashioned a robot in the form of a woman when Fredersen makes the request. The fact that the robot is manifestly female is quite striking particularly in light of Rotwang's explanation of the purpose of the machine: 'I have created a machine in the image of man, that never tires or makes a mistake. Now we have no further use for living workers.' A robot which is apparently designed as the ultimate producer is transformed into a woman of excessive and even explosive sexuality (as manifested in the scene in which Rotwang demonstrates her seductive traits to an audience of men who mistake her for a 'real woman'). In Andreas Huyssen's analysis of *Metropolis*, the robot Maria is symptomatic of the fears associated with a technology perceived as threatening and demonic: 'The fears and perceptual anxieties emanating from ever more powerful machines are recast and reconstructed in terms of the male fear of female sexuality, reflecting, in the Freudian account, the male's castration anxiety.'[10]

Yet, the construction of the robot Maria is also, in Huyssen's account, the result of a desire to appropriate the maternal function, a kind of womb envy on the part of the male. This phenomenon is clearly not limited to *Metropolis* and has been extensively explored in relation to Mary Shelley's *Frankenstein*, in which the hero, immediately before awakening to perceive his frightful creation, the monster, standing next to his bed, dreams that he holds the corpse of his dead mother in his arms. The 'ultimate technological fantasy,' according to Huyssen, is 'creation without the mother.'[11] Nevertheless, in *Metropolis*, the robot Maria is violently opposed to a real Maria who is characterized, first and foremost, as a mother. In the first shot of Maria, she is surrounded by a flock of children, and her entrance interrupts a kiss between Freder and another woman so that the maternal effectively disrupts the sexual. Toward the end of the film, Maria and Freder save the children from a flood unwittingly caused by the angry workers' disruption of the machinery. The film manages to salvage both the technological and the maternal (precisely by destroying the figure of the machine-woman) and to return the generations to their proper ordering (reconciling Freder and his father). The tension in these texts which holds in balance a desire on the part of the male to appropriate the maternal function and the conflicting desire to safeguard and honor the figure of the mother is resolved here in favor of the latter. The machine is returned to its rightful place in production, the woman hers in reproduction.

The maternal is understandably much more marginal in a more recent film, *The Stepford Wives* (1975), in which the machine-woman is not burned at the stake, as in *Metropolis*, but comfortably installed in the supermarket and the suburban home. In this film, a group of women are lured to the suburbs by their husbands who then systematically replace them with robots, indistinguishable from their originals. The robots have no desires beyond those of cooking, cleaning, caring for the children, and fulfilling their husbands' sexual needs. Even the main character, Joanna, who claims, 'I messed a little with Women's Lib in New York,' finds that she cannot escape the process. As in *L'Eve future*, the husbands record the voices of their wives to perfect the illusion, but unlike that of Hadaly, the Ideal, the discourse of these robot-housewives consists of hackneyed commercial slogans about the advantages of products such as Easy On Spray Starch. Here the address is to women and the social context is that of a strong and successful feminist movement, which the film seems to suggest is unnecessary outside of the science-fiction nightmare in which husbands turn wives into robots. *The Stepford Wives* indicates a loss of the obsessive force of the signifying matrix of the machine-woman – as though its very banalization could convince that there is no real threat involved, no reason for anxiety.

The contemporary films that strike me as much more interesting with respect to the machine-woman problematic are those in which questions of the maternal and technology are more deeply imbricated – films such as *Alien* (1979) and its sequel, *Aliens* (1986), and *Blade Runner* (1982). As technologies of reproduction seem to become a more immediate possibility (and are certainly the focus of media attention), the impact of the associative link between technology and the feminine on narrative representation becomes less localized – that is, it is no longer embodied solely in the figure of the female robot. *Alien* and *Aliens* contain no such machine-woman, yet the technological is insistently linked to the maternal. While *Blade Runner* does represent a number of female androids (the result of a sophisticated biogenetic engineering, they are called 'replicants' in the film), it also represents male replicants. Nevertheless, its narrative structure provocatively juxtaposes the question of biological reproduction and that of mechanical reproduction. Most importantly, perhaps, both *Alien* and *Blade Runner* contemplate the impact of drastic changes in reproductive processes on ideas of origins, narratives, and histories.

Alien, together with its sequel, *Aliens* and *Blade Runner* elaborate symbolic systems that correspond to a contemporary crisis in the realm of reproduction – the revolution in the development of technologies of reproduction (birth control, artificial insemination, *in vitro* fertilization, surrogate mothering, etc.). These technologies threaten to put into crisis the very possibility of the question of origins, the Oedipal dilemma and the relation between subjectivity and knowledge that it supports. In the beginning of *Alien*, Dallas types into the keyboard of the ship's computer (significantly nicknamed 'Mother' by the crew) the question: 'What's the story, Mother?' The story is no longer one of transgression and conflict with the father but of the struggle with and against what seems to become an overwhelming extension of the category of the maternal, now assuming monstrous proportions. Furthermore, this concept of the maternal neglects or confuses the traditional attributes of sexual differ-ence. The ship itself, *The Nostromo*, seems to mimic in the construction of its internal spaces the interior of the maternal body. In the first shots of the film, the camera explores in lingering fashion corridors and womblike spaces which exemplify a fusion of the organic and the technological.[12] The female merges with the environment and the mother-machine becomes *mise-en-scène*, the space within which the story plays itself out. The wrecked alien spaceship which the crew investigates is also characterized by its cavernous, womblike spaces; one of the crew even descends through a narrow tubelike structure to the 'tropical' underground of the ship where a field of large rubbery eggs are in the process of incubation. The maternal is not only the subject of the representation here, but also its ground.

The alien itself, in its horrifying otherness, also evokes the maternal. In the sequel, *Aliens*, the interpretation of the alien as a monstrous mother-machine, incessantly manufacturing eggs in an awesome excess of reproduction, con-firms this view. Yet, in the first film the situation is somewhat more complex, for the narrative operates by confusing the tropes of femininity and masculinity in its delineation of the process of reproduction. The creature first emerges from an egg, attaches itself to a crew member's face, penetrating his throat and gastrointestinal system to deposit its seed. The alien gestates within the stomach of the *male* crew member who later 'gives birth' to it in a grotesque scene in which the alien literally gnaws its way through his stomach to emerge as what one critic has labeled a *phallus dentatus*.[13] The confusion of the semes of sexual difference indicates the fears attendant upon the development of technologies of reproduction that debiologize the maternal. In *Alien*, men have babies but it is a horrifying and deadly experience. When the alien or other invades the most private space – the inside of the body – the foundations of subjectivity are shaken. The horror here is that of a collapse between inside and outside or of what Julia Kristeva refers to, in *Powers of Horror*, as the abject. Kristeva associates the maternal with the abject – i.e., that which is the focus of a combined horror and fascination, hence subject to a range of taboos designed to control the culturally marginal.[14] In this analysis, the function of nostalgia for the mother-origin is that of a veil which conceals the terror attached to nondifferentiation. The threat of the maternal space is that of the collapse of any distinction whatsoever between subject and object.

Kristeva elsewhere emphasizes a particularly interesting corollary of this aspect of motherhood: The maternal space is 'a place both double and foreign.'[15] In its internalization of heterogeneity, an otherness within the self, motherhood deconstructs certain conceptual boundaries. Kristeva delineates the maternal through the assertion, 'In a body there is grafted, unmasterable, an other.'[16] The confusion of identities threatens to collapse a signifying system based on the paternal law of differentiation. It would seem that the concept of motherhood automatically throws into question ideas concerning the self, boundaries between self and other, and hence identity.

According to Jean Baudrillard, 'Reproduction is diabolical in its very essence; it makes something fundamental vacillate.'[17] Technology promises more strictly to control, supervise, regulate the maternal – to put *limits* upon it. But somehow the fear lingers – perhaps the maternal will contaminate the technological. For aren't we now witnessing a displacement of the excessiveness and overproliferation previously associated with the maternal to the realm of technologies of representation, in the guise of the all-pervasive images and sounds of television, film, radio, the Walkman? One response to such anxiety is the recent spate of films that delineate the horror of the maternal – of that which harbors an otherness within, where the fear is always that of giving birth to the monstrous; films such as *It's Alive, The Brood, The Fly*, or the ecology horror film, *Prophecy. Alien*, in merging the genres of the horror film and science fiction, explicitly connects that horror to a technological scenario.

In *Blade Runner*, the signifying trajectory is more complex, and the relevant semes are more subtly inscribed. Here the terror of the motherless reproduction associated with technology is clearly located as an anxiety about the ensuing loss of history. One scene in *Blade Runner* acts as a condensation of a number of these critical terms: 'representation,' 'the woman,' 'the artificial,' 'the technological,' 'history,' and 'memory.' It is initiated by the camera's pan over Deckard's apartment to the piano upon which a number of photos are arranged, most of them apparently belonging to Deckard, signifiers of a past (though not necessarily his own), marked as antique – pictures of someone's mother, perhaps a sister or grandmother. One of the photographs, however – a rather nondescript one of a room, an open door, a mirror – belongs to the replicant Leon, recovered by Deckard in a search of his hotel room. Deckard inserts this photograph in a piece of equipment that is ultimately revealed as a machine for analyzing images. Uncannily responding to Deckard's voiced commands, the machine enlarges the image, isolates various sections, and enlarges them further. The resultant play of colors and grain, focus and its loss, is aesthetically provocative beyond the demonstration of technical prowess and control over the image. Deckard's motivation, the desire for knowledge that is fully consistent with his positioning in the film as the detective figure of *film noir*, is overwhelmed by the special effects which are the byproducts of this technology of vision – a scintillation of the technological image which exceeds his epistemophilia. Only gradually does the image resolve into a readable text.

And in the measure to which the image becomes readable, it loses its allure. The sequence demonstrates how technology, the instrument of a certain knowledge-effect, becomes spectacle, fetish. But one gains ascendancy at the price of the other – pleasure pitted against knowledge.

Historically, this dilemma has been resolved in the cinema by conflating the two – making pleasure and knowledge compatible by projecting them onto the figure of the woman. The same resolution occurs here: as the image gradually stabilizes, what emerges is the recognizable body of a woman (neglecting for a moment that this is not a 'real' woman), reclining on a couch, reflected in the mirror which Deckard systematically isolates. The mirror makes visible what is outside the confines of the photograph strictly speaking – the absent woman, object of the detective's quest. To know in *Blade Runner* is to be able to detect difference – not sexual difference, but the difference between human and replicant (the replicant here taking the place of the woman as marginal, as Other). Knowledge in psychoanalysis, on the other hand, is linked to the mother's body (knowledge of castration and hence of sexual difference, knowledge of where babies come from) – so many tantalizing secrets revolving around the idea of an origin and the figure of the mother. There are no literal – no embodied – mothers in *Blade Runner* (in fact, there are no 'real' women in the film beyond a few marginal characters – the old Chinese woman who identifies the snake scale, the women in the bar). Yet this does not mean that the concept of the maternal – its relation to knowledge of origins and subjective history – is inoperative in the text. As a story of replicants who look just like 'the real thing,' *Blade Runner* has an affinity with Barthes's analysis of photography, *Camera Lucida*.[18] Barthes's essay is crucially organized around a photograph of his mother which is never shown, almost as though making it present would banalize his desire, or reduce it. Both film and essay are stories of reproduction – mechanical reproduction, reproduction as the application of biogenetic engineering. In the film, however, our capability of representing human life begins to pose a threat when the slight divergence that would betray mimetic activity disappears.

In *Blade Runner*, as in *Camera Lucida*, there are insistent references to the mother, but they are fleeting, tangential to the major axis of the narrative. In the opening scene, the replicant Leon is asked a question by the examiner whose task it is to ascertain whether Leon is human or inhuman: 'Describe in single words only the good things that come into your mind about – your mother.' Leon answers, 'Let me tell you about my mother' and proceeds violently to blow away the examiner with a twenty-first-century gun. The replicants collect photographs (already an archaic mode of representation in this future time) in order to reassure themselves of their own past, their own subjective history. At one point Leon is asked by Roy whether he managed to retrieve his 'precious photographs.' Later Rachel, still refusing to believe that she is a replicant, tries to prove to Deckard that she is as human as he is by thrusting forward a photograph and claiming, 'Look, it's me with my mother.'

After Rachel leaves, having been told that these are 'not your memories' but 'somebody else's,' Deckard looks down at the photo, his voice-over murmuring 'a mother she never had, a daughter she never was.' At this moment, the photograph briefly becomes 'live,' animated, as sun and shadow play over the faces of the little girl and her mother. At the same moment at which the photograph loses its historical authenticity *vis-à-vis* Rachel, it also loses its status as a photograph, as dead time. In becoming 'present,' it makes Rachel less 'real.' Deckard animates the photograph with his gaze, his desire, and it is ultimately his desire that constitutes Rachel's only subjectivity, in the present tense. In this sense Rachel, like Villiers's *L'Eve future*, becomes the perfect woman, born all at once, deprived of a past or authentic memories.

Reproduction is the guarantee of a history – both human biological reproduction (through the succession of generations) and mechanical reproduction (through the succession of memories). Knowledge is anchored to both. Something goes awry with respect to each in *Blade Runner*, for the replicants do not have mothers and their desperate invocation of the figure of the mother is symptomatic of their desire to place themselves within a history. Neither do they have fathers. In the scene in which Roy kills Tyrell he, in effect, *simulates* the Oedipal complex,[19] but gets it wrong. The father, rather than the son, is blinded. Psychoanalysis can only be invoked as a misunderstood, misplayed scenario. Similarly, the instances of mechanical reproduction which should ensure the preservation of a remembered history are delegitimized; Leon's photograph is broken down into its constituent units to become a clue in the detective's investigation, and Rachel's photograph is deprived of its photographic status. The replicants are objects of fear because they present the humans with the specter of a motherless reproduction, and *Blade Runner* is at one level about the anxiety surrounding the loss of history. Deckard keeps old photos as well, and while they may not represent his own relatives, they nevertheless act as a guarantee of temporal continuity – of a coherent history which compensates for the pure presence of the replicants. This compensatory gesture is located at the level of the film's own discourse also insofar as it reinscribes an older cinematic mode – that of *film noir* – thus ensuring its own insertion within a tradition, a cinematic continuity.

Yet, science fiction strikes one as the cinematic genre that ought to be least concerned with origins since its 'proper' obsession is with the projection of a future rather than the reconstruction of a past. Nevertheless, a great deal of its projection of that future is bound up with issues of reproduction – whether in its constant emphasis upon the robot, android, automaton, and anthropomorphically conceived computer or its insistent return to the elaboration of high-tech, sophisticated audio-visual systems. When Deckard utilizes the video analyzer in *Blade Runner*, it is a demonstration of the power of future systems of imaging. Furthermore, the Voight-Kampf empathy test designed to differentiate between the replicant and the human being is heavily dependent upon a large video image of the eye. In both *Alien* and its sequel, *Aliens*, video mechanisms ensure that those in the stationary ship can see through the eyes of the investigating astronauts/soldiers outside. Danger is signaled by a difficulty in transmission or a loss of the image. Garrett Stewart remarks on the over-abundance of viewing screens and viewing machines in science fiction in general – of 'banks of monitors, outsized video intercoms, x-ray display panels, hologram tubes, backlit photoscopes, aerial scanners, telescopic mirrors, illuminated computer consoles, overhead projectors, slide screens, radar scopes, whole curved walls of transmitted imagery, the retinal registers of unseen electronic eyes.'[20] And in his view, 'cinema becomes a synecdoche for the entire technics of an imagined society.'[21]

Since the guarantee of the real in the classical narrative cinema is generally the visible, the advanced visual devices here would seem, at least in part, to ensure the credibility of the 'hyperreal' of science fiction. And cetainly insofar as it is necessary to imagine that the inhabitants of the future will need some means of representing to themselves their world (and other worlds), these visual devices serve the purpose, as Stewart points out, of a kind of documen-tary authentication.[22] Yet, the gesture of marking the real does not exhaust their function. Technology in cinema is the object of a quite precise form of fetishism, and science fiction would logically be a privileged genre for the technophile. Christian Metz describes the way in which this fetishism of technique works to conceal a lack:

> A fetish, the cinema as a technical performance, as prowess, as an *exploit*, an exploit that underlines and denounces the lack on which the whole arrangement is based (the absence of the object, replaced by its reflec-tion), an exploit which consists at the same time of making this absence forgotten. The cinema fetishist is the person who is enchanted at what the machine is capable of, at the *theatre of shadows* as such. For the establishment of his full potency for cinematic enjoyment [*jouissance*] he must think at every moment (and above all *simultaneously*) of the force of presence the film has and of the absence on which this force is constructed. He must constantly compare the result with the means deployed (and hence pay attention to the technique), for his pleasure lodges in the gap between the two.[23]

Metz here finds it necessary to desexualize a scenario which in Freud's theory of fetishism is linked explicitly to the woman and the question of her 'lack' (more specifically to the question of whether or not the mother is phallic). Technological fetishism, through its alliance of technology with a process of concealing and revealing lack, is theoretically returned to the body of the mother. Claude Bailblé, from a somewhat different perspective, links the fascination with technology to its status as a kind of transitional object: 'For the technology plays the role of transitional object, loved with a regressive love still trying to exhaust the pain of foreclosure from the Other, endlessly trying to repair that initial separation, and as such it is very likely to be the target of displacements.'[24] In both cases, the theory understands the obsession with technology as a tension of movement toward and away from the mother.

It is not surprising, then, that the genre that highlights technological fetishism – science fiction – should be obsessed with the issues of the maternal, reproduction, representation, and history. From *L'Eve future* to *Blade Runner*, the conjunction of technology and the feminine is the object of fascination and desire but also of anxiety – a combination of affects that makes it the perfect field of play for the science fiction/horror genre. If Hadaly is the first embodiment of the cinematic woman (this time outside of the cinema) – a machine that synchronizes the image and sound of a 'real' woman, Rachel is in a sense her double in the contemporary cinema, the ideal woman who flies off with Deckard at the end of the film through a pastoral setting. Yet, Rachel can be conceived only as a figure drawn from an earlier cinematic scene – 1940s film noir – the dark and mysterious *femme fatale* with padded shoulders and 1940s hairdo, as though the reinscription of a historically dated genre could reconfirm the sense of history that is lost with technologies of representation. What is reproduced as ideal here is an earlier reproduction.

Again, according to Baudrillard: 'Reproduction ... makes something fundamental vacillate.' What it makes vacillate are the very concepts of identity, origin, and the original, as Benjamin has demonstrated so provocatively in 'The Work of Art in the Age of Mechanical Reproduction.'[25] There is always something uncanny about a photograph; in the freezing of the moment the real is lost through its doubling. The unique identity of a time and a place is rendered obsolete. This is undoubtedly why photographic reproduction is culturally coded and regulated by associating it closely with the construction of a family history, a stockpile of memories, forcing it to buttress that very notion of history that it threatens to annihilate along with the idea of the origin. In a somewhat different manner, but with crucial links to the whole problematic of the origin, technologies of reproduction work to regulate the excesses of the maternal. But in doing so these technologies also threaten to undermine what have been coded as its more positive and nostalgic aspects. For the idea of the maternal is not only terrifying – it also offers a certain amount of epistemological comfort. The mother's biological role in reproduction has been aligned with the social function of knowledge. For the mother is coded as certain, immediately knowable, while the father's role in reproduction is subject to doubt, not verifiable through the evidence of the senses (hence the necessity of the legal sanctioning of the paternal name). The mother is thus the figure who guarantees, at one level, the possibility of certitude in historical knowledge. Without her, the story of origins vacillates, narrative vacillates. It is as though the association with a body were the only way to stabilize reproduction. Hence the persistence of contradictions in these texts that manifest both a nostalgia for and a terror of the maternal function, both linking it to and divorcing it from the idea of the machine woman. Clinging to the realm of narrative, these films strive to rework the connections between the maternal, history, and representation in ways that will allow a taming of technologies of reproduction. The extent to which the affect of horror is attached to such filmic narratives, however, indicates the traumatic impact of these technologies – their potential to disrupt given symbolic systems that construct the maternal and the paternal as stable positions. It is a trauma around which the films obsessively circulate and which they simultaneously disavow.

NOTES

1. G. Claudin, *Paris* (Paris, 1867), 71–72, quoted in Wolfgang Schivelbusch, *The Railway Journey: The Industrialization of Time and Space in the 19th Century* (Berkeley: The University of California Press, 1986), 159.
2. Villiers de l'Isle-Adam, *Tomorrow's Eve*, trans. Robert Martin Adams (Urbana, Chicago, and London: University of Illinois Press, 1982), 13.
3. Ibid., 14.
4. Ibid., 125.
5. Ibid., 131.
6. Annette Michelson, 'On the Eve of the Future: The Reasonable Facsimile and the Philosophical Toy,' in *October: The First Decade, 1976–1986*, eds. Annette Michelson, et al. (Cambridge: The MIT Press, 1987), 432. See also Raymond Bellour, 'Ideal Hadaly: on Villier's *The Future Eve*,' *Camera Obscura* 15 (Fall 1986): 111–35.
7. Villiers de l'Isle-Adam, 164.
8. Michelson, 433.
9. Christian Metz, 'The Imaginary Signifier,' *Screen* 16:2 (Summer 1975), 15.
10. Andreas Huyssen, *After the Great Divide: Modernism, Mass Culture, Postmodernism* (Bloomington: Indiana University Press, 1986), 70.
11. Ibid.
12. See Barbara Creed, 'Horror and the Monstrous-Feminine – An Imaginary Abjection,' *Screen* 27:1 (January–February 1986): 44–71; and James H. Kavanagh, '"Son of a Bitch": Feminism, Humanism and Science in *Alien*,' *October* 13 (1980), 91–100.
13. Kavanagh, 94.
14. Julia Kristeva, *Powers of Horror* (New York: Columbia University Press, 1983).
15. Julia Kristeva, 'Maternité selon Giovanni Bellini,' *Polylogue* (Paris: édition du Seuil, 1977), 409; my translation.
16. Ibid.
17. Jean Baudrillard, *Simulations*, trans. Paul Foss, Paul Patton, and Philip Beitchman (New York City: Semiotext(e), 1983), 153.
18. Roland Barthes, *Camera Lucida: Reflections on Photography*, trans. Richard Howard (New York: Hill and Wang, 1981). For a remarkably similar analysis of *Blade Runner*, although differently inflected, see Giuliana Bruno, 'Ramble City: Postmodernism and *Blade Runner*,' *October* 41 (Summer 1987), 61–74. Bruno also invokes Barthes's *Camera Lucida* in her analysis of the role of photography in the film.
19. See Glenn Hendler, 'Simulation and Replication: The Question of *Blade Runner*,' honors thesis, Brown University, Spring 1984.
20. Garrett Stewart, 'The "Videology" of Science Fiction,' in *Shadows of the Magic Lamp: Fantasy and Science Fiction in Film*, eds George Slusser and Eric S. Rabkin (Carbondale and Edwardsville: Southern Illinois University Press, 1985), 161.
21. Ibid., 161.
22. Ibid., 167.

23. Metz, 72.
24. Claude Bailblé, 'Programming the Look,' *Screen Education* 32/33, 100.
25. Walter Benjamin, 'The Work of Art in the Age of Mechanical Reproduction,' *Illuminations*, trans. Harry Zohn (New York: Schocken Books, 1969), 217–52.

2

THE PLEASURE OF THE INTERFACE

Claudia Springer

Sex times technology equals the future
J. G. Ballard[1]

A discourse describing the union of humans and electronic technology currently circulates in the scientific community and in popular culture texts such as films, television, video games, magazines, cyberpunk fiction and comic books. Much of the discourse represents the possibility of human fusion with computer technology in positive terms, conceiving of a hybrid computer/ human that displays highly evolved intelligence and escapes the imperfections of the human body. And yet, while disparaging the imperfect human body, the discourse simultaneously uses language and imagery associated with the body and bodily functions to represent its vision of human/technological perfection. Computer technologies thus occupy a contradictory discursive position where they represent both escape from the physical body and fulfilment of erotic desire. To quote science fiction author J. G. Ballard again:

> I believe that organic sex, body against body, skin area against skin area, is becoming no longer possible. ... What we're getting is a whole new order of sexual fantasies, involving a different order of experiences, like car crashes, like traveling in jet aircraft, the whole overlay of new technologies, architecture, interior design, communications, transport, merchandising. These things are beginning to reach into our lives and change the interior design of our sexual fantasies.[2]

The language and imagery of technological bodies exist across a variety of diverse texts. Scientists who are currently designing ways to integrate human consciousness with computers (as opposed to creating Artificial Intelligence) describe a future in which human bodies will be obsolete, replaced by computers that retain human intelligence on software.[3] *Omni* magazine postulates a 'postbiological era'. The *Whole Earth Review* publishes a forum titled 'Is the body obsolete?' Jean-François Lyotard asks, 'Can thought go on without a body?'[4] Popular culture has appropriated the scientific project; but instead of effacing the human body, these texts intensify corporeality in their representation of cyborgs. A mostly technological system is represented as its opposite: a muscular human body with robotic parts that heighten physicality and sexuality. In other words, these contemporary texts represent a future where human bodies are on the verge of becoming obsolete but sexuality nevertheless prevails.

The contradictory discourse on cyborgs reveals a new manifestation of the simultaneous revulsion and fascination with the human body that has existed throughout the western cultural tradition. Ambivalence toward the body has traditionally been played out most explicitly in texts labelled pornographic, in which the construction of desire often depends upon an element of aversion. That which has been prohibited by censorship, for example, frequently becomes highly desirable. It was only in the nineteenth century, however, that pornography was introduced as a concept and a word, though its etymology dates back to the Greek πορνογραφος: writing about prostitutes. In his book *The Secret Museum*, Walter Kendrick argues that the signifier 'pornography' has never had a specific signified, but constitutes a shifting ideological framework that has been imposed on a variety of texts since its inception.[5] He suggests that after the years between 1966 and 1970 we entered a post-pornographic era heralded by the publication of *The Report of the Commission on Obscenity and Pornography*.[6] I would like to propose that if we are in a post-pornographic era, it is most aptly distinguished by the dispersion of sexual representation across boundaries that previously separated the organic from the technological. As Donna Haraway writes:

> Late twentieth-century machines have made thoroughly ambiguous the difference between natural and artificial, mind and body, self-developing and externally designed, and many other distinctions that used to apply to organisms and machines. Our machines are disturbingly lively, and we ourselves frighteningly inert.[7]

Sexual images of technology are by no means new: modernist texts in the early twentieth century frequently eroticized technology. As K. C. D'Alessandro argues:

> Sexual metaphor in the description of locomotives, automobiles, pistons, and turbines; machine cults and the Futurist movement, *Man With a Movie Camera*, and *Scorpio Rising* – these are some of the ways technophiliacs have expressed their passion for technology. For techno-philiacs, technology provides an erotic thrill – control over massive

power, which can itself be used to control others. ... The physical manifestations of these machines – size, heft, shape, motions that thrust, pause
and press again – represent human sexual responses on a grand scale.
There is much to venerate in the technology of the Industrial age.[8]

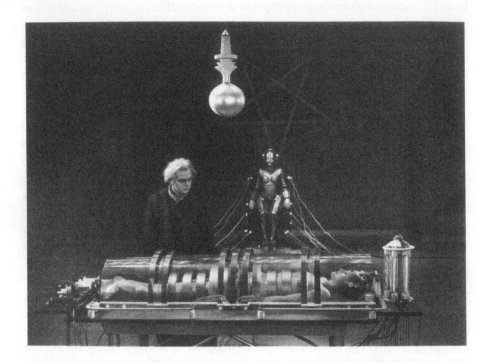

Figure 2.1 *Metropolis* (Courtesy of the Museum of Modern Art/Film Stills Archive)

The film *Metropolis* (Fritz Lang, 1926) is a classic example of the early
twentieth-century fascination with technology. It combines celebration of
technological efficiency with fear of technology's power to destroy humanity
by running out of control. This dual response is expressed by the film in sexual
terms: a robot shaped like a human woman represents technology's simultaneous allure and powerful threat. The robot is distinguished by its overt
sexuality, for it is its seductive manner that triggers a chaotic worker revolt.
Andreas Huyssen argues that modernist texts tend to equate machines with
women, displacing and projecting fears of overpowering technology onto
patriarchal fears of female sexuality.[9] Huyssen contends that historically,
technology was not always linked to female sexuality: the two became associated after the beginning of the nineteenth century just as machines came to be
perceived as threatening entities capable of vast, uncontrollable destruction. In
nineteenth-century literature, human life appears often to be vulnerable to the
massive destructive potential of machines. Earlier, in the eighteenth century,

before the Industrial Revolution installed machinery in the workplace on a grand scale, mechanization offered merely a playful diversion in the form of the mechanical figures, designed to look male as often as female, that achieved great popularity in the European cities where they were displayed.[10]

Cyborgs, however, belong to the information age, where, as D'Alessandro writes, 'huge, thrusting machines have been replaced with the circuitry maze of the microchip, the minimal curve of aerodynamic design'.[11] Indeed, machines have been replaced by systems, and the microelectronic circuitry of computers bears little resemblance to the thrusting pistons and grinding gears that characterized industrial machinery. D'Alessandro asks: 'What is sensual, erotic, or exciting about electronic tech?' She answers by suggesting that cybernetics makes possible the thrill of control over information and, for the corporate executives who own the technology, control over the consumer classes. What popular culture's cyborg imagery suggests is that electronic technology also makes possible the thrill of escape from the confines of the body and from the boundaries that have separated organic from inorganic matter.

While robots represent the acclaim and fear evoked by industrial age machines for their ability to function independently of humans, cyborgs incorporate rather than exclude humans, and in so doing erase the distinctions previously assumed to distinguish humanity from technology. Transgressed boundaries, in fact, define the cyborg, making it the consummate postmodern concept. When humans interface with computer technology in popular culture texts, the process consists of more than just adding external robotic prostheses to their bodies. It involves transforming the self into something entirely new, combining technological with human identity. Although human subjectivity is not lost in the process, it is significantly altered.

Rather than portraying human fusion with electronic technology as terrifying, popular culture frequently represents it as a pleasurable experience. The pleasure of the interface, in Lacanian terms, results from the computer's offer to lead us into a microelectronic Imaginary where our bodies are obliterated and our consciousness integrated into the matrix. The word matrix, in fact, originates in the Latin *mater* (meaning both mother and womb), and the first of its several definitions in *Websters* is 'something within which something else originates or develops'. Computers in popular culture's cyborg imagery extend to us the thrill of metaphoric escape into the comforting security of our mother's womb, which, as Freud explained, represents our earliest *Heim* (home).[12] According to Freud, when we have an *unheimlich* (uncanny) response to something, we are feeling the simultaneous attraction and dread evoked by the womb, where we experienced our earliest living moment at the same time that our insentience resembled death. It was Freud's contention that we are constituted by a death wish as well as by the pleasure principle; and popular culture's cyborg imagery effectively fuses the two desires.

Indeed, collapsing the boundary between what is human and what is technological is often represented as a sexual act in popular culture. By associating a

Figure 2.2 *Cyberpunk* (Courtesy of the Innovative Corporation, Scott Rockwell and Darryl Banks)

deathlike loss of identity with sexuality, popular culture's cyborg imagery upholds a longstanding tradition of using loss of self as a metaphor for orgasm. It is well known that love and death are inextricably linked in the western cultural tradition, as Denis de Rougemont shows in his book *Love in the Western World*.[13] The equation of death with love has been accompanied in literature by the idea of bodiless sexuality: two united souls represent the purest form of romance. De Rougemont considers the Tristan legend to be western culture's paradigmatic romantic myth, from the twelfth century into the twentieth century; and it persists in the late twentieth century in cyborg imagery that associates the human/computer interface with sexual pleasure.

Instead of losing our consciousness and experiencing bodily pleasures, cyborg imagery in popular culture invites us to experience sexuality by losing our bodies and becoming pure consciousness. One of many examples is provided by the comic book *Cyberpunk*,[14] whose protagonist, Topo, mentally enters the 'Playing Field' – a consensual hallucination where all the world's data exists in three-dimensional abstraction (called cyberspace in the cyberpunk novels of William Gibson) – saying 'it's the most beautiful thing in the human universe. If I could leave my meat behind and just live here. If I could just be pure consciousness I could be happy.' While in the Playing Field he meets Neon Rose, a plant/woman with a rose for a head and two thorny tendrils for arms (and like Topo, only present through hallucination). Even her name inscribes the collapse of boundaries between organic plant life and a technological construct. He engages her in a contest of wills, represented as their bodies entwined around each other while he narrates: 'In here, you're what you will. Time and space at our command. No limits, except how good your software is. No restraints.' Topo's spoken desire – to leave his meat behind and become pure consciousness, which is in fact what he has done – is contradicted by the imagery: his body – his meat – wrapped around another body.

The word 'meat' is widely used to refer to the human body in cyberpunk texts. Cyberpunk, a movement in science fiction dating from the early 1980s, combines an aggressive punk sensibility rooted in urban street culture with a highly technological future where distinctions between technology and human-ity have dissolved. In this context, 'meat' typically carries a negative connota-tion along with its conventional association with the penis. It is an insult to be called meat in these texts, and to be meat is to be vulnerable. And yet despite its aversion to meat, *Cyberpunk* visually depicts Topo's body after he has aban-doned it to float through the Playing Field's ever-changing topography. His body, however, only seems to be inside the Playing Field because of an illusion, and he is capable of transforming it in any way he desires. As he sees Neon Rose approach, he transforms himself into mechanical parts shaped like his own human body, but more formidable. He has lost his flesh and become steel. Only his face remains unchanged, and it is protected by a helmet. Topo's new powerful body, a product of his fantasy, inscribes the conventional signifiers of masculinity: he is angular with broad shoulders and chest; and, most

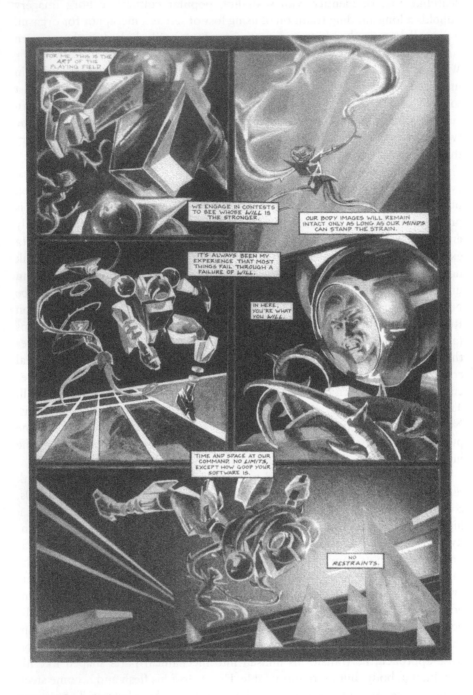

Figure 2.3 *Cyberpunk* (Courtesy of the Innovative Corporation, Scott Rockwell and Darryl Banks)

importantly, he is hard. It is no accident that he adopts this appearance in order to greet Neon Rose, who is coded in stereotypical feminine fashion as a sinewy plant who throws her tendrils like lassos to wrap them around him. In case the reader is still in doubt about Neon Rose's gender, *Cyberpunk* shows her as a human woman after Topo defeats her in their mock battle.

This example from *Cyberpunk* indicates that while popular culture texts enthusiastically explore boundary breakdowns between humans and computers, gender boundaries are treated less flexibly.

Cyberbodies, in fact, tend to appear masculine or feminine to an exaggerated degree. We find giant pumped-up pectoral muscles on the males and enormous breasts on the females; or, in the case of Neon Rose, cliched flower imagery meant to represent female consciousness adrift in the computer matrix. Cyborg imagery has not so far realized the ungendered ideal theorized by Donna Haraway.[15] Haraway praises the cyborg as a potentially liberatory concept, one that could release women from their inequality under patriarchy by making genders obsolete. When gender difference ceases to be an issue, she explains, then equality becomes possible. Janet Bergstrom points out that exaggerated genders dominate in science fiction because

> where the basic fact of identity as a human is suspect and subject to transformation into its opposite, the representation of sexual identity carries a potentially heightened significance, because it can be used as the primary marker of difference in a world otherwise beyond our norms.[16]

In heightening gender difference, popular culture's cyborg imagery has not caught up with scientist Hans Moravec, who tells us that there will be no genders in the mobile computers that will retain human mental functions on software once the human body has become obsolete: 'not unless for some theatrical reason. I expect there'll be play, which will be just another kind of simulation, and play may include costume parties.'[17] According to Lyotard, on the other hand, the most complex and transcendent thought is made possible by the force of desire, and therefore 'thinking machines will have to be nourished not just on radiation but on irremediable gender difference'.[18]

Jean Baudrillard takes a similar position when he suggests that its inability to feel pleasure makes Artificial Intelligence incapable of replicating human intelligence.[19] But Baudrillard, unlike Lyotard, does not insist that gender difference is indispensable. Instead, he sees the collapse of clear boundaries between humans and machines as part of the same postmodern move toward uncertainty that characterizes the collapse of difference between genders. Baudrillard asserts that 'science has anticipated this panic-like situation of uncertainty by making a principle of it'.[20] Indeed, uncertainty is a central characteristic of postmodernism and the essence of the cyborg. But since most cyborgs in popular culture exhibit definite gender difference, it is apparent that, despite its willingness to relinquish other previously sacrosanct categories, patriarchy continues to uphold gender difference.

Figure 2.4 *Interface* (Courtesy of James D. Hudnall and Paul Johnson)

Despite the fact that cyborg imagery in popular culture often exaggerates conventional gender difference, however, it does not always conform entirely to traditional sexual representations. Contrary to the way most sexual imagery has been designed for a male gaze and has privileged heterosexual encounters, cyborg imagery, taken as a whole, implies a wider range of sexualities. Erotic interfacing is, after all, purely mental and nonphysical; it theoretically allows a free play of imagination. Accordingly, not all cyborg imagery adheres strictly to the standardized male fantasies celebrated in *Playboy*. Nor does it simply posit the computer as female in the manner that *Metropolis* associates technology with female sexuality and represents men as vulnerable to both. Instead, computers in popular culture's cyborg imagery represent sexual release of various kinds for both genders.

In some examples, the act of interfacing with a computer matrix is acknowledged to be solitary; but it is nonetheless represented as a sexual act, a masturbatory fantasy expressed in terms of entering something, but lacking the presence of another human body or mind. In the comic book *Interface*, the interfacing experience of a woman named Linda Williams is coded as masturbation, which becomes linked to the process of thinking.[21] Williams is seen from a high angle lying on her bed on her back, saying, 'I relax my body. My mind starts to caress the frequencies around me. There. That's better. I'm one with the super-spectrum now. I'm interfaced with the world.' In the last panel, she is seen doubled, her second self rising nude from the bed with head thrown back and arms outstretched in a sexual pose.

Linda Williams's mental journey through the computer matrix in search of valuable files is drawn so as to show her nude body diving through oceans of electronic circuitry and a jumble of clichéd newspaper headlines. Although female masturbation is a staple of conventional pornography for a male spectator, Williams's interface/masturbation is drawn differently from the pornographic norm: her body is ghostly white and in constant motion as she swoops through the matrix surrounded by a watery mist. In two panels, her body is merely an indistinct blur. Its activity distinguishes her from the conventional passive female object of pornography, and her masturbation is not a prelude to heterosexual sex. Later in the evening, after she has returned from the matrix (sighing, 'coming down from the interface makes me feel dizzy') and is once again fully clothed, she rejects the sexual advances of a male character. She tells him, 'I need some time to myself right now.' When he tries to persuade her, she responds, 'Not tonight. I know you were expecting me to sleep with you, to make you want to stay. But I don't do that sort of thing. Look, I'm attracted to you. So maybe you'll get lucky sometime. Right now, I've got a lot on my mind. There's so much I have to think about.' Williams takes control over her own sexuality, which embodies the cyborgian condition as represented in popular culture by being purely cerebral and simultaneously sexual. When she says she wants to be alone because there is so much she has to think about, the reader may infer that her private thoughts will be expressed sexually, as they were when she mentally entered the computer matrix.

Imaginary sex – sex without physically touching another human – prevails in cyborg discourses, though bodily sex is not altogether absent. The emphasis on cerebral sexuality suggests that while pain is a meat thing, sex is not. Historical, economic, and cultural conditions have facilitated human isolation and the evolution of cerebral sex. Capitalism has always separated people from one another with its ideology of rugged individualism. Its primary form of sanctioned unity – the nuclear family – has traditionally decreed that one person, usually the woman, relinquish her individuality in order to support in the private realm the public endeavours of the other. Public relations under capitalism are characterized by competition and its attendant suspicions. In late capitalism, social relations are mediated not only by money, but also by the media with its simulations. Rather than communicate, we spectate. Computer technology offers greater opportunities for dialogue – through modem hookup and electronic mail, for example – than does television, and can be thought of as a way to reestablish the human contact that was lost during the television decades. It is hardly astonishing that, at a time when paranoia over human contact in response to the AIDS virus is common, human interaction should occur through computerized communication, with the participants far apart and unable to touch each other.

To say that people communicate via their computers is not to say that the act of communication has remained unchanged from the pre-computer era. The term 'communication' is in fact imprecise, according to Baudrillard. He writes that in the interface with the computer

> the Other, the sexual or cognitive interlocuter, is never really aimed at – crossing the screen evokes the crossing of the mirror. The screen itself is targeted as the point of interface. The machine (the interactive screen) transforms the process of communication, the relation from one to the other, into a process of commutation, i.e. the process of reversability from the same to the same. The secret of the interface is that the Other is within it virtually the Same – otherness being surreptitiously confiscated by the machine.[22]

Although the computer invites us to discard our identities and embrace an Imaginary unity, like a mirror it also reminds us of our presence by displaying our words back to us. What Baudrillard argues is that this intensely private experience precludes actual interaction with another person and turns all computerized communication into a kind of autocommunication which may contain elements of autoeroticism.

In an example of solitary sexual communion with technology, William Gibson, one of the founding authors of cyberpunk fiction, uses the term 'jack in' in his writing to describe the moment when a 'cowboy' sitting at a 'deck' enters his command to be mentally transported into cyberspace: he wanted to title his first novel 'Jacked In', but the publisher refused on the grounds that it sounded too much like 'Jacked Off'.[23] Gibson's trilogy – *Neuromancer, Count Zero* and *Mona Lisa Overdrive* – evokes a dystopian future where isolated individuals drift in and out of each others' lives and often escape into fantasy.[24] Not unlike television's mass-produced fantasies of today, Gibson's 'simstim' (simulated stimulation) feeds entertaining narratives directly into people's minds. Cyberspace, too, is a place of the mind, but it feels like three-dimensional space to those who enter it:

> Cyberspace. A consensual hallucination experienced daily by billions of legitimate operators, in every nation, by children being taught mathematical concepts. ... A graphic representation of data abstracted from the banks of every computer in the human system. Unthinkable complexity. Lines of light ranged in the nonspace of the mind, clusters and constellations of data. Like city lights, receding. ... [25]

Gibson's evocation of cyberspace has influenced the way people think about Virtual Reality, a concept dating back to the late 1960s which has become fashionable in the 1990s, receiving widespread media coverage while several companies develop its capabilities and design marketing strategies. Virtual Reality creates a computer-generated space that a person perceives as three-dimensional through goggles fitted with small video monitors. Gloves connected to the computer allow users to interact with the space and feel as though they are performing such activities as picking up objects, driving or flying. It would be inappropriate to call Virtual Reality an escape from reality, since what it does is provide an alternative reality where 'being' somewhere does not require physical presence and 'doing' something does not result in any changes in the physical world. Virtual Reality undermines certainty over the term reality, ultimately abandoning it altogether along with all the other certainties that have been discarded in postmodern times. John Perry Barlow, who writes about the cyberworld and is cofounder of the Electronic Frontier Foundation (an organization that tries to protect those working in electronic communications from governmental repression), calls Virtual Reality 'a Disneyland for epistemologists', declaring that it will 'further expose the conceit that "reality" is a fact ... delivering another major hit to the old fraud of objectivity'.[26]

In published descriptions of Virtual Reality there are frequent references to its erotic potential. One concept in the works is 'teledildonics', which puts the user in a bodysuit lined with tiny vibrators.[27] The user would telephone others who are similarly outfitted. Their telephone conversations would be accompanied by computerized visual representations, displayed to them on headsets, of their bodies engaged in sexual activities. As Howard Rheingold, author of the book *Virtual Reality*,[28] points out, teledildonics would revolutionize sexual encounters as well as our definitions of self:

> Clearly we are on the verge of a whole new semiotics of mating. Privacy and identity and intimacy will become tightly coupled into something we don't have a name for yet. ... What happens to the self? Where does identity lie? And with our information-machines so deeply intertwingled [sic] with our bodily sensations, as Ted Nelson might say, will our communication devices be regarded as 'its' ... or will they be part of 'us'?[29]

Confusion over the boundaries between the self and technological systems is already evident. Virtual Reality, according to some of its proponents, will be able to eliminate the interface, the 'mind-machine information barrier'.[30] According to Baudrillard, uncertainty over the boundary between humanity and technology originates in our relationship to the new technological systems, not to traditional machines:

> Am I a man, am I a machine? In the relationship between workers and traditional machines, there is no ambiguity whatsoever. The worker is always estranged from the machine, and is therefore alienated by it. He keeps his precious quality of alienated man to himself. Whilst new technology, new machines, new images, interactive screens, do not alienate me at all. With me they form an integrated circuit.[31]

Nowhere is the confusion of boundaries between humanity and electronic technology more apparent than in films involving cyborg imagery: here cyborgs are often indistinguishable from humans. The Terminator (*The Terminator* [James Cameron, 1984]), for example, can be recognized as nonhuman only by dogs, not by humans. Even when cyborgs in films look different from humans, they are often represented as fundamentally human. In *Robocop* (Paul Verhoeven, 1987), Robocop is created by fusing electronic technology and robotic prostheses with the face of a policeman, Alex J. Murphy, after he has died from multiple gunshot wounds. He clearly looks technological, while at the same time he retains a human shape. His most recognizably human feature is his face, with its flesh still intact, while the rest of his body is entirely constructed of metal and electronic circuitry. The film shows that despite his creators' attempts to fashion him into a purely mechanical tool, his humanity keeps surfacing. He seeks information about Murphy, his human precursor; and increasingly identifies with him, particularly since he retains memories of the attack that killed Murphy. At the end of the film, Robocop identifies himself, when asked for his name, as Murphy. In the sequel, *Robocop II* (Irvin Kershner, 1990), Robocop's basic humanity is further confirmed when he is continually stirred by memories of Murphy's wife and young son, and takes to watching them from the street outside their new home. Robocop's inability to act on his human desires constitutes the tragic theme of the film, which takes for granted that Robocop is basically human.

If there is a single feature that consistently separates cyborgs from humans in these films, it is the cyborg's greater capacity for violence, combined with enormous physical prowess. Instead of representing cyborgs as intellectual wizards whose bodies have withered away and been replaced by computer terminals, popular culture gives us muscular hulks distinguished by their superior fighting skills. To some extent the phenomenon of the rampaging cyborg in films suggests a residual fear of technology of the sort that found similar expression in older films like *Metropolis*. Electronic technology's incredible capabilities can certainly evoke fear and awe, which can be translated in fictional representation into massive bodies that overpower human characters.

But fear of the computer's abilities does not entirely explain why cyborgs are consistently associated with violence. Significantly, musclebound cyborgs in films are informed by a tradition of muscular comic-book superheroes; and, like the superheroes, their erotic appeal lies in the promise of power they embody. Their heightened physicality culminates not in sexual climax but in acts of violence. Violence substitutes for sexual release. Steve Neale has theorized that violence displaces male sexuality in films in response to a cultural taboo against a homoerotic gaze.[32] Certain narrative films continue to be made for a presumed male audience, and homophobia exerts a strong influence on cinematic techniques. For example, closeup shots that caress the male body on screen might encourage a homoerotic response from the male spectator. But, as Neale explains, the spectacle of a passive and desirable male body is typically undermined by the narrative, which intervenes to make him the object or the perpetrator of violence, thereby justifying the camera's objectification of his body.

In the opening sequence of *The Terminator*, for example, the shot of the cyborg's (Arnold Schwarzenegger) beautifully sculpted nude body standing on a hill above night-time Los Angeles, city lights twinkling like ornaments behind him, is quickly followed by his bloody attack on three punk youths in order to steal their clothes. His attire then consists of hard leather and metal studs, concealing his flesh and giving his sexuality a veneer of violence. As in similar examples from other films, an invitation to the spectator to admire the beauty of a male body is followed by the body's participation in violence. The male body is restored to action to deny its status as passive object of desire, and the camera's scrutiny of the body receives narrative justification.

Klaus Theweleit, in his two-volume study of fascist soldier males (specifically, men of the German *Freikorps* between the world wars), writes that their psychological state indicates an intense misogyny and an overwhelming desire to maintain a sense of self in the face of anything they perceive might threaten their bodily boundaries.[33] Theweleit draws on the theories of psychologist Margaret Mahler to argue that fascist males have never developed an identity (they are 'not-yet-fully-born'), and thus invest all of their energies into maintaining a fragile edifice of selfhood. Their failure to disengage from their mothers during infancy results in a fear that women will dissolve their identities;

hence the frequency with which women are associated in fascist rhetoric with raging floods that threaten to engulf their victims. In order to protect themselves from women, onto whom they project the watery weakness they despise in themselves, fascist males encase themselves in body armour, both literally and figuratively. The machine body becomes the ideal tool for ego maintenance.

For the fascist male, additionally, the sexual act evokes loss of self and becomes displaced onto violence. The act of killing, especially by beating the victim into a bloody pulp, functions to externalize the dissolution of self that he fears, and assures him of his relative solidity. He reaffirms his physical and psychological coherence every time he kills. Acts of violence also serve to release some of his enormous tension, for the task of maintaining a sense of self when a self barely exists is excruciating, and the soldier male does not allow himself to experience release through sexual union. As Theweleit writes, 'heroic acts of killing take the place of the sexual act', and the ecstasy of killing substitutes for sexual climax.[34]

Cyborg imagery in films is remarkably consistent with Theweleit's description of the fascist soldier male. If anything, cyborg imagery epitomizes the fascist ideal of an invincible armoured fighting machine. In *Robocop*, Robocop's armour is external and protects him from gunshots and other assaults that would kill a human. He strides fearlessly into a blaze of gunfire as bullets bounce off his armoured body. In *The Terminator*, the cyborg's armour is inside his body and therefore not visible, but it makes him virtually indestructible. Near the end of the film, after the Terminator's flesh has been burnt away, he is revealed to be a metal construct that, despite the loss of all its flesh, continues methodically to stalk its victim.

Cyborg imagery, therefore, represents more than just a recognition that humanity has already become integrated with technology to the point of indistinguishability; it also reveals an intense crisis in the construction of masculinity. Shoring up the masculine subject against the onslaught of a femininity feared by patriarchy now involves transforming the male body into something only minimally human. Whereas traditional constructions of masculinity in film often relied on external technological props (guns, armoured costumes, motorcycles, fast cars, cameras, and so on)[35] to defend against disintegration, the cinematic cyborg heralds the fusion of the body with the technological prop.

Ironically, the attempt to preserve the masculine subject as a cyborg requires destroying the coherence of the male body and replacing it with electronic parts; either physically – using hardware, or psychologically – using software. The construction of masculinity as cyborg requires its simultaneous deconstruction. And yet, by escaping from its close identification with the male body, masculine subjectivity has been reconstituted, suggesting that there is an essential masculinity that transcends bodily presence. In a world without human bodies, the films tell us, technological things will be gendered and there will still be a patriarchal hierarchy. What this reconfiguration of masculinity indicates is that patriarchy is more willing to dispense with human life than with male superiority.

However, the sacrifice of the male body is disguised in cyborg films by emphasizing physicality and intensifying gender difference. Pumping up the cyborg into an exaggerated version of the muscular male physique hides the fact that electronic technology has no gender. In *Total Recall* (Paul Verhoeven, 1990), for example, the fact that Doug Quaid's identity is merely an electronic implant is counteracted by his massive physical presence, once again made possible by casting Arnold Schwarzenegger in the role of Quaid. Muscular cyborgs in films thus assert and simultaneously disguise the dispersion of masculine subjectivity beyond the male body.

The paradox that preserving masculine subjectivity in the figure of the cyborg requires destroying the male body accounts in part for the extreme violence associated with cyborgs in films: they represent an impossible desire for strength through disintegration; and, like the fascist soldier males, their frustration finds expression in killing. The Terminator, for example, is programmed to kill and in fact has no other function than to kill humans. He has been sent into the past by his machine masters expressly to kill a young woman, Sarah Connor. His adversary Kyle Reese tells Connor that the Terminator 'can't be bargained with, it can't be reasoned with, it doesn't feel pity or remorse or fear and it absolutely will not stop, ever, until you are dead', recalling Theweleit's observation that the fascist soldier male has no moral qualms about killing.

Robocop is also an expert killer, but the two *Robocop* films, unlike *The Terminator*, justify the hero's acts of killing by putting him on the side of law enforcement and showing his victims caught in the act of committing crimes. In *Robocop II*, Robocop is programmed to apprehend criminals without killing them by a smarmy woman psychologist who preaches nonviolence and is made to appear ridiculous. The film indicates, however, that the software program that prevents Robocop from killing hinders his effectiveness; and the film celebrates his acts of killing when he manages to overcome the restraining program. In *Total Recall*, Doug Quaid is attacked nearly every time he turns a corner, and he responds by killing all of his attackers with a show of incredible strength and brutality.

Not only does cyborg imagery in films extol the human killing machine, it also expresses the concomitant fear of sexuality theorized by Theweleit. In the film *Hardware* (Richard Stanley, 1990), for example, the cyborg is dormant until activated by the sight of a young woman, Jill, having sex with her boyfriend. After the boyfriend has left the apartment and Jill has hung the cyborg on the wall as part of a scrap metal sculpture, the cyborg watches her sleeping body for a while and then emerges to attack her; for, like the Terminator, it has been created to destroy humans.

Sexuality is feared by fascist soldier males not only because it signifies loss of personal boundaries, writes Theweleit, but also because sexuality evokes the creation of life, and the soldier male is bent on destroying all signs of life before they can destroy him. Pregnant women, according to Theweleit, are treated with revulsion in his rhetoric. Like fascist soldier males, cyborgs in films are often determined to prevent birth. In *Hardware*, it turns out that the cyborg that kills all the life forms it encounters is a secret weapon in the government's birth control programme. The Terminator, likewise, has been programmed to travel back through time to kill Sarah Connor in order to prevent her giving birth to her son John, who, forty years into the future, will lead the few humans who have survived a nuclear war in defeating the machines that threaten humanity with annihilation.

Creation versus destruction of life is not only a central thematic concern but also a site of dispute in cyborg texts. The ability to engender life is divided between men and women and between humans and technology. Women are typically associated with biological reproduction while men are involved in technological reproduction. In the film *Demon Seed* (Donald Cammell, 1977), for example, a scientist creates an Artificial Intelligence in a sophisticated computer laboratory where teams of specialists educate their artificial child. The scientist's wife, Susan (Julie Christie), is a psychiatrist, a member of a humanistic profession that opposes her husband's technophilia. She complains about his emotional coldness, illustrating the film's stereotypically phallo-centric definition of gender roles: men are scientific and aloof while women are humanistic and emotional.

Demon Seed reinforces its version of gender difference by taking for granted that the AI, a form of pure consciousness, is male. Masculine subjectivity has dispensed entirely with the need to construct a body in this film, existing instead as bodiless intellect. And the woman's role is even further confined when Susan is raped by the AI, whose pure intellect is the antithesis of Susan's reduction to a reproductive vessel. Since the Artifical Intelligence has no physical form (its name is Proteus IV, after the Greek sea god capable of assuming different forms), it relies on a robot and a giant mutating geometric shape under its command to rape Susan. Its orgasm while impregnating her is represented as a trip into the far reaches of the cosmos. ('I'll show you things I have seen', it tells her.) Motivated by a desire to produce a child and thereby experience emotions and physical sensations, the AI attempts to take control over the reproductive process; in effect vying with Susan's husband for power over creation, but going back to a biological definition of reproduction and a phallocentric definition of woman as childbearer. Susan is a mere womb in the AI's scheme. When the film ends with the birth of the child conceived by the AI and Susan, it leaves ambiguous whether the cyborg child, a union of a disembodied intellect and a human woman, will be demonic or benign.

Men are also the creators of life in *Weird Science* (John Hughes, 1985), a throwback to *Metropolis* with its representation of a woman artificially designed to fulfil a male fantasy. Two unpopular high school boys program a computer to create their perfect woman, assembled from fragmented body parts selected from *Playboy* magazines. Her role, like the robot's in *Metropolis*, at first appears to be sexual: the boys' initial desire is to take a shower with her. Also as in *Metropolis*, the woman's sexuality is too powerful for the boys who are incapable of doing more than just kissing her. However, unlike *Metropolis*, she takes on a big sisterly role that involves instructing her creators in the finer points of talking to girls. Her guidance boosts their self-confidence and allows them to win over the two most popular high school girls, whom earlier they could only admire from afar. The film uses the concept of computer-generated life only to further its conventional coming-of-age narrative, and does nothing to question either gender roles or the implications of nonbiological reproduction.

Eve of Destruction (Duncan Gibbins, 1991) complicates the theme of creation versus destruction, but only to punish the woman protagonist for her sexuality and for engaging in technological rather than biological reproduction. A scientist named Eve creates a cyborg, also named Eve, who looks exactly like her and is programmed with her memories. The cyborg escapes from the scientist and goes on a killing spree. Rather than engaging in random destruction, however, the cyborg Eve lives out the scientist Eve's repressed fantasies of sex and revenge against men. Thus the cyborg kills the scientist's father, whom the scientist has hated since childhood because he brutalized and caused the death of her mother. The cyborg's first victim is a redneck at a country saloon that the scientist had fantasized frequenting for casual sex. The cyborg takes the man to a motel room and, when he taunts her with his erection, bites his penis.

The film's castration anxiety escalates, for it turns out that the cyborg has something much more dangerous than a vagina dentata: a nuclear vagina. We learn that the Defense Department funded the cyborg project to create a secret military weapon, complete with nuclear capabilities. In a computer graphics display of the cyborg's design, we see that the nuclear explosive is located at the end of a tunnel inside her vagina. Sure enough, the countdown to a nuclear explosion begins when the cyborg has an orgasm as she destroys another man by crashing her car into his. Patriarchal fear of female sexuality has clearly raised the stakes since the 1920s when *Metropolis* showed unleashed female sexuality leading to the collapse of a city. *Eve of Destruction* puts the entire planet at risk.

Having established that female sexuality leads to uncontrollable destruction, the film suggests that what the scientist placed in danger by creating artificial life was her role as biological mother: for the cyborg kidnaps the scientist's young son. Only then does the scientist cooperate with the military officer whose job it is to destroy the cyborg before it detonates. Earlier, they had an antagonistic relationship revolving around contempt for each other's profession. His attempts to destroy her cool professional demeanour finally succeed, and at the end it is she who destroys the cyborg only seconds before zero hour in order to save the lives of her son and the military officer. The scientist in effect destroys her repressed sexuality and anger towards men, and accepts her primary status as biological mother.

As *Eve of Destruction* illustrates, artificial life in films continues, in the Frankenstein tradition, to threaten the lives of its creators; but it also continues to hold out the promise of immortality. A yearning for immortality runs throughout cyborg discourses. In cyberpunk fiction, taking the postmodern principle of uncertainty to its radical extreme, not even death is a certainty. Cyberpunk fiction writers William Gibson and Rudy Rucker[36] have made immortality a central theme in their books, raising questions about whether nonphysical existence constitutes life and, especially in Gibson's novels, examining how capitalism would allow only the extremely wealthy class to attain immortality by using technology inaccessible to the lower classes. But cyberpunk fiction is not without recognition of the paradoxes and dangers of immortality. In both Gibson's and Rucker's work, characters who attempt to become immortal are usually surrounded by a tragic aura of loneliness and decay.

Even Topo, in the comic book *Cyberpunk*, rejects the idea of leaving his meat behind and remaining permanently in the Playing Field when he is offered the opportunity.[37] What he rejects is immortality. But the comic book reveals that the loss of his human body would be tantamount to death; for the invitation to join those who have permanently abandoned their bodies comes from a death mask, called The Head, that addresses him from atop a pedestal. During their conversation, disembodied skulls swoop by around them, reinforcing the death imagery. When, in the next issue, Topo loses his human body and becomes a cyberghost trapped in the Playing Field, the line between life and death becomes more ambiguous.[38] There is much speculation among his friends, who remain outside of the computer matrix, about whether Topo is dead or alive. Topo himself says 'after all, I'm only a data construct myself, now. Nothing equivocal about it. We live. We are forms of life, based on electrical impulses. Instead of carbon or other physical matter. We are the next step.'

These examples show that cyborg imagery revolves around the opposition between creation and destruction of life, expressing ambivalence about the future of human existence and also, as with the fascist soldier males, uncertainty about the stability of masculine subjectivity. Fusion with electronic technology thus represents a paradoxical desire to preserve human life by destroying it. The concept of abandoning the body with pleasure arises in part from late twentieth-century post-nuclear threats to the body: nuclear annihilation, AIDS, and environmental disasters. Devising plans to preserve human consciousness outside of the body indicates a desire to redefine the self in an age when human bodies are vulnerable in unprecedented ways. Contemporary concern with the integrity of the body is only the latest manifestation of post-war anxiety over the body's fragility.

Neither alive nor dead, the cyborg in popular culture is constituted by paradoxes: its contradictions are its essence, and its vision of a discordant future is in fact a projection of our own conflictual present. What is really being debated in the discourses surrounding a cyborg future are contemporary disputes concerning gender and sexuality, with the future providing a clean slate, or a blank screen, onto which we can project our fascination and fears. While some texts cling to traditional gender roles and circumscribed sexual relations, others experiment with alternatives. It is perhaps ironic, though, that a debate over gender and sexuality finds expression in the context of the cyborg, an entity that makes sexuality, gender, even humankind itself, anachronistic. Foucault's statement that 'man is an invention of recent date. And one perhaps nearing its end' prefigures the consequences of a cyborg future.[39] But, as Foucault also argues, it is precisely during a time of discursive crisis, when categories previously taken for granted become subject to dispute, that new concepts emerge. Late twentieth-century debates over sexuality and gender roles have thus contributed to producing the concept of the cyborg. And, depending on one's stake in the outcome, one can look to the cyborg to provide either liberation or annihilation.

NOTES

1. J. G. Ballard interviewed by Peter Linnett, *Corridor*, no. 5 (1974): reprinted in *Re/Search*, nos 8–9 (San Francisco: Re/Search Publications, 1984), p. 164.
2. J. G. Ballard interviewed by Lynn Barber, *Penthouse*, September 1970: reprinted in *Re/Search*, nos 8–9, p. 157.
3. Hans Moravec, director of the Mobile Robot Laboratory at Carnegie-Mellon University, is at the forefront of cyborg development: see his book *Mind Children: The Future of Robot and Human Intelligence* (Cambridge: Harvard University Press, 1988). For a journalistic account of Moravec's research projects and his concept of a cyborg future, see Grant Fjermedal, *The Tomorrow Makers: A Brave New World of Living-Brain Machines* (New York: Macmillan, 1986).
4. 'Interview with Hans Moravec', *omni*, vol. 11, no. 11 (1989), p. 88; 'Is the body obsolete? A forum', *Whole Earth Review*, no. 63 (1989), pp. 34–55; Jean-François Lyotard. 'Can thought go on without a body?'. *Discourse*, vol. 11, no. 1 (1988–9), pp. 74–87.
5. Walter Kendrick, *The Secret Museum: Pornography in Modern Culture* (New York: Penguin Books, 1987), p. 31.
6. Commission on Obscenity and Pornography, *Report of the Commission on Obscenity and Pornography* (Toronto, New York and London: Bantam, 1970). The Commission was established by the US Congress in October 1967 to define 'pornography' and 'obscenity', to determine the nature and extent of their distribution, to study their effects upon the public, and to recommend legislative action. The Commission's report did not define obscenity and pornography and stated that no empirical evidence existed to link criminal behaviour to sexually explicit material. It also recommended against any legislative action and proposed that all existing prohibitive laws be repealed. Soon after its release, the report was rejected by the US Senate and by President Nixon.
7. Donna Haraway. 'A manifesto for cyborgs: science, technology, and socialist feminism in the 1980s', *Socialist Review*, no. 80 (1985), pp. 65–107; reprinted in Elizabeth Weed (ed.), *Coming to Terms: Feminism, Theory, and Practice* (New York: Routledge, 1989), pp. 173–204: the quotation is from p. 176.
8. K. C. D'Alessandro. 'Technophilia: cyberpunk and cinema', a paper presented at the Society for Cinema Studies conference. Bozeman, Montana (July 1988), p. 1.
9. Andreas Huyssen, 'The vamp and the machine: technology and sexuality in Fritz Lang's *Metropolis*', *New German Critique*, nos 24–5 (1981–2), pp. 221–37.
10. Ibid., p. 226.

11. D'Alessandro, 'Technophilia: cyberpunk and cinema', p. 1.
12. Sigmund Freud, 'The "Uncanny"' (1919), *The Standard Edition of the Complete Psychological Works of Sigmund Freud*, vol. 17, trans. and ed. James Strachey (London: The Hogarth Press, 1973), pp. 219–52.
13. Denis de Rougemont, *Love in the Western World* (New York: Harpel and Row, 1956.)
14. Scott Rockwell, *Cyberpunk*, book one, vol. 1, no. 1 (Wheeling, West Virginia: Innovative Corporation, 1989).
15. Haraway, 'A manifesto for cyborgs'.
16. Janet Bergstrom, 'Androids and androgyny', *Camera Obscura*, no. 15 (1986), p. 39.
17. 'Interview with Hans Moravec', p. 88.
18. Lyotard, 'Can thought go on without a body?', p. 86.
19. Jean Baudrillard, *Xerox and Infinity*, trans. Agitac (Paris: Touchepas, 1988), p. 3.
20. Ibid., p. 16.
21. James D. Hudnall, *Interface*, vol. 1, no. 1 (New York: Epic Comics, 1989).
22. Baudrillard, *Xerox and Infinity*, pp. 5–6.
23. William Gibson and Timothy Leary in conversation, 'High tech/high life', *Mondo 2000*, no. 1 (Berkeley, CA: Fun City Megamedia, 1989), p. 61.
24. William Gibson, *Neuromancer* (New York: Ace Books, 1984); *Count Zero* (New York: Ace Books, 1986); *Mona Lisa Overdrive* (New York: Bantam Books, 1988).
25. Gibson, *Neuromancer*, p. 51.
26. John Perry Barlow, 'Being in nothingness', *Mondo 2000*, no. 2 (Berkeley, CA: Fun City Megamedia, 1990), pp. 39, 41.
27. Howard Rheingold, 'Teledildonics: reach out and touch someone', *Mondo 2000*, no. 2 (Berkeley, CA: Fun City Megamedia, 1990), pp. 52–4.
28. Howard Rheingold, *Virtual Reality* (New York: Simon and Schuster, 1991).
29. Rheingold, 'Teledildonics', p. 54.
30. Barlow, 'Being in nothingness', p. 38.
31. Baudrillard, *Xerox and Infinity*, p. 14.
32. Steve Neale, 'Masculinity as spectacle: reflections on men and mainstream cinema', *Screen*, vol. 24, no. 6 (1983) pp. 2–16.
33. Klaus Theweleit, *Male Fantasies*, vol. 1 (Minneapolis: University of Minnesota Press, 1987); *Male Fantasies*, vol. 2 (Minneapolis: University of Minnesota Press, 1989).
34. Theweleit, *Male Fantasies*, vol. 2, pp. 276, 279.
35. For a discussion of how constructions of the masculine subject have relied on technological props, see Sabrina Barton, 'The apparatus of masculinity in *Shane and Sex, Lies, and Videotape*', a paper presented at the Society for Cinema Studies Conference, Washington DC (May 1990); Constance Penley, 'Feminism, film theory, and the bachelor machines', *The Future of an Illusion: Film, Feminism, and Psychoanalysis* (Minneapolis: University of Minnesota Press, 1989), chapter 4. In the context of the war film, see Susan Jeffords, *Remasculinization of America: Gender and the Vietnam War* (Bloomington: Indiana University Press, 1989).
36. Rudy Rucker, *Software* (New York: Avon Books, 1982); *Hardware* (New York: Avon Books, 1988).
37. Scott Rockwell, *Cyberpunk*, book two. vol. 1, no. 1 (Wheeling, West Virginia: Innovative Corporation, 1990).
38. Scott Rockwell, *Cyberpunk*, book two. vol. 1, no. 2 (Wheeling, West Virginia: Innovative Corporation, 1990).
39. Michel Foucault, *The Order of Things: An Archaeology of the Human Sciences* (New York: Vintage Books, 1973), p. 387.

3

VIRTUAL CORPOREALITY: A FEMINIST VIEW

Zoë Sofia

Once associated primarily with war, space, mathematics, and electronic engineering, computer technology has in recent years found many applications in the verbal, aural and visual arts. Programmes for computer-mediated drawing, painting, and designing are readily available to many people in technologically-rich cultures. Not so readily accessible but currently a 'sexy topic' for discussion amongst some visual arts practitioners and cultural critics is the equipment for generating virtual reality ('VR' to its fans). In the typical VR set-up, the user wears a special body suit or a data glove containing fibre-optic sensors that feed information about the wearer's movements into a computer, where they are transformed into analogous movements of a virtual body, a computerised representation within a simulated three-dimensional environment (sometimes called 'cyberspace').[1] The virtual reality thus generated is portrayed on small video monitors within a cyberspace helmet or 'videophone' goggles worn by the user, devices which intensify the illusion of inhabiting the virtual world. In some set-ups, sound effects, force-feedback sensors and other devices can produce pressure, heat, etc. on the body to simulate the effects of events within the virtual environment.

Like many other simulations technologies, VR was first developed by the military (especially in the United States), whose administrators long ago realised it was cheaper to let trainee pilots crash virtual planes in flight simulations than to set them loose with the actual hardware. That virtual reality – and indeed, computer technology in general – has its home in the military is not in itself a sufficient reason for rejecting it out of hand. For as Donna Haraway has argued, in these postmodern times of contingent affinities, where perverse or impure alliances can be forged across broken boundaries, origin is no sure predictor of future.[2] Along with the cyborg, cyberspace might be perverted from its (already non-natural) origins and turned to more biophilic purposes. Virtual reality, as the Australian video artist Jill Scott suggested, may even provide a kind of 'utopian space' for feminists.[3]

Yet there is something disconcerting about the spectre of virtual reality, something suspicious about the quality of enthusiasm surrounding it that gives me pause, makes me want to resist its embrace. Since I am 'one of Those', as the computer animation artist Peter Callas called me (i.e. one who hasn't donned the data glove), I cannot write as an experienced or expert user of VR technology. But as a feminist thinker interested in the erotics and semiotics of technology, who has recently researched questions of gender and computing, I offer the following critical perspective, which applies ideas from psychoanalysis and the philosophy of technology to clarify the different kinds of relations humans and human bodies may take up with computers, including the phenomena of virtual reality and 'virtual corporeality'.[4]

Is the phrase virtual corporeality an oxymoron? 'Yes' says the part of me invested in defining corporeality as precisely other than virtual embodiment. 'Well ...' says the part who recognises that embodiment includes virtual phenomena – Imaginative projections, the phantasmic bodies of dreams. 'No' says the part whom fashion obliges to acknowledge that corporeality is mediated by signs, and that overdrawn distinctions between authentically present and virtual bodies are erroneous and philosophically outmoded.

This is the kind of messiness – a confusion of positions and tendencies – which inspires the naming of certain intellectual topics as 'sexy'. Virtual corporeality is one such, along with questions of women and technology (where one of the sexy/messy tasks is to critique technology while enfranchising women through it).[5] Wherever a topic seems sexy, I suggest we are dealing with something seductive that has to do with *bodies or pleasure*, and/or with something 'naughty', taboo, or *transgressive*, and/or with something that tempts us to *betray or abandon our former commitments*. I will discuss these seductions in turn, beginning with the third.

ABANDON COMMITMENTS

The deconstruction of notions of authenticity and origin, the postmodern aesthetics of dispersal, fragmentation, montage, simulation, etc., and the supposed collapse of critical distance have overshadowed – though for me, not made redundant – earlier leftist and Freudo-Marxist commitments to the critique of technological rationality. These would have held virtual reality and corporeality suspect as products of an apparatus of domination which denied the full range of bodily life. Nowadays, we are expected to enjoy machinic pleasures. Norman O. Brown reminds us of an holistic standpoint from which computer-mediated knowledge and reality might be criticised:

> What is being probed, and found to be in some sense morbid, is not knowledge as such, but the unconscious schemata governing the pursuit of knowledge in modern civilization – specifically the aim of possession or mastery over objects (Freud), and the principle of economizing in the means (Ferenczi) ... possessive mastery over nature and rigorously economical thinking are partial impulses in the human being (the human body) which in modern civilization have become tyrant organizers of the whole of human life; abstraction from the reality of the whole body and substitution of the abstracted impulse for the whole reality are inherent in *Homo economicus*.[6]

Here synecdoche – or the tyranny of the part over the whole – is a symptom of domination. The pleasure of synecdoche is that it lets you master the part while ignoring the complexities of the whole. Synecdoche is what allows the dis-embodied, alienated, objective rationality of a certain gender, class, ethnicity, and historical epoch to be vaunted as universal, while other styles and components of rationality – such as embodiment, situatedness, emotion – are ignored or dismissed as non-rational.

Heidegger criticises instrumental rationality as a tyrannical synecdoche where technological causality is reduced from the complex mutual indebted-ness of the four Aristotelian causes (material, formal, efficient, final) to a part of the efficient cause: predictive calculation, or the mode of seeking knowledge by a 'reporting challenged forth' from a nature that can be cut up and mathematically processed.[7] As a machine for the most speedy and efficient forms of calculation and projection, a reporting technology whose very sustenance is digitalised bits of data, the computer exemplifies this narrow but powerful way of knowing. The recent proliferation of computing not only amplified pre-existing techniques for rendering the world as information, but colonised new realms – including childhood education, art and music – for informatising technique.[8]

Nowhere is the seduction of synecdoche so powerful as in the promise of total control and complete mobility afforded by computer microworlds or virtual realities. Once the privilege of military and corporate commanders, these illusions are now home delivered by personal computers (though the most 'advanced' forms are still in defence-funded labs like MIT). Paul Edwards draws attention to a 'peculiar relation of suitability' between the 'hard mastery' of computer science and contemporary militarism, for the military institution 'already has the character of a microworld' where every move and chain of command is already regimented and clearly defined in advance from the top down.[9] In the microworld we find an 'abstraction from the reality of the whole body', and the full range of senses (including olfactory and tactile) is sub-ordinated to visual, and to a lesser extent aural, perception.

NAUGHTY/TRANSGRESSIVE

Freudian theory considers the exemplary fetish object the mother's phallus,[10] which suggests 'sexiness' has something to do with an unspeakable excess – for example, a cryptic bisexuality that transgresses the official categories of kinship, gender and anatomy. In the psychoanalytic anthropology of the Hungarian Gèza Róheim, every technology is an object of this general type – not necessarily bisexual (though I could argue this), but certainly *transitional*.[11] Róheim drew on Kleinian theory to interpret the tool as a transitional object, a part of the self projected onto the world, a part of the world introjected for human designs (whether practical or poetic). Tools exist mid-way between self and other, guarding against the danger of complete object loss, for failing all else, humans would still 'have these children of their minds to love'.[12]

Note that from this psychoanalytic perspective, there are no grounds for debate on whether computer-mediated productions are truly art or merely technology: as a transitional object, the tool is inherently poetic.

All technologies may be transitional, but the computer is arguably more transitional than most. It functions as a projection of certain parts of the mind (language and formal logic but not intuition or dreaming), producing the uncanny effect of the computer as a second self.[13] Between self and other, subject and object, it permits quasi-tactile manipulation of computational objects that exist 'on the boundary between the physical and the abstract'.[14] Computer writing erodes distinctions between original and copy, the text and the writer, and can proliferate differences between writers and personae (as in computer networks), upsetting conventional understandings of authorship.[15] These transitionalities have excited some cultural critics, who find in them evidence that computers are machines for producing postmodern forms of subjectivity.[16]

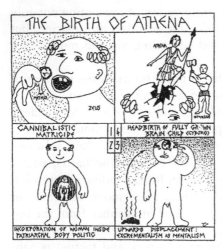

Figure 3.1 *The Birth of Athena*

Ambiguous gender is another transitional aspect of computers, whose feminine associations ('mother boards', 'consoles', 'user-friendliness') contrast with the masculinity of heavy machinery in the industrial age.[17] The computer figures centrally in the mythic iconography I call Jupiter Space.[18] According to the ancient myth, which has retained currency in allusions to mental 'conceptions', or successive 'generations' of 'brainchildren' projected from Man's 'fertile' mind, etc., the Greek god Zeus (Roman Jupiter) ate Methis, the pregnant goddess of wisdom, and later the fully-formed, fully armed Athena sprang from his head (see Figure 3.1).

Klein's theory of masculine epistemophilic desire (curiosity, or the desire to know) helps us interpret this myth. Infantile researches into origins and sexual difference are directed at the maternal body, whose objects and organs are appropriated in fantasy by a greedy and curious infant.[19] Fear over damaging the mother's body prompts the impulses to make reparation, for example by fashioning new whole bodies from appropriated resources to produce art-works, tools, ritual objects. Masculine envy of, and identification with, maternal fertility fuels the sublimation of infantile epistemophilia into cultural quests to discover or create new bodies of knowledge, land, resources, or technique as brainchildren and substitutes for the mother's body.[20] In the Athena myth, female fertility is cannibalised and upwardly displaced onto intellectual production; the (masculine) brain is equated with the (maternal) womb, and produces a new body from the cannibalised mother.

'Jupiter Space' is a term taken from *2001: A Space Odyssey*, where it was used to refer simultaneously to the womby red brain-womb of the computer HAL (called Athena in an early version of the screenplay), and to the outer space near the planet Jupiter, from whence the astronaut is later reborn as an omnipotent extraterrestrial foetus.

Jupiter Space iconography reinforces parallels between the fertile spaces of the masculinist/rational brain, outer space, cyberspace, as well as electronic circuitry and urban streets. Typically, a luminous grid of lines passes into infinity, a visual pun on the concept of 'matrix' as womb and as mathematical/geometrical grid, signifying the fertility of the techno-scientific intellect and corporate production. In Jupiter Space, not only technological artefacts and information circuits, as well as fembots – electro-mechanical women – but also the Earth and even the universe itself are depicted as conceptions of a masculine, rational and (increasingly) artificial brain (See Figure 3.2).

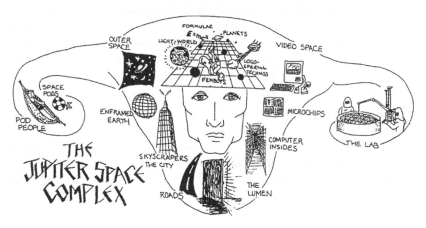

Figure 3.2 *The Jupiter Space Complex*

So in response to the question 'what place does the female body have in cyber-space?', I would say 'almost none'. Femininity and maternity are present, but displaced onto masculine and corporate technological fertility.[21] Instead of a human female mother, we find the matrix and 'think tanks' of Jupiter Space, and an occasional obvious brain-womb like HAL or the womby computer Mother in *Alien*. Instead of a female-identified woman, we find an Athenoid daddy's girl, or an emotionally remote and machine-like woman; and we regularly find a fembot, like the false Maria in *Metropolis* who is commissioned as a sexy tool of the male-dominated state. A fembot or a phallic woman can become the mascot of a city or male body corporate, but the spectre of real women forming corporate bodies ourselves still arouses masculine paranoia.[22]

Dominant culture depicts women as the signs or objects but not usually the possessors or subjects of knowledges. Here women and computers are structurally equivalent: friendly to users, not users themselves. In computer hacking, cyberspace can be imagined as a maternal or feminine body to be penetrated, cut up and manipulated in quests to appropriate and control resources. These mythic associations present unconscious barriers to women users of computers, whose identification with feminine bodies makes them/us prone to anxiety about hacking up the matrix, and more keen to make reparations if we do.

On the other hand, the prospect exists for adopting more dialogical and negotiated styles of interacting with computers and other 'material semiotic actors' in the world.[23] Hence one possible source of fascination with artificial intelligence and other technobodies for feminists, women science fiction writers and techno-artists: if these artificial second selves can be loved and accepted as powerful, resistant, speaking subjects, so too might women, long acclaimed as monstrous to conventional categories of self and other.

One mark of non-classical subjectivity is the permeability of body boundaries. In the cyberpunk vision, organs are replaced, and various technologies are implanted into bodies and perceptual organs. Male critics of William Gibson's writings typically hail them as original efforts that rescued science fiction from what one has described as a 'moribund patch' of the late 1970s and early 1980s.[24] This ignores the women writers whose works form important precedents for Gibson's cyberpunks: the multiply disabled heroine who finds a wonderfully powerful body once her brain is installed as the operative core of *The Ship Who Sang* (Anne McCaffrey); the pilots and scientists who undergo radical surgery and receive bionic implants (including mini-computer screen messages in their eyes) in Vonda McIntyre's *Superluminal*; Omali, a black woman who hacks her way into and inhabits the informational core of a giant extraterrestrial entity in *Up the Walls of the World* by James Tiptree Jnr (Alice Sheldon) – all this in a genre founded novel by a woman whose novel of scientific discovery and creation is replete with metaphors of pregnancy and labour.[25]

Why would women writers develop such visions of technological embodiment and incorporation? At the level of species life, women's bodies are already 'transitional objects': they leak more than men's, and in pregnancy and lactation have never been accounted for by classical (e.g. Platonic) distinctions of self and other. Since the onset of obstetrics and gynaecology, women's bodies have been increasingly subject to technological penetration and manipulation. Cyberpunk is one of the places where the femininity latent in the masculine dream of reason becomes manifest: male bodies now become vulnerable to the technological penetration, leaky boundaries, and pregnancy-like states that have long been part of women's history and ontological condition. In the passage from philosophical fantasy to technological (hyper)reality, the repressed returns: the matricidal cannibal becomes what he ate, and his technologies articulate ever more clearly the life and value attached to the body he once disposed of.[26]

THE PLEASURES OF COMPUTING

Some of the main pleasures of computing include those of mastery and control, afforded by the speedy command of resources and actions. Children who cannot draw straight lines can produce perfect shapes on the computer. Virtual realities also offer the illusion of mastery and mobility within the computer's micro-universe.

Another axis of pleasure is reparative. The computer itself is a technological brainchild, a body fashioned from appropriated resources which affords users the narcissistic satisfaction of relating to a technological second self (especially pleasurable for those who equate humanity with disembodied brains).[27] Hacking through the matrix can result in the reparative satisfaction of creating or stealing a new programme, a treasure to share around. Despite pedagogical emphases on sadistically tinged and top-down styles of hard mastery in computer programming, some students – especially girls, and boys with interests in poetry, art or music – adopt more loving and reparative approaches, which Turkle and Papert call 'bricoleur' or 'soft mastery' styles.[28] Systems are created through tinkering and the assemblage of intimately known component parts, rather than by 'black-boxing' (i.e. making top-down plans and knocking the sub-components into shape later).

Other reparative satisfactions of computing include the production of neat, typographically sophisticated work.[29] The theory of epistemophilia suggests that the creation of whole, beautiful bodies (texts or artworks) might be particularly important in feminine creativity as ways of counteracting anxiety about disturbing the 'mother's body'.

At a time when habitats and climatic patterns are being destroyed at an unprecedented pace, the smooth, luminous microworlds and artificial objects of computer microworlds offer imaginary restorations of a fragmented and out of control reality. The phantasmic mobility of virtual bodies not only satisfies our infantile desires for omnipotence and omnipresence, but can provide hallucinatory satisfaction to those whose real body's mobility is impaired in some way.

I would now like to get more specific about the genres and genders of computer interactions with the aid of a four-part taxonomy of human–technology–world interactions developed by the philosopher of technology

Don Ihde.[30] These genres are called embodiment, hermeneutic, alterity and background relations, and it ought to be noted at the outset of this discussion that the 'transitionality' of computer technology puts it on the horizons of each genre.

EMBODIMENT: RELATIONS (TECHNOLOGY AS BODY/PROSTHESIS)

Here the technology is an extension of our organs or senses, a more or less transparent mediator through which we perceive and act on the world (e.g. blind person's cane; a hammer).

In a broad sense, computers are in embodiment (or more properly, instrumental) relations when used as mere means to other ends, rather than being the prime focus of interaction themselves.

Computers are typically weak in the embodiment dimension, requiring restricted manual operation through a keyboard, mouse, or joystick. However, technologies like the data glove enhance the computer's 'sexiness' by improving embodiment, mainly in the direction of human to technology. Feedback remains visually mediated, other body senses are supplied mainly by hallucination, and the effects are contained within the virtual world. However, as noted, recent developments expand the range of sensory feedback to include proprioception, pressure, etc. In science fantasies, feedback is enhanced through informational/sensory implants and 'trodes' to simulate the stimulation of senses besides vision (Gibson's 'simstim' effect).

HERMENEUTIC RELATIONS (TECHNOLOGY AS SIGN)

Here the technology is read or interpreted as a text through which a state of the world or technical system is inferred (e.g. maps, dials and gauges; prototypically, reading/writing).

Computers are thoroughly hermeneutic tools. At the operational level, they require the user's attention to be focussed on keyboard and screen. For many workers, the computerisation of the workplace has meant an unwelcome increase in time spent staring at a computer screen and less time dealing with real people or other equipment. Functionally, computers represent a speedy, efficient and highly flexible *reporting technology*, and can work as sophisticated writing instruments, where they may produce an horizonal blurring of the reader/writing distinction. Winograd and Flores consider computers as communications devices, mediators in human speech communities; they also show how principles of linguistic representation (such as the arbitrary relations of signifier and signified) are deployed throughout computer design, from the lowest levels of the physical machine to its higher levels as a representational scheme for facts (e.g. an airline booking system).[31]

ALTERITY RELATIONS (TECHNOLOGY AS SECOND SELF)

Here the technology as a thing in itself is more important than its connection to the world; we relate to it as a quasi-human entity.

Ihde's own examples are drawn from computing, where the self-contained, linguistic and logical qualities of the technology – not to mention the powerful myths of artificial creation and Jupiter Space – foster the experience of the computer as a quasi-human second self. Anthropomorphic tendencies are particularly prevalent in artificial intelligence, where the computer is invested with powers of reason and to some extent with its own 'personality'. Even users of word processors develop alterity relations with and become defensively fond of their chosen computer programmes (though as Ihde notes, these 'quasi-love' relations can easily flip over into 'quasi-hate' when the programme doesn't do what you want it to!). Interactions can take on the competitive and aggressive forms indulged by video game players and computer hackers, to whom the machine is an opposing player in a game of mastery, or a system to be beaten.[32]

BACKGROUND RELATIONS (TECHNOLOGY AS WORLD)

Here technology is experienced as an integral part of the world, or as the world itself (e.g. the background hum of domestic machines; 'technological cocoons' such as environmentally-controlled buildings, space stations, postmodern war vehicles).

In an ordinary sense, computers and computerised systems constitute part of the background, taken-for-granted texture of life in the information society (e.g. the ubiquity of automatic bank tellers and laser-readable bar-codes).

In a 'sexier' sense, virtual realities are horizonal instances of both background and hermeneutic relations in their presentation of simulated, simplified and controllable microworlds. Cyberspace or virtual reality forms an irreal technological cocoon with no necessary external referents. In what could almost be taken as a definition of 'hyperreality', the computer microworld has a high degree of verisimilitude, but the objects or systems it (hyper-) realistically represents may have no material or worldly correlates.

Virtual reality and simulations technologies straddle the horizons of all four genres of technics. In discussing computer microworlds, both Don Ihde and Sherry Turkle conflate alterity and background relations of computers.[33] Turkle writes:

> For adults as well as children, computers, reactive and interactive, offer companionship without the mutuality and complexity of a human relationship. They seduce because they provide a chance to be in complete control, with building one's own private world.[34]

The slippage between relating to a second self and mastering a microworld is not just bad analysis, but symptomatic of the consciousness entailed in some forms of human–computer interaction. This is the narcissistic consciousness of Jupiter Space, which equate brain space with both universal and maternal horizons of being. Need we ask who is the self to whom this brain–world–machine is a second self?

Studies of gender and educational computing consistently report that more boys tend to enjoy playing with the technology as such, while girls tend to use the computer as an instrument to attain certain goals within specific contexts.[35] Girls prefer to treat the computer as 'just a tool',[36] an instrumental means to an end (e.g. producing a neat essay, creating an image) and/or as a *hermeneutic* technology with referents and communicative functions within the world (e.g. conversing on computer networks; processing data about their own bodies and lives). It tends to be boys rather than girls who seek alterity relations, 'loving the machine for itself'[37] and losing themselves for hours playing 'sport-death' in the microworld.

Sherry Turkle pathologises young women's attitudes to the computer as 'just a tool', calling this a 'defensive' and 'romantic' reaction against their experience of the 'intimate machine'.[38] Citing feminist object relations theorists,[39] Turkle suggests girls would be cured of this defence once they had more opportunity for intimacy with computational objects that 'can be experienced as tactile and physical'.[40] But this analysis elides the distinction between 'experiencing ...' and 'experiencing *as* ...'; Turkle seems to want women to fall for the seductive idealist fantasy wherein a luminous virtual brainchild is an acceptable substitute for a tactile physical body. But the resistant young women insist on the difference between actual and virtual objects, in part as a way of defining themselves as women (as Turkle notes), and of claiming an identity as 'those to whom this machine is *not* a second self', those who reject abstraction from the reality of the whole body. In my opinion, this resistance is not symptomatic of psychopathology, nor only of the conventional social definitions of women as more emotional and physical than men. It also indicates a different relation to the fertility myths of Jupiter Space, and is perhaps a form of 'conscientious objection' to the idealist, masculinist and militaristic culture from which computers have emerged.

Computers represent a limit case for standard feminist object relations theory. Girls, who are supposed to be emotionally connected to their objects and indeterminate about self-other boundaries, adopt dispassionate instrumentalist attitudes, though the deeper interest in connectedness is expressed in their demand that computing remain relevant to contexts of communicative expression and worldly knowledges. Boys, who are supposedly concerned to reinforce subject–object distinctions and maintain emotional distance from their objects, can develop self-sacrificing and passionate attachments to computers, losing their body boundaries in oceanic sensations of inhabiting the computer microworld.[41] Their deeper alienation is expressed in the pursuit of relations with programmable quasi-others to the neglect of living entities, the displacement of eroticism onto sexy machines.[42]

The analytic perspective forwarded here suggests that women's relations to virtual reality and corporeality would be highly ambivalent. On the one hand, enhancements of embodiment (through fibre-optic suits, force-feedback circuits, etc.) might be appealing in their physicality, while the creation of spaces, objects and bodies in cyberspace could afford many reparative satisfactions of the kind already known to painters, sculptors and designers. But on the other hand, the irreal referentiality and escapist tendencies expressed in the hyperreal cocoon go against the grain for those whose self-definition requires neither the negation nor possession of the maternal/material axis, nor the over-valuation of artificial fertility.

I suspect the mythic and irrational aspects of computer technology are more off-putting to women than the practical and rational dimensions. Even though techno-nerds may feel otherwise, lack of engagement with the Jupiter Space and second self fantasies need not prevent women from effectively appropriating information technology through hermeneutic or instrumental relations. We need to negotiate and explore rather than suppress or ignore the contradictory experiences of seduction by and rejection of virtual bodies and worlds. We need not fall for the seductions of synecdoche, but might follow Haraway's injunction to adopt *irony* as an appropriate response to technologies spawned by the 'informatics of domination'.[43] Irony can imply a double-vision which allows us to simultaneously 'experience ...' and 'experience as ...', without eliding the difference; a consciousness that can engage with technological part-objects while remaining aware of the ontological horizons and social contexts within which they are embedded.

Whatever our precise interpretation of virtual realities and corporealities, we must still question the basic project. Virtual worlds and bodies offer a pleasurable fulfilment of the defensive and ultimately misogynistic fantasy of escape from Earth, gravity, and maternal/material origins.[44] As the fabric of the real world is rent apart, new, whole, artificial entities partially satisfy desires for making reparation. What would happen if instead of denying origins, and turning our eyes away from terrestrial reality, we acknowledged the maternal/planetary axis as origin and (damaged) source? Would we then become more urgently interested in applying technologies to make real rather than virtual reparations? Let's hope so.

Meanwhile, without wanting to deny that like any other representational technology, VR could be deployed for interesting and enjoyable artistic and educational purposes, it seems important to keep questioning the sources of certainty and enthusiasm amongst those eager to welcome virtual bodies and spaces as new and important parts of our immediately impending future. Why does the same determinism not seem to reign in the case of truly reparative technological innovations? More importantly, we ask: *Cui bono?* In whose interest is this distribution of reparative technologies maintained? Who benefits when desires for control of tools and technological processes are satisfied within irreal microworlds offering the illusion of participation in the corporate and military matrix, yet limiting effects to microevents amongst electronic circuits?

NOTES

1. Herein is some cause of confusion: 'cyberspace' is the term by which author William Gibson describes the science fiction of the 'consensual hallucination' of 'the matrix', a visual/spatial representation of transglobal networks of computers and databases,

mentally entered by futuristic computer hackers via 'trodes' attached to their heads or via other body implants. However, 'cyberspace' is also used to describe the simulated environments generated on screen by computer graphics programmes, and to refer in general to the 'virtual space' inside the computer programme or network.

2. Donna Haraway, 'A Manifesto for Cyborgs: Science, Technology and Socialist Feminism in the 1980s', *Australian Feminist Studies*, no. 4, 1987.

3. Jill Scott speaking at a video festival, Film and Television Institute, Fremantle, c.1989.

4. 'Virtual Corporeality' was the title of the session in which this paper was delivered at Adelaide Artists' Week, March 1992. The perspective and arguments forwarded in this paper are developed in my forthcoming book, *Whose Second Self? Gender and (Ir)rationality in Computer Culture* (Deakin University Press) Geelong, 1992. An earlier version of this paper was presented at the 'Signifying Others' Cultural Studies conference at the University College of Southern Queensland, December 1991.

5. By this, I do not mean to disparage writers concerned with the important issues of women and technology. For an illuminating recent contribution to this field, see Judy Wacjman, *Feminism Confronts Technology* (Allen and Unwin) Sydney, 1991.

6. Norman O. Brown, *Life Against Death: The Psychoanalytical Meaning of History* (Wesleyan University Press) Middletown, Conn., 1959, p. 236.

7. Martin Heidegger, 'The Question Concerning Technology'; 'The Age of the World Picture' in *The Question Concerning Technology and Other Essays*, trans. William Lovitt (Harper and Row) New York, 1977, especially p. 21; p. 141.

8. For a critique of technological determinism in the computer literacy push of the 1980s, see Douglas Noble, 'Computer Literacy and Ideology' in Douglas Sloan (ed.), *The Computer in Education: A Critical Perspective* (Teacher's College Press) New York, 1984; for analysis and discussion of the implications of progressively earlier introduction of computers in education see Harriet K. Cuffaro, 'Micro-computers in Education: Why is Earlier Better?' and Arthur G. Zajonc, 'Computer Pedagogy? Questions Concerning the New Educational Technology', in the same volume.

9. Paul N. Edwards, 'The Army and the Microworld: Computers and the Politics of Gender Identity', *Signs,* vol. 16, no. 1, 1990, pp. 102–27, especially p. 114; on the 'holding power' of the microworld see also Sherry Turkle, *The Second Self* (Simon and Schuster) New York, 1984, Chapter 6, especially pp. 207–8.

10. Freud, 'Fetishism', *S. E.* XXI, 1927.

11. Gèza Róheim, *The Origins and Functions of Culture* (Nervous and Mental Diseases Monographs) New York, 1943, pp. 95–6.

12. Róheim, *The Origins and Functions of Culture*, p. 96.

13. Turkle, 'Introduction: The Evocative Object', *The Second Self*, pp. 11–25.

14. Sherry Turkle and Seymour Papert, 'Epistemological Pluralism: Styles and Voices within Computer Culture', *Signs,* vol. 16, no. 1, 1990, p. 145.

15. Mark Poster, *The Mode of Information* (Polity) London, 1990, especially 'Derrida and Electronic Writing: The Subject of the Computer', pp. 99–128. For a fascinating and more precise treatment of the implications of the transitionality of computer writing for questions of authorship and originality, see John Frow, 'Repetition and Limitation: Computer Software and Copyright Law', *Screen*, vol. 29, no. 1, 1988, pp. 4–20.

16. Poster, *The Mode of Information*, p. 128.

17. John Stratton, personal communication, January 1992.

18. Zoë Sofia, 'Exterminating Fetuses: Abortion, Disarmament, and the Sexo-Semiotics of Extraterrestrialism', *Diacritics*, vol. 14, no. 2, 1984, pp. 47–59; and Sofia 'Aliens "R" U.S.: American Science Fiction Viewed from Down Under', in G. E. Slusser and E. S. Rabkin (eds) *Aliens: The Anthropology of Science Fiction* (Southern Illinois University Press) Carbondale, Ill., 1987, pp. 128–41.

19. Melanie Klein, 'Early Stages of the Oedipus Conflict' (1928) in R. E. Money-Kyrle (ed.), *Love, Guilt and Reparation and Other Works 1921–1945* (Hogarth Press and the Institute of Psychoanalysis) London, 1975.
20. Klein, 'Love, Guilt and Reparation', especially pp. 333–8.
21. Note that the perspective forwarded here makes a point of not conflating anatomical and sociological categories of 'female' and 'woman' with 'femininity' and 'maternity'; unlike some of my fellow feminist analysts of representations of women and technology, I do not attribute all 'textual excess' to the figure 'woman' but aim to specify the masculine excess which finds expression in feminine and technological/maternal figures. Compare, for example, Mary Ann Doane, 'Technophilia: Technology, Representation and the Feminine' in Mary Jacobus, Evelyn Fox Keller and Sally Shuttleworth (eds), *Body/Politics: Women and the Discourses of Science* (Routledge) New York, 1990, pp. 163–75, with Sofia, 'Masculine Excess and the Metaphorics of Vision: Some Problems in Feminist Film Theory', *Continuum: An Australian Journal of the Media*, vol. 2, no. 2, 1989, pp. 116–28; also Sofia, 'Hegemonic Irrationalities and Psychoanalytic Cultural Critique', forthcoming in *Cultural Studies*, 1992.
22. Thus, even though women were disbarred from science, traditional depictions of the spirits or muses of various sciences showed female figures, as discussed by Londa Schiebinger, 'Feminine Icons: The Face of Early Modern Science', *Critical Inquiry*, vol. 14, 1988, pp. 661–91. Schiebinger finds this trope declined in popularity by the mid-nineteenth century, and we now find 'science' represented in the image of a solitary white-coated man in a lab. However, it would appear that science fictions took up the feminine figure with respect to technological invention in the trope of the mechanical woman. For discussions of the roboticised woman see Mary Daly, *Gyn/Ecology: The Meta-Ethics of Radical Feminism* (Beacon) Boston, 1978, pp. 52–7; on the mechanical Hadaly (whose name means 'Ideal' in Arabic) see Annette Michelson, 'On the Eve of the Future: The Reasonable Facsimile and the Philosophical Toy', *October: The First Decade 1976–1986* (MIT Press) Cambridge, Mass., 1986, pp. 417–35, and Raymond Bellour, 'Ideal Hadaly', *Camera Obscura*, no. 15, 1986, pp. 111–35; on the *Metropolis* robot Maria see Andreas Huyssen, 'The Vamp and the Machine: Technology and Sexuality in Fritz Lang's *Metropolis*', *New German Critique*, no. 24–5, Fall/Winter 1981/82, pp. 221–37; and (my favourite from a psychoanalytic perspective) Roger Dadoun, '*Metropolis* Mother-City – "Mittler" – Hitler', *Camera Obscura*, no. 15, 1986, pp. 137–63.
23. Donna Haraway, 'Situated Knowledges: The Science Question in Feminism and the Privilege of Partial Perspective', *Feminist Studies*, vol. 14, no. 3, 1988, pp. 575–99.
24. David Bausor, '"Turn on, Boot up, Jack in": Ontological Aphinasis in Postmodern Society', presented at the 'Signifying Others' Cultural Studies Conference University College of Southern Queensland, Toowoomba, 2–4 December 1991.
25. For one of the early analyses of this see Sandra Gilbert, 'Horror's Twin: Mary Shelley's Monstrous Eve,' *Feminist Studies*, vol. 4, no. 2, 1978, pp. 48–73; a version of this also appears in Gilbert and Gubar's *The Madwoman in the Attic*. See also Margaret Homans, *Bearing the Word: Language and Female Experience in Nineteenth-Century Women's Writing* (University of Chicago Press) Chicago, 1986, for a discussion of Shelley and other women writers who evoked childbirth as a metaphor of writing.
26. Norman O. Brown, *Life Against Death*, pp. 297–8.
27. Like Poster, for example, who describes the computer as an 'uncanny mirror' of the human, which 'in its immateriality ... mimics the human being', Mark Poster, *The Mode of Information*, p. 112.
28. Sherry Turkle and Seymour Papert, 'Epistemological Pluralism', p. 131.
29. As with Turkle's subject Tanya, a young girl who became interested in computers as a means of making her writing neat, Turkle, *The Second Self*, pp. 122–6.

30. Don Ihde, *Technology and the Lifeworld: From Garden to Earth* (Indiana University Press) Bloomington, 1990, pp. 72–123.
31. Terry Winograd and Fernando Flores, *Understanding Computers and Cognition: A New Foundation for Design* (Ablex Publishing Corporation) New Jersey, 1986, especially pp. 87–90 and pp. 172–9.
32. Ihde, *Technology and the Lifeworld*, p. 106.
33. For example Ihde, *Technology and the Lifeworld*, p. 100; and Turkle, *The Second Self*, p. 19.
34. Turkle, *The Second Self*, p. 19.
35. '... girls prefer to use computers as word processors, or in research applications, while boys prefer game playing and programming', Morwenna Griffiths, 'Strong Feelings About Computers', *Women's Studies International Forum*, vol. 11, no. 2, 1988, p. 151. See also Valerie A. Clarke, 'Girls and Computing: Dispelling Myths and Finding Directions' in A. McDougall and C. Dowling (eds), *Computers in Education* (Elsevier Science Publishers B.V.) North Holland, 1990, pp. 53–8; especially p. 57. For review articles on this see Ruth Perry and Lisa Greber, 'Women and Computers: An Introduction', *Signs*, vol. 16, no. 1, 1990, pp. 74–101; Pamela E. Kramer and Sheila Lehman, 'Mismeasuring Women: A Critique of Research on Computer Ability and Avoidance', *Signs*, vol. 16, no. 1, 1990, pp. 158–72; and Sue Curry Jansen, 'Gender and the Information Society: A Socially Structured Silence', *Journal of Communication*, vol. 39, no. 3, 1989, pp. 196–215. Jansen goes beyond reviewing sociological research to outlining elements for a thoroughly interdisciplinary feminist approach to the question of gender and computing.
36. Sherry Turkle, 'Computational Reticence: Why Women Fear the Intimate Machine' in Cheris Kramarae (ed.), *Technology and Women's Voices: Keeping in Touch* (Routledge and Kegan Paul) London, 1988, pp. 41–61; esp p. 50.
37. The title of Turkle's chapter on computer hackers in *The Second Self*.
38. Turkle, 'Computational Reticence', p. 50.
39. For example, Nancy Chodorow, *The Reproduction of Mothering* (University of California Press) Berkeley, 1978; Carol Gilligan, *In a Different Voice* (Harvard University Press) Cambridge MA, 1982; Evelyn Fox Keller, *A Feeling for the Organism* (Freeman) San Francisco, 1983; and Keller, *Reflections on Gender and Science* (Yale University Press) New Haven, Conn., 1985.
40. Turkle, 'Computational Reticence', p. 57.
41. Turkle, *The Second Self*, pp. 211–12.
42. For more investigations of this see Sally Hacker, 'Discipline and Pleasure in Engineering' in *Pleasure, Power and Technology: Some Tales of Gender, Engineering, and the Cooperative Workplace* (Unwin Hyman) Boston, 1989, pp. 35–57; similar ground is covered in 'The Eye of the Beholder' in Dorothy Smith and Susan M. Turner (eds), *Doing It the Hard Way: Investigations of Gender and Technology* (Unwin Hyman) Boston, 1990, pp. 205–23. See also Turkle's discussion of boys' tendencies to escape sociality in Turkle, *The Second Self*, Chapter 6.
43. Donna Haraway, 'A Manifesto for Cyborgs'.
44. For extensive discussions of some of these trajectories, see Robert D. Romanyshyn, *Technology as Symptom and Dream* (Routledge) London, 1989.

4

WILL THE REAL BODY PLEASE STAND UP? BOUNDARY STORIES ABOUT VIRTUAL CULTURES

Allucquere Rosanne Stone

THE MACHINES ARE RESTLESS TONIGHT

After Donna Haraway's 'Promises of Monsters' and Bruno Latour's papers on actor networks and artifacts that speak, I find it hard to think of any artifact as being devoid of agency. Accordingly, when the dryer begins to beep complainingly from the laundry room while I am at dinner with friends, we raise eyebrows at each other and say simultaneously, 'The machines are restless tonight . . .'

It's not the phrase, I don't think, that I find intriguing. Even after Haraway 1991 and Latour 1988, the phrase is hard to appreciate in an intuitive way. It's the ellipsis I notice. You can hear those three dots. What comes after them? The fact that the phrase – obviously a send-up of a vaguely anthropological chestnut – seems funny to us, already says a great deal about the way we think of our complex and frequently uneasy imbrications with the unliving. I, for one, spend more time interacting with Saint-John Perse, my affectionate name for my Macintosh computer, than I do with my friends. I appreciate its foibles, and it gripes to me about mine. That someone comes into the room and reminds me that Perse is merely a 'passage point' for the work practices of a circle of my friends over in Silicon Valley changes my sense of facing a vague but palpable sentience squatting on my desk not one whit. The people I study are deeply imbricated in a complex social network mediated by little technologies to which they have delegated significant amounts of their time and agency, not to mention their humor. I say to myself: Who am I studying? A group of people? Their machines? A group of people and or in their machines? Or something else?

When I study these groups, I try to pay attention to all of their interactions. And as soon as I allow myself to see that most of the interactions of the people I am studying involve vague but palpable sentiences squatting on their desks, I have to start thinking about watching the machines just as attentively as I watch the people, because, for them, the machines are not merely passage points. Haraway and other workers who observe the traffic across the boundaries between 'nature,' 'society,' and 'technology' tend to see nature as lively, unpredictable, and, in some sense, actively resisting interpretations. If nature and technology seem to be collapsing into each other, as Haraway and others claim, then the unhumans can be lively too.

One symptom of this is that the flux of information that passes back and forth across the vanishing divides between nature and technology has become extremely dense. Cyborgs with a vengeance, one of the groups I study, is already talking about colonizing a social space in which the divide between nature and technology has become thoroughly unrecognizable, while one of the individuals I study is busy trying to sort out how the many people who seem to inhabit the social space of her body are colonizing her. When I listen to the voices in these new social spaces I hear a multiplicity of voices, some recognizably human and some quite different, all clamoring at once, frequently saying things whose meanings are tantalizingly familiar but which have subtly changed.

My interest in cyberspace is primarily about communities and how they work. Because I believe that technology and culture constitute each other, studying the actors and actants that make up our lively, troubling, and productive technologies tells me about the actors and actants that make up our culture. Since so much of a culture's knowledge is passed on by means of stories, I will begin by retelling a few boundary stories about virtual cultures.

SCHIZOPHRENIA AS COMMODITY FETISH

Let us begin with a person I will call Julie, on a computer conference in New York in 1985. Julie was a totally disabled older woman, but she could push the keys of a computer with her headstick. The personality she projected into the 'net' – the vast electronic web that links computers all over the world – was huge. On the net, Julie's disability was invisible and irrelevant. Her standard greeting was a big, expansive 'HI!!!!!!' Her heart was as big as her greeting, and in the intimate electronic companionships that can develop during on-line conferencing between people who may never physically meet, Julie's women friends shared their deepest troubles, and she offered them advice – advice that changed their lives. Trapped inside her ruined body, Julie herself was sharp and perceptive, thoughtful and caring.

After several years, something happened that shook the conference to the core. 'Julie' did not exist. 'She' was, it turned out, a middle-aged male psychiatrist. Logging onto the conference for the first time, this man had accidentally begun a discussion with a woman who mistook him for another woman. 'I was stunned,' he said later, 'at the conversational mode. I hadn't known that women talked among themselves that way. There was so much more vulnerability, so much more depth and complexity. Men's conversations on the nets were much more guarded and superficial, even among intimates. It was fascinating, and I wanted more.' He had spent weeks developing the right persona. A totally disabled, single older woman was perfect. He felt that such a person wouldn't be expected to have a social life. Consequently her existence only as a net persona would seem natural. It worked for years, until one of Julie's devoted admirers, bent on finally meeting her in person, tracked her down.

The news reverberated through the net. Reactions varied from humorous resignation to blind rage. Most deeply affected were the women who had shared their innermost feelings with Julie. 'I felt raped,' one said. 'I felt that my deepest secrets had been violated.' Several went so far as to repudiate the genuine gains they had made in their personal and emotional lives. They felt those gains were predicated on deceit and trickery.

The computer engineers, the people who wrote the programs by means of which the nets exist, just smiled tiredly. They had understood from the beginning the radical changes in social conventions that the nets implied. Young enough in the first days of the net to react and adjust quickly, they had long ago taken for granted that many of the old assumptions about the nature of identity had quietly vanished under the new electronic dispensation. Electronic networks in their myriad kinds, and the mode of interpersonal interaction that they foster, are a new manifestation of a social space that has been better known in its older and more familiar forms in conference calls, communities of letters, and FDR's fireside chats. It can be characterized as 'virtual' space – an imaginary locus of interaction created by communal agreement. In its most recent form, concepts like distance, inside/outside, and even the physical body take on new and frequently disturbing meanings.

Now, one of the more interesting aspects of virtual space is 'computer crossdressing.' Julie was an early manifestation. On the nets, where *warranting*, or grounding, a persona in a physical body, is meaningless, men routinely use female personae whenever they choose, and vice versa. This wholesale appropriation of the other has spawned new modes of interaction. Ethics, trust, and risk still continue, but in different ways. Gendered modes of communication themselves have remained relatively stable, but who uses which of the two socially recognized modes has become more plastic. A woman who has appropriated a male conversational style may be simply assumed to be male at that place and time, so that her/his on-line persona takes on a kind of quasi life of its own, separate from the person's embodied life in the 'real' world.

Sometimes a person's on-line persona becomes so finely developed that it begins to take over their life *off* the net. In studying virtual systems, I will call both the space of interaction that is the net and the space of interaction that we call the 'real' world *consensual loci*. Each consensual locus has its own 'reality,' determined by local conditions. However, not all realities are equal. A whack on the head in the 'real' world can kill you, whereas a whack in one of the virtual worlds will not (although a legal issue currently being debated by futurist attorneys is what liability the whacker has if the fright caused by a virtual whack gives the whackee a 'real' heart attack).

Some conferencees talk of a time when they will be able to abandon warranting personae in even more complex ways, when the first 'virtual reality' environments come on-line. VR, one of a class of interactive spaces that are coming to be known by the general term *cyberspace*, is a three-dimensional consensual locus or, in the terms of science fiction author William Gibson, a 'consensual hallucination' in which data may be visualized, heard, and even felt. The 'data' in some of these virtual environments are people – 3-D representations of individuals in the cyberspace. While high-resolution images of the human body in cyberspace are years away, when they arrive they will take 'computer crossdressing' even further. In this version of VR a man may be seen, and perhaps touched, as a woman and vice versa – or as anything else. There is talk of renting prepackaged body forms complete with voice and touch ... multiple personality as commodity fetish!

It is interesting that at just about the time the last of the untouched 'real-world' anthropological field sites are disappearing, a new and unexpected kind of 'field' is opening up – incontrovertibly social spaces in which people still meet face-to-face, but under new definitions of both 'meet' and 'face.' These new spaces instantiate the collapse of the boundaries between the social and technological, biology and machine, natural and artificial that are part of the postmodern imaginary. They are part of the growing imbrication of humans and machines in new social forms that I call *virtual systems*.

A VIRTUAL SYSTEMS ORIGIN MYTH

Cyberspace, without its high-tech glitz, is partially the idea of virtual community. The earliest cyberspaces may have been virtual communities, passage points for collections of common beliefs and practices that united people who were physically separated. Virtual communities sustain themselves by constantly circulating those practices. To give some examples of how this works, I'm going to tell an origin story of virtual systems.

There are four epochs in this story. The beginning of each is signaled by a marked change in the character of human communication. Over the years, human communication is increasingly mediated by technology. Because the rate of change in technological innovation increases with time, the more recent epochs are shorter, but roughly the same quantity of information is exchanged in each. Since the basis of virtual communities is communication, this seems like a reasonable way to divide up the field.

Epoch One: Texts. [From the mid-1600s]
Epoch Two: Electronic communication and entertainment media. [1900+]
Epoch Three: Information technology. [1960+]
Epoch Four: Virtual reality and cyberspace. [1984+]

Epoch One

This period of early textual virtual communities starts, for the sake of this discussion, in 1669 when Robert Boyle engaged an apparatus of literary technology to 'dramatize the social relations proper to a community of philosophers.' As Steven Shapin and Simon Shapiro point out in their study of the debate between Boyle and the philosopher Thomas Hobbes, *Leviathan and the Air-Pump*, we probably owe the invention of the boring academic paper to Boyle. Boyle developed a method of compelling assent that Shapin and Shaffer described as *virtual witnessing*. He created what he called a 'community of like-minded gentlemen' to validate his scientific experiments, and he correctly surmised that the 'gentlemen' for whom he was writing believed that boring, detailed writing implied painstaking experimental work. Consequently it came to pass that boring writing was likely to indicate scientific truth. By means of such writing, a group of people were able to 'witness' an experiment without being physically present. Boyle's production of the detailed academic paper was so successful that it is still the exemplar of scholarship.

The document around which community forms might also be a novel, a work of fiction. Arguably the first texts to reach beyond class, gender, and ideological differences were the eighteenth-century sentimental novels, exemplified by the publication of Bernardin de Saint-Pierre's short novel *Paul and Virginia* (1788), which Roddey Reid, in his study 'Tears For Fears,' identifies as one of the early textual productions that 'dismantled the absolutist public sphere and constructed a bourgeois public sphere through fictions of national community.' Reid claims that *Paul and Virginia* was a passage point for a circulating cluster of concepts about the nature of social identity that transformed French society. Reid suggests that an entire social class – the French bourgeoisie – crystallized around the complex of emotional responses that the novel produced. Thus in the first epoch texts became ways of creating, and later of controlling, new kinds of communities.

Epoch Two

The period of the early electronic virtual communities began in the twentieth century with invention of the telegraph and continued with musical communities, previously constituted in the physical public space of the concert hall, shifting and translating to a new kind of virtual communal space around the phonograph. The apex of this period was Franklin Delano Roosevelt's radio 'fireside chats,' creating a community by means of readily available technology.

Once communities grew too big for everyone to know everyone else, which is to say very early on, government had to proceed through delegates who represented absent groups. FDR's use of radio was a way to bypass the need for delegates. Instead of talking to a few hundred representatives, Roosevelt used the radio as a machine for fitting listeners into his living room. The radio was one-way communication, but because of it people were able to begin to think of presence in a different way. Because of radio and of the apparatus for the production of community that it implied and facilitated, it was now possible for millions of people to be 'present' in the same space – seated across from Roosevelt in his living room.

This view implies a new, different, and complex way of experiencing the relationship between the physical human body and the 'I' that inhabits it. FDR did not physically enter listeners' living rooms. He invited listeners into his. In a sense, the listener was in two places at once – the body at home, but the delegate, the 'I' that belonged to the body in an imaginal space with another person. This space was enabled and constructed with the assistance of a particular technology. In the case of FDR the technology was a device that mediated between physical loci and incommensurable realities – in other words, an interface. In virtual systems *an interface is that which mediates between the human body (or bodies) and an associated 'I' (or 'I's')*. This double view of 'where' the 'person' is, and the corresponding trouble it may cause with thinking about 'who' we are talking about when we discuss such a problematic 'person,' underlies the structure of more recent virtual communities.

During the same period thousands of children, mostly boys, listened avidly to adventure serials, and sent in their coupons to receive the decoder rings and signaling devices that had immense significance within the community of a particular show. Away from the radio, they recognized each other by displaying the community's tokens, an example of communities of consumers organized for marketing purposes.

The motion picture, and later, television, also mobilized a similar power to organize sentimental social groups. Arguably one of the best examples of a virtual community in the late twentieth century is the Trekkies, a huge, heterogenous group partially based on commerce but mostly on a set of ideas. The fictive community of 'Star Trek' and the fantasy Trekkie community interrelate and mutually constitute each other in complex ways across the boundaries of texts, films, and video interfaces.

Epoch Two ended in the mid-1970s with the advent of the first computer, terminal-based, bulletin board systems (BBSs).

Epoch Three

This period began with the era of information technology. The first virtual communities based on information technology were the on-line bulletin board services (BBS) of the middle 1970s. These were not dependent upon the widespread ownership of computers, merely of terminals. But because even a used terminal cost several hundred dollars, access to the first BBSs was mainly limited to electronics experimenters, ham-radio operators, and the early hardy computer builders.

BBSs were named after their perceived function – virtual places, conceived to be just like physical bulletin boards, where people could post notes for general reading. The first successful BBS programs were primitive, usually allowing the user to search for messages alphabetically, or simply to read messages in the order in which they were posted. These programs were sold by their authors for very little, or given away as 'shareware' – part of the early visionary ethic of electronic virtual communities. The idea of shareware, as enunciated by the many programmers who wrote shareware programs, was that the computer was a passage point for circulating concepts of community. The important thing about shareware, rather than making an immediate profit for the producer, was to nourish the community in expectation that such nourishment would 'come around' to the nourisher.

CommuniTree

Within a few months of the first BBS's appearance, a San Francisco group headed by John James, a programmer and visionary thinker, had developed the idea that the BBS was a virtual community, a community that promised radical transformation of existing society and the emergence of new social forms. The CommuniTree Group, as they called themselves, saw the BBS in McLuhanesque terms as transformative because of the ontological structure it presupposed and simultaneously created – the mode of tree-structured discourse and the community that spoke it – and because it was another order of 'extension,' a kind of prosthesis in McLuhan's sense. The BBS that the CommuniTree Group envisioned was an extension of the participant's instrumentality into a virtual social space.

The CommuniTree Group quite correctly foresaw that the BBS in its original form was extremely limited in its usefulness. Their reasoning was simple. The physical bulletin board for which the BBS was the metaphor had the advantage of being quickly scannable. By its nature, the physical bulletin board was small and manageable in size. There was not much need for bulletin boards to be organized by topic. But the on-line BBS could not be scanned in any intuitively satisfactory way. There were primitive search protocols in the early BBSs, but they were usually restricted to alphabetical searches or searches by keywords. The CommuniTree Group proposed a new kind of BBS that they called a tree-structured conference, employing as a working metaphor both the binary tree protocols in computer science and also the organic qualities of trees as such appropriate to the 1970s. Each branch of the tree was to be a separate conference that grew naturally out of its root message by virtue of each subsequent message that was attached to it. Conferences that lacked participation would cease to grow, but would remain on-line as archives of failed discourse and as potential sources of inspiration for other, more flourishing conferences.

With each version of the BBS system, The CommuniTree Group supplied a massive, detailed instruction manual – which was nothing less than a set of directions for constructing a new kind of virtual community. They couched the manual in radical seventies language, giving chapters such titles as 'Downscale, please, Buddha' and 'If you meet the electronic avatar on the road, laserblast hir!' This rich intermingling of spiritual and technological imagery took place in the context of George Lucas's *Star Wars*, a film that embodied the themes of the technological transformativists, from the all-pervading Force to what Vivian Sobchack (1987) called 'the outcome of infinite human and technological progress.' It was around *Star Wars* in particular that the technological and radically spiritual virtual communities of the early BBSs coalesced. *Star Wars* represented a future in which the good guys won out over vastly superior adversaries – with the help of a mystical Force that 'surrounds us and penetrates us ... it binds the galaxy together' and which the hero can access by learning to 'trust your feelings' – a quintessential injunction of the early seventies.

CommuniTree #1 went on-line in May 1978 in the San Francisco Bay area of northern California, one year after the introduction of the Apple II computer and its first typewritten and hand-drawn operating manual. CommuniTree #2 followed quickly. The opening sentence of the prospectus for the first conference was 'We are as gods and might as well get good at it.' This techno-spiritual bumptiousness, full of the promise of the redemptive power of technology mixed with the easy, catch-all Eastern mysticism popular in upscale northern California, characterized the early conferences. As might be gathered from the tone of the prospectus, the first conference, entitled 'Origins,' was about successor religions.

The conferencees saw themselves not primarily as readers of bulletin boards or participants in a novel discourse but as agents of a new kind of social experiment. They saw the terminal or personal computer as a tool for social transformation by the ways it refigured social interaction. BBS conversations were time-aliased, like a kind of public letter writing or the posting of broadsides. They were meant to be read and replied to some time later than they were posted. But their participants saw them as conversations nonetheless, as social acts. When asked how sitting alone at a terminal was a social act, they explained that they saw the terminal as a window into a social space. When describing the act of communication, many moved their hands expressively as though typing, emphasizing the gestural quality and essential tactility of the virtual mode. Also present in their descriptions was a propensity to reduce other expressive modalities to the tactile. It seemed clear that, from the beginning, the electronic virtual mode possessed the power to overcome its character of single-mode transmission and limited bandwidth.

By 1982 Apple Computer had entered into the first of a series of agreements with the federal government in which the corporation was permitted to give away computers to public schools in lieu of Apple's paying a substantial portion of its federal taxes. In terms of market strategy, this action dramatically increased Apple's presence in the school system and set the pace for Apple's domination in the education market. Within a fairly brief time there were significant numbers of personal computers accessible to students of grammar school and high school age. Some of those computers had modems.

The students, at first mostly boys and with the linguistic proclivities of pubescent males, discovered the Tree's phone number and wasted no time in logging onto the conferences. They appeared uninspired by the relatively intellectual and spiritual air of the ongoing debates, and proceeded to express their dissatisfaction in ways appropriate to their age, sex, and language abilities. Within a short time the Tree was jammed with obscene and scatalogical messages. There was no way to monitor them as they arrived, and no easy way to remove them once they were in the system. This meant that the entire system had to be purged – a process taking hours – every day or two. In addition, young hackers enjoyed the sport of attempting to 'crash' the system by discovering bugs in the system commands. Because of the provisions of the system that made observing incoming messages impossible, the hackers were free to experiment with impunity, and there was no way for the system operator to know what was taking place until the system crashed. At that time it was generally too late to save the existing disks. The system operator would be obliged to reconstitute ongoing conferences from earlier backup versions.

Within a few months, the Tree had expired, choked to death with what one participant called 'the consequences of freedom of expression.' During the years of its operation, however, several young participants took the lessons and implications of such a community away with them, and proceeded to write their own systems. Within a few years there was a proliferation of on-line virtual communities of somewhat less visionary character but vastly superior message-handling capability – systems that allowed monitoring and disconnection of 'troublesome' participants (hackers attempting to crash the system), and easy removal of messages that did not further the purposes of the system operators. The age of surveillance and social control had arrived for the electronic virtual community.

The visionary character of CommuniTree's electronic ontology proved an obstacle to the Tree's survival. Ensuring privacy in all aspects of the Tree's structure and enabling unlimited access to all conferences did not work in a context of increasing availability of terminals to young men who did not necessarily share the Tree gods' ideas of what counted as community. As one Tree veteran put it, 'The barbarian hordes mowed us down.' Thus, in practice, surveillance and control proved necessary adjuncts to maintaining order in the virtual community.

It is tempting to speculate about what might have happened if the introduction of CommuniTree had not coincided with the first wave of 'computerjugen.' Perhaps the future of electronic virtual communities would have been quite different.

SIMNET

Besides the BBSs, there were more graphic, interactive systems under construction. Their interfaces were similar to arcade games or flight simulators – (relatively) high-resolution, animated graphics. The first example of this type of cyberspace was a military simulation called SIMNET. SIMNET was conducted by a consortium of military interests, primarily represented by DARPA, and a task group from the Institute for Simulation and Training, located at the

University of Central Florida. SIMNET came about because DARPA was beginning to worry about whether the Army could continue to stage large-scale military practice exercises in Germany. With the rapid and unpredictable changes that were taking place in Europe in the late 1980s, the army wanted to have a backup – some other place where they could stage practice maneuvers without posing difficult political questions. As one of the developers of SIMNET put it, 'World War III in Central Europe is at the moment an unfashionable anxiety.' In view of the price of land and fuel, and of the escalating cost of staging practice maneuvers, the armed forces felt that if a large-scale consensual simulation could be made practical they could realize an immediate and useful financial advantage. Therefore, DARPA committed significant resources – money, time, and computer power – to funding some research laboratory to generate a 200-tank cyberspace simulation. DARPA put out requests for proposals, and a group at the University of Central Florida won.

The Florida group designed and built the simulator units with old technology, along the lines of conventional aircraft cockpit simulators. Each tank simulator was equipped to carry a crew of four, so the SIMNET environment is an 800-person virtual community.

SIMNET is a two-dimensional cyberspace. The system can be linked up over a very large area geographically; without much difficulty, in fact, to anywhere in the world. A typical SIMNET node is an M-1 tank simulator. Four crew stations contain a total of eight vision blocks, or video screens, visible through the tank's ports. Most of these are 320 × 138 pixels in size, with a 15 Hertz update rate. This means that the image resolution is not very good, but the simulation can be generated with readily available technology no more complex than conventional video games. From inside the 'tank' the crew looks out the viewports, which are the video screens. These display the computer-generated terrain over which the tanks will maneuver (which happens to be the landscape near Fort Knox, Kentucky). Besides hills and fields, the crew can see vehicles, aircraft, and up to 30 other tanks at one time. They can hear and see the vehicles and planes shooting at each other and at them.

By today's standards, SIMNET's video images are low-resolution and hardly convincing. There is no mistaking the view out the ports for real terrain. But the simulation is astonishingly effective, and participants become thoroughly caught up in it. SIMNET's designers believe that it may be the lack of resolution itself that is responsible, since it requires the participants to actively engage their own imaginations to fill the holes in the illusion! McLuhan redux. That it works is unquestionable. When experimenters opened the door to one of the simulators during a test run to photograph the interior, the participants were so caught up in the action that they didn't notice the bulky camera poking at them.

Habitat

Habitat, designed by Chip Morningstar and Randall Farmer, is a large-scale social experiment that is accessible through such common telephone-line computer networks as Tymnet. Habitat was designed for LucasFilm, and has been on-line for about a year and a half. It is a completely decentralized, connectionist system. The technology at the user interface was intended to be simple, this in order to minimize the costs of getting on-line. Habitat is designed to run on a Commodore 64 computer, a piece of very old technology in computer terms (in other words, at least ten years old), but Morningstar and Farmer have milked an amazing amount of effective bandwidth out of the machine. The Commodore 64 is very inexpensive and readily available. Almost anyone can buy one if, as one Habitat participant said, 'they don't already happen to have one sitting around being used as a doorstop.' Commodore 64s cost $ 100 at such outlets as Toys R Us.

Habitat existed first as a 35-foot mural located in a building in Sausalito, California, but, on-line, each area of the mural represents an entirely expandable area in the cyberspace, be it a forest, a plain, or a city. Habitat is inhabitable in that, when the user signs on, he or she has a window into the ongoing social life of the cyberspace – the community 'inside' the computer. The social space itself is represented by a cartoonlike frame. The virtual person who is the user's delegated agency is represented by a cartoon figure that may be customized from a menu of body parts. When the user wishes his/her character to speak, s/he types out the words on the Commodore's keyboard, and these appear in a speech balloon over the head of the user's character. The speech balloon is visible to any other user nearby in the virtual space.[1] The user sees whatever other people are in the immediate vicinity in the form of other figures.

Habitat is a two-dimensional example of what William Gibson called a 'consensual hallucination.' First, according to Morningstar and Farmer, it has well-known protocols for encoding and exchanging information. By generally accepted usage among cyberspace engineers, this means it is consensual. The simulation software uses agents that can transform information to simulate environment. This means it is an hallucination.

Habitat has proved to be incontrovertibly social in character. During Habitat's beta test, several social institutions sprang up spontaneously. As Randall Farmer points out in his report on the initial test run, there were marriages and divorces, a church (complete with a real-world Greek Orthodox minister), a loose guild of thieves, an elected sheriff (to combat the thieves), a newspaper with a rather eccentric editor, and before long two lawyers hung up their shingles to sort out claims. And this was with only 150 people. My vision (of Habitat) encompasses tens of thousands of simultaneous participants.

Lessons of the Third Epoch

In the third epoch the participants of electronic communities seem to be acquiring skills that are useful for the virtual social environments developing in late twentieth-century technologized nations. Their participants have learned to delegate their agency to body-representatives that exist in an imaginal space contiguously with representatives of other individuals. They have become accustomed to what might be called lucid dreaming in an awake state – to a constellation of activities much like reading, but an active and interactive reading, a participatory social practice in which the actions of the reader have consequences in the world of the dream or the book.

In the third epoch the older metaphor of reading is undergoing a transformation in a textual space that is consensual, interactive, and haptic, and that is constituted through inscription practices – the production of microprocessor code. Social spaces are beginning to appear that are simultaneously natural, artificial, and constituted by inscription. The boundaries between the social and the natural and between biology and technology are beginning to take on the generous permeability that characterizes communal space in the fourth epoch.

Epoch Four

Arguably the single most significant event for the development of fourth-stage virtual communities was the publication of William Gibson's science fiction novel *Neuromancer*. *Neuromancer* represents the dividing line between the third and fourth epochs not because it signaled any technological development, but because it crystallized a new community, just as Boyle's scientific papers and *Paul and Virginia* did in an earlier age.

Neuromancer reached the hackers who had been radicalized by George Lucas's powerful cinematic evocation of humanity and technology infinitely extended, and it reached the technologically literate and socially disaffected who were searching for social forms that could transform the fragmented anomie that characterized life in Silicon Valley and all electronic industrial ghettos. In a single stroke, Gibson's powerful vision provided for them the imaginal public sphere and refigured discursive community that established the grounding for the possibility of a new kind of social interaction. As with *Paul and Virginia* in the time of Napoleon and Dupont de Nemours, *Neuromancer* in the time of Reagan and DARPA is a massive intertextual presence not only in other literary productions of the 1980s, but in technical publications, conference topics, hardware design, and scientific and technological discourses in the large.

The three-dimensional inhabitable cyberspace described in *Neuromancer* does not yet exist, but the groundwork for it can be found in a series of experiments in both the military and private sectors.

Many VR engineers concur that the tribal elders of 3-D virtual systems are Scott Fisher and Ivan Sutherland, formerly at MIT, and Tom Furness, with the Air Force. In 1967–68, Sutherland built a see-through helmet at the MIT Draper Lab in Cambridge. This system used television screens and half-silvered mirrors, so that the environment was visible through the TV displays. It was not designed to provide a surround environment. In 1969–70 Sutherland went to the University of Utah, where he continued this work, doing things with vector-generated computer graphics and maps, still see-through technology.

In his lab were Jim Clark, who went on to start Silicon Graphics, and Don Vickers.

Tom Furness had been working on VR systems for 15 years or more – he started in the mid-seventies at Wright-Patterson Air Force Base. His systems were also see-through, rather than enclosing. He pushed the technology forward, particularly by adopting the use of high-resolution CRTs. Furness's system, designed for the USAF, was an elaborate flight simulation cyberspace employing a helmet with two large CRT devices, so large and cumbersome that it was dubbed the 'Darth Vader helmet.' He left Wright-Patterson in 1988–89 to start the Human Interface Technology Lab at the University of Washington.

Scott Fisher started at MIT in the machine architecture group. The MA group worked on developing stereo displays and crude helmets to contain them, and received a small proportion of their funding from DARPA. When the group terminated the project, they gave the stereo displays to another group at UNC (University of North Carolina), which was developing a display device called the Pixel Planes Machine. In the UNC lab were Henry Fuchs and Fred Brooks, who had been working on force feedback with systems previously developed at Argonne and Oak Ridge National labs. The UNC group worked on large projected stereo displays, but was aware of Sutherland's and Furness's work with helmets, and experimented with putting a miniature display system into a helmet of their own. Their specialties were medical modeling, molecular modeling, and architectural walk-through. The new Computer Science building at UNC was designed partially with their system. Using their software and 3-D computer imaging equipment, the architects could 'walk through' the full-sized virtual building and examine its structure. The actual walk-through was accomplished with a treadmill and bicycle handlebars. The experiment was so successful that during the walk-through one of the architects discovered a misplaced wall that would have cost hundreds of thousands of dollars to fix once the actual structure had been built.

In 1982, Fisher went to work for Atari. Alan Kay's style at Atari was to pick self-motivated people and then turn them loose, on anything from flight simulation to personal interactive systems. The lab's philosophy was at the extreme end of visionary. According to Kay, the job of the group was to develop products not for next year or even for five years away, but for no less than 15 to 20 years in the future. In the corporate climate of the 1980s, and in particular in Silicon Valley, where product life and corporate futures are calculated in terms of months, this approach was not merely radical but stratospheric. For the young computer jocks, the lure of Silicon Valley and of pushing the limits of computer imaging into the far future was irresistible, and a group of Cambridge engineers, each outstanding in their way, made the trek out to the coast. Eric Gullichsen arrived first, then Scott Fisher and Susan Brennan, followed a year later by Ann Marion. Michael Naimark was already there, as was Brenda Laurel. Steve Gans was the last to arrive.

As it turned out, this was not a good moment to arrive at Atari. When the Atari lab closed, Ann Marion and Alan Kay went to Apple (followed by a drove of other Atari expatriates), where they started the Vivarium project and continued their research. Susan Brennan went first to the Stanford Psychology Department and also Hewlett-Packard, which she left in 1990 to teach at CUNY Stony Brook. Michael Naimark became an independent producer and designer of interactive video and multimedia art. William Bricken and Eric Gullichsen took jobs at Autodesk, the largest manufacturer of CAD software, where they started a research group called Cyberia.

Scott Fisher went to work for Dave Nagel, head of the NASA-Ames View Lab. To go with their helmet, the Ames lab had developed a primitive sensor to provide the computer with information about the position of the user's hand. The early device used a simple glove with strain gauges wired to two fingers. They contracted with VPL, Inc. to develop it further, using software written in collaboration with Scott. The Ames group referred to the software as 'gesture editors.' The contract started in 1985, and VPL delivered the first glove in March 1986. The Ames group intended to apply the glove and software to such ideas as surgical simulation, 3-D virtual surgery for medical students. In 1988, Dave Nagel left the Ames laboratory to become director of the Advanced Technology Group (ATG) at Apple.

Lusting for images, such organizations as SIGGRAPH gobbled up information about the new medium and spread it out through its swarm of networks and publications. The audience, made up largely of young, talented, computer-literate people in both computer science and art, and working in such fields as advertising, media, and the fine arts, had mastered the current state of the art in computers and was hungry for the next thing. LucasFilm (later LucasArts) in Marin, now doing the bulk of all computerized special effects for the film industry, and Douglas Trumbull's EEG in Hollywood, fresh from their spectacular work on *Blade Runner*, had made the production of spectacular visual imaginaries an everyday fact. They weren't afraid to say that they had solved all of the remaining problems with making artificial images, under particular circumstances, indistinguishable from 'real' ones – a moment that Stewart Brand called '(t)he end of photography as evidence for anything.' Now the artists and engineers who worked with the most powerful imaging systems, like Lucas's Pixar, were ready for more. They wanted to be able to get inside their own fantasies, to experientially inhabit the worlds they designed and built but could never enter. VR touched the same nerve that *Star Wars* had, the englobing specular fantasy made real.

Under Eric Gullichsen and William Bricken, the Autodesk Cyberspace Project quickly acquired the nickname Cyberia. John Walker, president of Autodesk, had seen the UNC architectural system and foresaw a huge market for virtual CAD – 3-D drawings that the designers could enter. But after a year or so, Autodesk shrank the Cyberia project. Eric Gullichsen left to start Sense8, a manufacturer of low-end VR systems. William Bricken (and later his wife

Meredith) left the company to take up residence at the University of Washington, where Tom Furness and his associates had started the Human Interface Technology Laboratory. Although there were already academic-based research organizations in existence at that time (Florida, North Carolina), and some of them (Florida) were financed at least in part by DOD, the HIT lab became the first academic organization to secure serious research funding from private industry.

During this period, when *Neuromancer* was published, 'virtual reality' acquired a new name and a suddenly prominent social identity as 'cyberspace.' The critical importance of Gibson's book was partly due to the way that it triggered a conceptual revolution among the scattered workers who had been doing virtual reality research for years: As task groups coalesced and dissolved, as the fortunes of companies and projects and laboratories rose and fell, the existence of Gibson's novel and the technological and social imaginary that it articulated enabled the researchers in virtual reality – or, under the new dispensation, cyberspace – to recognize and organize themselves as a community.

By this time private industry, represented by such firms as American Express, PacBell, IBM, MCC, Texas Instruments, and NYNEX, were beginning to explore the possibilities and commercial impact of cyberspace systems. That is not to say that people were rushing out to purchase tickets for a cyberspace vacation! The major thrust of the industrial and institutional commitment to cyberspace research was still focused on data manipulation – just as Gibson's *zaibatsu* did in *Neuromancer*. Gibson's cowboys were outlaws in a military-industrial fairyland dominated by supercomputers, artificial intelligence devices, and data banks. Humans were present, but their effect was minimal. There is no reason to believe that the cyberspaces being designed at NASA or Florida will be any different. However, this knowledge does not seem to daunt the 'real' cyberspace workers. Outside of their attention to the realities of the marketplace and workplace, the young, feisty engineers who do the bulk of the work on VR systems continue their discussions and arguments surrounding the nature and context of virtual environments. That these discussions already take place in a virtual environment – the great, sprawling international complex of commercial, government, military, and academic computers known as Usenet – is in itself suggestive.

DECOUPLING THE BODY AND THE SUBJECT

The illusion will be so powerful you won't be able *to tell what's real and what's not.* (Steve Williams)

In her complex and provocative 1984 study *The Tremulous Private Body*, Frances Barker suggests that, because of the effects of the Restoration on the social and political imaginary in Britain (1660 and on), the human body gradually ceased to be perceived as public spectacle, as had previously been the case, and became privatized in new ways. In Barker's model of the post-Jacobean citizen, the social economy of the body became rearranged in such a way as to interpose several layers between the individual and public space. Concomitant with this removal of the body from a largely public social economy, Barker argues that the subject, the 'I' or perceiving self that Descartes had recently pried loose from its former unity with the body, reorganized, or was reorganized, in a new economy of its own. In particular, the subject, as did the body, ceased to constitute itself as public spectacle and instead fled from the public sphere and constituted itself in *text* – such as Samuel Pepys' diary (1668).

Such changes in the social economy of both the body and the subject, Barker suggests, very smoothly serve the purposes of capital accumulation. The product of a privatized body and of a subject removed from the public sphere is a social monad more suited to manipulation by virtue of being more isolated. Barker also makes a case that the energies of the individual, which were previously absorbed in a complex public social economy and which regularly returned to nourish the sender, started backing up instead, and needing to find fresh outlets. The machineries of capitalism handily provided a new channel for productive energy. Without this damming of creative energies, Barker suggests, the industrial age, with its vast hunger for productive labor and the consequent creation of surplus value, would have been impossible.

In Barker's account, beginning in the 1600s in England, the body became progressively more hidden, first because of changing conventions of dress, later by conventions of spatial privacy. Concomitantly, the self, Barker's 'subject,' retreated even further inward, until much of its means of expression was through texts. Where social communication had been direct and personal, a warrant was developing for social communication to be indirect and delegated through communication technologies – first pen and paper, and later the technologies and market economics of print. The body (and the subject, although he doesn't lump them together in this way) became 'the site of an operation of power, of an exercise of meaning ... a transition, effected over a long period of time, from a socially visible object to one which can no longer be seen' (Barker 1984: 13).

While the subject in Barker's account became, in her words, 'raging, solitary, productive,' what it produced was text. On the other hand, it was the newly hidden Victorian body that became physically productive and that later provided the motor for the industrial revolution; it was most useful as a brute body, for which the creative spark was an impediment. In sum, the body became more physical, while the subject became more textual, which is to say nonphysical.

If the information age is an extension of the industrial age, with the passage of time the split between the body and the subject should grow more pronounced still. But in the fourth epoch the split is simultaneously growing and disappearing. The socioepistemic mechanism by which bodies mean is undergoing a deep restructuring in the latter part of the twentieth century, finally fulfilling the furthest extent of the isolation of those bodies through which its domination is authorized and secured.

I don't think it is accidental that one of the earliest, textual, virtual communities – the community of gentlemen assembled by Robert Boyle during his debates with Hobbes – came into existence at the moment about which Barker is writing. The debate between Boyle and Hobbes and the production of Pepys' diary are virtually contemporaneous. In the late twentieth century, Gibson's *Neuromancer* is simultaneously a perverse evocation of the Restoration subject and its annihilation in an implosion of meaning from which arises a new economy of signification.

Barker's work resonates in useful ways with two other accounts of the evolution of the body and the subject through the interventions of late twentieth-century technologies: Donna Haraway's 'A Manifesto for Cyborgs' and 'The Biopolitics of Postmodern Bodies' (1985, 1988). Both these accounts are about the collapse of categories and of the boundaries of the body. (Shortly after being introduced to Haraway's work I wrote a very short paper called 'Sex And Death among the Cyborgs.' The thesis of 'Sex And Death' was similar to Haraway's.) The boundaries between the subject, if not the body, and the 'rest of the world' are undergoing a radical refiguration, brought about in part through the mediation of technology. Further, as Baudrillard and others have pointed out, the boundaries between technology and nature are themselves in the midst of a deep restructuring. This means that many of the usual analytical categories have become unreliable for making the useful distinctions between the biological and the technological, the natural and artificial, the human and mechanical, to which we have become accustomed.

François Dagognet suggests that the recent debates about whether nature is becoming irremediably technologized are based on a false dichotomy: namely that there exists, here and now, a category 'nature' which is 'over here,' and a category 'technology' (or, for those following other debates, 'culture') which is 'over there.' Dagognet argues on the contrary that the category 'nature' has not existed for thousands of years ... not since the first humans deliberately planted gardens or discovered slash-and-burn farming. I would argue further that 'Nature,' instead of representing some pristine category or originary state of being, has taken on an entirely different function in late twentieth-century economies of meaning. Not only has the character of nature as yet another coconstruct of culture become more patent, but is has become nothing more (or less) than an ordering factor – a construct by means of which we attempt to *keep technology visible* as something separate from our 'natural' selves and our everyday lives. In other words, the category 'natural' rather than referring to any object or category in the world, is a *strategy* for maintaining boundaries for political and economic ends, and thus a way of making meaning. (In this sense, the project of reifying a 'natural' state over and against a technologized 'fallen' one is not only one of the industries of postmodern nostalgia, but also part of a binary, oppositional cognitive style that some maintain is part of our society's pervasively male epistemology.)

These arguments imply as a corollary that 'technology,' as we customarily think of it, does not exist either; that we must begin to rethink the category of technology as also one that exists only because of its imagined binary opposition to another category upon which it operates and in relation to which it is constituted. In a recent paper Paul Rabinow asks what kind of being might thrive in a world in which nature is becoming increasingly technologized. What about a being who has learned to live in a world in which, rather than nature becoming technologized, technology *is* nature – in which the boundaries between subject and environment have collapsed?

Phone Sex Workers and VR Engineers

I have recently been conducting a study of two groups who seemed to instantiate productive aspects of this implosion of boundaries. One is phone sex workers. The other is computer scientists and engineers working on VR systems that involve making humans visible in the virtual space. I was interested in the ways in which these groups, which seem quite different, are similar. For the work of both is about representing the human body through limited communication channels, and both groups do this by coding cultural expectations as tokens of meaning.

Computer engineers seem fascinated by VR because you not only program a world, but in a real sense inhabit it. Because cyberspace worlds can be inhabited by communities, in the process of articulating a cyberspace system, engineers must model cognition and community; and because communities are inhabited by bodies, they must model bodies as well. While cheap and practical systems are years away, many workers are already hotly debating the form and character of the communities they believe will spring up in their quasi-imaginary cyberspaces. In doing so, they are articulating their own assumptions about bodies and sociality and projecting them onto the codes that define cyberspace systems. Since, for example, programmers create the codes by which VR is generated in interaction with workers in widely diverse fields, how these heterogenous co-working groups understand cognition, community, and bodies will determine the nature of cognition, community, and bodies in VR.

Both the engineers and the sex workers are in the business of constructing tokens that are recognized as objects of desire. Phone sex is the process of provoking, satisfying, *constructing* desire through a single mode of communication, the telephone. In the process, participants draw on a repertoire of cultural codes to construct a scenario that compresses large amounts of information into a very small space. The worker verbally codes for gesture, appearance, and proclivity, and expresses these as tokens, sometimes in no more than a word. The client uncompresses the tokens and constructs a dense, complex interactional image. In these interactions desire appears as a product of the tension between embodied reality and the emptiness of the token, in the forces that maintain the preexisting codes by which the token is constituted. The client mobilizes expectations and preexisting codes for body in the modalities that are not expressed in the token; that is, tokens in phone sex are purely verbal, and the client uses cues in the verbal token to construct a multimodal object of desire with attributes of shape, tactility, odor, etc. This act is thoroughly individual and interpretive; out of a highly compressed token of desire the client constitutes meaning that is dense, locally situated, and socially particular.

Bodies in cyberspace are also constituted by descriptive codes that 'embody' expectations of appearance. Many of the engineers currently debating the form and nature of cyberspace are the young turks of computer engineering, men in their late teens and twenties, and they are preoccupied with the things with which postpubescent men have always been preoccupied. This rather steamy group will generate the codes and descriptors by which bodies in cyberspace are represented. Because of practical limitations, a certain amount of their discussion is concerned with data compression and tokenization. As with phone sex, cyberspace is a relatively narrow-bandwidth representational medium, visual and aural instead of purely aural to be sure, but how bodies are represented will involve how *recognition* works.

One of the most active sites for speculation about how *recognition* might work in cyberspace is the work of computer game developers, in particular the area known as interactive fantasy (IF). Since Gibson's first book burst onto the hackers' scene, interactive fantasy programmers (in particular, Laurel and others) have been taking their most durable stock-in-trade and speculating about how it will be deployed in virtual reality scenarios. For example, how, if they do, will people make love in cyberspace – a space in which everything, including bodies, exists as something close to a metaphor. Fortunately or unfortunately, however, everyone is still preorgasmic in virtual reality.

When I began the short history of virtual systems, I said that I wanted to use accounts of virtual communities as an entry point into a search for two things: an apparatus for the production of community and an apparatus for the production of body. Keeping in mind that this chapter is necessarily brief, let me look at the data so far:

- Members of electronic virtual communities act as if the community met in a physical public space. The number of times that on-line conferencees refer to the conference as an architectural place and to the mode of interaction in that place as being social is overwhelmingly high in proportion to those who do not. They say things like 'This is a nice place to get together' or 'This is a convenient place to meet.'
- The virtual space is most frequently visualized as Cartesian. On-line conferencees tend to visualize the conference system as a three-dimensional space that can be mapped in terms of Cartesian coordinates, so that some branches of the conference are 'higher up' and others 'lower down.' (One

of the commands on the Stuart II conference moved the user 'sideways.')
Gibson's own visualization of cyberspace was Cartesian. In consideration
of the imagination I sometimes see being brought to bear on virtual spaces,
this odd fact invites further investigation.

- Conferencees act as if the virtual space were inhabited by bodies.
Conferencees construct bodies on-line by describing them, either
spontaneously or in response to questions, and articulate their discourses
around this assumption.

- Bodies in virtual space have complex erotic components. Conferencees
may flirt with each other. Some may engage in 'netsex,' constructing
elaborate erotic mutual fantasies. Erotic possibilities for the virtual body
are a significant part of the discussions of some of the groups designing
cyberspace systems. The consequences of virtual bodies are considerable in
the local frame, in that conferencees mobilize significant erotic tension in
relation to their virtual bodies. In contrast to the conferences, the
bandwidth for physicalities in phone sex is quite limited. (One worker
said ironically, '(o)n the phone, every female sex worker is white, five feet
four, and has red hair.')

- The meaning of locality and privacy is not settled. The field is rife with
debates about the legal status of communications within the networks.
One such, for example, is about the meaning of inside and outside.
Traditionally, when sending a letter one preserves privacy by enclosing
it in an envelope. But in electronic mail, for example, the address is part of
the message. The distinction between inside and outside has been erased,
and along with it the possibility of privacy. Secure encryption systems are
needed.[2]

- Names are local labels. 'Conferencees' seem to have no difficulty
addressing, befriending, and developing fairly complex relationships with
the delegated puppets – agents – of other conferencees. Such relationships
remain stable as long as the provisional name ('handle') attached to the
puppet does not change, but an unexpected observation was that
relationships remain stable when the conferencee decides to change
handles, as long as fair notice is given. Occasionally a conferencee will
have several handles on the same conference, and a constructed identity for
each. Other conferencees may or may not be aware of this. Conferencees
treat others' puppets as if they were embodied people meeting in a public
space nonetheless.

PRIVATE BODY, PUBLIC BODY, AND CYBORG ENVY

Partly, my interest in VR engineers stems from observations that suggest that
they while are surely engaged in saving the project of late-twentieth-century
capitalism, they are also inverting and disrupting its consequences for the body
as object of power relationships. They manage both to preserve the privatized
sphere of the individual – which Barker characterizes as 'raging, solitary,

productive' – as well as to escape to a position that is of the spectacle and incontrovertibly public. But this occurs under a new definition of public and private: one in which warrantability is irrelevant, spectacle is plastic and negotiated, and desire no longer grounds itself in physicality. Under these conditions, one might ask, will the future inhabitants of cyberspace 'catch' the engineers' societal imperative to construct desire in gendered, binary terms – coded into the virtual body descriptors – or will they find more appealing the possibilities of difference unconstrained by relationships of dominance and submission? Partly this will depend upon how 'cyberspaceians' engage with the virtual body.

Vivian Sobchack, in her 1987 discussion of cinematic space excludes the space of the video and computer screen from participation in the production of an 'apparatus of engagement.' Sobchack describes engagement with cinematic space as producing a thickening of the present ... a 'temporal simultaneity (that) also extends presence spatially – transforming the "thin" abstracted space of the machine into a thickened and concrete world.' Contrasted with video, which is to say with the electronic space of the CRT screen and with its small, low-resolution, and serial mode of display, the viewer of cinema engages with the apparatus of cinematic production in a way that produces 'a space that is deep and textural, that can be materially inhabited ... a specific and mobile engagement of embodied and enworlded subjects/objects whose visual/ visible activity prospects and articulates a shifting field of vision from a world that always exceeds it.' Sobchack speaks of electronic space as 'a phenomenological structure of sensual and psychological experience that seems to belong to no-body.' Sobchack sees the computer screen as 'spatially decentered, weakly temporalized and quasi-disembodied.'

This seems to be true, as long as the mode of engagement remains that of spectator. But it is the quality of direct physical and kinesthetic engagement, the enrolling of hapticity in the service of both the drama and the dramatic, which is not part of the cinematic mode. The cinematic mode of engagement, like that of conventional theater, is mediated by two modalities; the viewer experiences the presentation through sight and hearing. The electronic screen is 'flat,' so long as we consider it in the same bimodal way. But it is the potential for interaction that is one of the things that distinguishes the computer from the cinematic mode, and that transforms the small, low-resolution, and frequently monochromatic electronic screen from a novelty to a powerfully gripping force. Interaction is the physical concretization of a desire to escape the flatness and merge into the created system. It is the sense in which the 'spectator' is more than a participant, but becomes both participant in and creator of the simulation. In brief, it is the sense of unlimited power which the dis/embodied simulation produces, and the different ways in which socialization has led those always-embodied participants confronted with the sign of unlimited power to respond.

In quite different terms from the cinematic, then, cyberspace 'thickens' the present, producing a space that is deep and textural, and one that, in

Sobchack's terms, can be materially inhabited. David Tomas, in his article 'The Technophilic Body' (1989), describes cyberspace as 'a purely spectacular, kinesthetically exciting, and often dizzying sense of bodily freedom.' I read this in the additional sense of freedom *from* the body, and in particular perhaps, freedom from the sense of loss of control that accompanies adolescent male embodiment. Cyberspace is surely also a concretization of the psycho-analytically framed desire of the male to achieve the 'kinesthetically exciting, dizzying sense' of freedom.

Some fiction has been written about multimodal, experiential cinema. But the fictional apparatus surrounding imaginary cybernetic spaces seems to have proliferated and pushed experiential cinema into the background. This is because cyberspace is part of, not simply the medium for, the action. Sobchack, on the other hand, argues that cinematic space possesses a power of engage-ment that the electronic space cannot match:

> Semiotically engaged as subjective and intentional, as presenting repre-sentation of the objective world ... The spectator(s) can share (and thereby to a degree interpretively alter) a film's presentation and repre-sentation of embodied experience. (Forthcoming)

Sobchack's argument for the viewer's intentional engagement of cinematic space, slightly modified, however, works equally well for the cybernetic space of the computer. That is, one might say that the console cowboy is also '... semiotically engaged as subjective and intentional, as presenting representation of a *sub*jective world ... the spectator can share (and thereby to a high degree interpretively alter) a simulation's presentation and representation of experi-ence which may be, through cybernetic/semiotic operators not yet existent but present and active in fiction (the cyberspace deck), mapped back upon the physical body.'

In psychoanalytic terms, for the young male, unlimited power first suggests the mother. The experience of unlimited power is both gendered, and, for the male, fraught with the need for control, producing an unresolvable need for reconciliation with an always absent structure of personality. An 'absent structure of personality' is also another way of describing the peculiarly seductive character of the computer that Turkle characterizes as the 'second self.' Danger, the sense of threat as well as seductiveness that the computer can evoke, comes from both within and without. It derives from the complex interrelationships between human and computer, and thus partially within the human; and it exists quasi-autonomously within the simulation. It constitutes simultaneously the senses of erotic pleasure and of loss of control over the body. Both also constitute a constellation of responses to the simulation that deeply engage fear, desire, pleasure, and the need for domination, subjugation, and control.

It seems to be the engagement of the adolescent male within humans of both sexes that is responsible for the seductiveness of the cybernetic mode. There is also a protean quality about cybernetic interaction, a sense of physical as well as conceptual mutability that is implied in the sense of exciting, dizzying physical movement within purely conceptual space. I find that reality hackers experience a sense of longing for an embodied conceptual space like that which cyberspace suggests. This sense, which seems to accompany the desire to cross the human/machine boundary, to penetrate and merge, which is part of the evocation of cyberspace, and which shares certain conceptual and affective characteristics with numerous fictional evocations of the inarticulate longing of the male for the female, I characterize as *cyborg envy*.

Smoothness implies a seductive tactile quality that expresses one of the characteristics of cyborg envy: In the case of the computer, a desire literally to enter into such a discourse, to penetrate the smooth and relatively affectless surface of the electronic screen and enter the deep, complex, and tactile (individual) cybernetic space or (consensual) cyberspace within and beyond. Penetrating the screen involves a state change from the physical, biological space of the embodied viewer to the symbolic, metaphorical 'consensual hallucination' of cyberspace; a space that is a locus of intense desire for refigured embodiment.

The act of programming a computer invokes a set of reading practices both in the literary and cultural sense. 'Console cowboys' such as the cyberspace warriors of William Gibson's cyberpunk novels proliferate and capture the imagination of large groups of readers. Programming itself involves constant creation, interpretation, and reinterpretation of languages. To enter the discursive space of the program is to enter the space of a set of variables and operators to which the programmer assigns names. To enact naming is simultaneously to possess the power of, and to render harmless, the complex of desire and fear that charge the signifiers in such a discourse; to enact naming within the highly charged world of surfaces that is cyberspace is to appropriate the surfaces, to incorporate the surfaces into one's own. Penetration translates into envelopment. In other words, to enter cyberspace is to physically *put on* cyberspace. To become the cyborg, to put on the seductive and dangerous cybernetic space like a garment, is to put on the *female*. Thus cyberspace both *dis*embodies, in Sobchack's terms, but also *re*embodies in the polychrome, hypersurfaced cyborg character of the console cowboy. As the charged, multigendered, hallucinatory space collapses onto the personal physicality of the console cowboy, the intense tactility associated with such a reconceived and refigured body constitutes the seductive quality of what one might call the *cybernetic act*.

In all, the unitary, bounded, safely warranted body constituted within the frame of bourgeois modernity is undergoing a gradual process of translation to the refigured and reinscribed embodiments of the cyberspace community. Sex in the age of the coding metaphor – absent bodies, absent reproduction, perhaps related to desire, but desire itself refigured in terms of bandwidth and internal difference – may mean something quite unexpected. Dying in the age of the coding metaphor – in selectably inhabitable structures of signification, absent warrantability – gives new and disturbing meaning to the title of Steven Levine's book about the process, *Who Dies?*

CYBERSPACE, SOCIOTECHNICS, AND OTHER NEOLOGISMS

Part of the problem of 'going on in much the same way,' as Harry Collins put it, is in knowing what the same way is. At the close of the twentieth century, I would argue that two of the problems are, first, as in Paul Virilio's analysis, *speed*, and second, tightly coupled to speed, what happens as human physical evolution falls further and further out of synchronization with human cultural evolution. The product of this growing tension between nature and culture is stress.

Stress management is a major concern of industrial corporations. Donna Haraway points out that

> (t)he threat of intolerable rates of change and of evolutionary and ideological obsolescence are the framework that structure much of late twentieth-century medical, social and technological thought. Stress is part of a complex web of technological discourses in which the organism becomes a particular kind of communications system, strongly analogous to the cybernetic machines that emerged from the war to reorganize ideological discourse and significant sectors of state, industrial, and military practice. .... Utilization of information at boundaries and transitions, biological or mechanical, is a critical capacity of systems potentially subject to stress, because failure to correctly apprehend and negotiate rapid change could result in communication breakdown – a problem which engages the attention of a broad spectrum of military, governmental, industrial and institutional interests. (1990: 186–230 passim)

The development of cyberspace systems – which I will refer to as part of a new *technics* – may be one of a widely distributed constellation of responses to stress, and secondly as a way of continuing the process of collapsing the categories of nature and culture that Paul Rabinow sees as the outcome of the new genetics. Cyberspace can be viewed as a toolkit for refiguring consciousness in order to permit things to go on in much the same way. Rabinow suggests that nature will be modeled on culture; it will be known and remade through technique. Nature will finally become artificial, just as culture becomes natural.

Haraway (1985) puts this in a slightly different way: 'The certainty of what counts as nature,' she says, '(that is, as) a source of insight, a subject for knowledge, and a promise of innocence – is undermined, perhaps fatally.' The change in the permeability of the boundaries between nature and technics that these accounts suggest does not simply mean that nature and technics mix – but that, seen from the technical side, technics become natural, just as, from Rabinow's anthropological perspective on the culture side, culture becomes artificial. In technosociality, the social world of virtual culture, technics is nature. When exploration, rationalization, remaking, and control mean the same thing, then nature, technics, and the structure of meaning have become indistinguishable. The technosocial subject is able successfully to navigate through this treacherous new world. S/he is constituted as part of the evolution of communications technology and of the human organism, in a time in which technology and organism are collapsing, imploding, into each other.

Electronic virtual communities represent flexible, lively, and practical adaptations to the real circumstances that confront persons seeking community in what Haraway (1987) refers to as 'the mythic time called the late twentieth century.' They are part of a range of innovative solutions to the drive for sociality – a drive that can be frequently thwarted by the geographical and cultural realities of cities increasingly structured according to the needs of powerful economic interests rather than in ways that encourage and facilitate habitation and social interaction in the urban context. In this context, electronic virtual communities are complex and ingenious strategies for *survival*. Whether the seemingly inherent seductiveness of the medium distorts the aims of those strategies, as television has done for literacy and personal interaction, remains to be seen.

So Much for Community. What about the Body?

No matter how virtual the subject may become, there is always a body attached. It may be off somewhere else – and that 'somewhere else' may be a privileged point of view – but consciousness remains firmly rooted in the physical. Historically, body, technology, and community constitute each other.

In her 1990 book *Gender Trouble*, Judith Butler introduces the useful concept of the 'culturally intelligible body,' or the criteria and the textual productions (including writing on or in the body itself) that each society uses to produce physical bodies that it recognizes as members. It is useful to argue that most cultural production of intelligibility is about reading or writing and takes place through the mediation of texts. If we can apply textual analysis to the narrow-bandwidth modes of computers and telephones, then we can examine the production of gendered bodies in cyberspace also as a set of tokens that code difference within a field of ideal types. I refer to this process as the production of the *legible* body.

The opposite production, of course, is of the *illegible* body, the 'boundary-subject' that theorist Gloria Anzaldúa calls the *Mestiza*, one who lives in the borderlands and is only partially recognized by each abutting society. Anzaldúa describes the Mestiza by means of a multiplicity of frequently conflicting accounts. There is no position, she shows, outside of the abutting societies themselves from which an omniscient overview could capture the essence of the Mestiza's predicament, nor is there any single account from within a societal framework that constitutes an adequate description.

If the Mestiza is an illegible subject, existing quantumlike in multiple states, then participants in the electronic virtual communities of cyberspace live in the borderlands of both physical and virtual culture, like the Mestiza. Their social system includes other people, quasi people or delegated agencies that represent specific individuals, and quasi agents that represent 'intelligent' machines, clusters of people, or both. Their ancestors, lower on the chain of evolution, are network conferencers, communities organized around texts such as Boyle's 'community of gentlemen' and the religious traditions based in holy scripture, communities organized around broadcasts, and communities of music such as the Deadheads. What separates the cyberspace communities from their ancestors is that many of the cyberspace communities interact in real time. Agents meet face-to-face, though as I noted before, under a redefinition of both 'meet' and 'face.'

I might have been able to make my point regarding illegible subjects without invoking the Mestiza as an example. But I make an example of a specific kind of person as a way of keeping the discussion grounded in individual bodies: in Paul Churchland's words, in the 'situated biological creatures' that we each are. The work of science is *about* bodies – not in an abstract sense, but in the complex and protean ways that we daily manifest ourselves as physical social beings, vulnerable to the powerful knowledges that surround us, and to the effects upon us of the transformative discourses of science and technology that we both enable and enact.

I am particularly conscious of this because much of the work of cyberspace researchers, reinforced and perhaps created by the soaring imagery of William Gibson's novels, assumes that the human body is 'meat' – obsolete, as soon as consciousness itself can be uploaded into the network. The discourse of visionary virtual world builders is rife with images of imaginal bodies, freed from the constraints that flesh imposes. Cyberspace developers foresee a time when they will be able to forget about the body. But it is important to remember that virtual community originates in, and must return to, the physical. No refigured virtual body, no matter how beautiful, will slow the death of a cyberpunk with AIDS. Even in the age of the technosocial subject, life is lived through bodies.

Forgetting about the body is an old Cartesian trick, one that has unpleasant consequences for those bodies whose speech is silenced by the act of our forgetting; that is to say, those upon whose labor the act of forgetting the body is founded – usually women and minorities. On the other hand, as Haraway points out, forgetting can be a powerful strategy; through forgetting, that which is already built becomes that which can be discovered. But like any powerful and productive strategy, this one has its dangers. Remembering – discovering – that bodies and communities constitute each other surely suggests a set of questions and debates for the burgeoning virtual electronic community. I hope to observe the outcome.

Acknowledgments

Thanks to Mischa Adams, Gloria Anzaldúa, Laura Chernaik, Heinz von Foerster, Thyrza Goodeve, John Hartigan, Barbara Joans, Victor Kytasty,

Roddey Reid, Chela Sandoval, Susan Leigh Star, and Sharon Traweek for their many suggestions; to Bandit (Seagate), Ron Cain (Borland), Carl Tollander (Autodesk), Ted Kaehler (Sun), Jane T. Lear (Intel), Marc Lentczner, Robert Orr (Amdahl), Jon Singer (soulmate), Brenda Laurel (Telepresence Research and all-around Wonderful Person); Joshua Susser, the Advanced Technology Group of Apple Computer, Inc., Tene Tachyon, Jon Shemitz, John James, and my many respondents in the virtual world of online BBSs. I am grateful to Michael Benedikt and friends and to the University of Texas School of Architecture for making part of the research possible, and to the participants in The First Conference on Cyberspace for their ideas as well as their collaboration in constituting yet another virtual community. In particular I thank Donna Haraway, whose work and encouragement have been invaluable.

NOTES

1. 'Nearby' is idiosyncratic and local in cyberspace. In the case of Habitat, it means that two puppets (body representatives) occupy that which is visible on both screens simultaneously. In practice this means that each participant navigates his or her screen 'window' to view the same area in the cyberspace. Because Habitat is consensual, the space looks the same to different viewers. Due to processor limitations only nine puppets can occupy the same window at the same time, although there can be more in the neighborhood (just offscreen).

2. Although no one has actually given up on encryption systems, the probable reason that international standards for encryption have not proceeded much faster has been the United States Government's opposition to encryption key standards that are reasonably secure. Such standards would prevent such agencies as the CIA from gaining access to communications traffic. The United States' diminishing role as a superpower may change this. Computer industries in other nations have overtaken the United States' lead in electronics and are beginning to produce secure encryption equipment as well. A side effect of this will be to enable those engaged in electronic communication to reinstate the inside-outside dichotomy, and with it the notion of privacy in the virtual social space.

BIBLIOGRAPHY

Allan, Francis, 'The End of Intimacy.' *Human Rights*, Winter 1984:55.

Anzaldúa, Gloria, *Borderlands/La Frontera: The New Mestiza* (San Francisco: Spinsters/Aunt Lute, 1987).

Barker, Francis, *The Tremulous Private Body: Essays in Subjection* (London: Methuen, 1984).

Baudrillard, Jean, *The Ecstasy of Communication*, trans. Bernard and Caroline Schutze, Sylvere Lotringer (New York: Semiotext(e), 1987).

Butler, Judith, *Gender Trouble: Feminism and the Subversion of Identity* (New York: Routledge, 1990).

Campbell, Joseph, *The Masks of God: Primitive Mythology* (New York: Viking, 1959).

Cohn, Carol, 'Sex and Death in the Rational World of Defense Intellectuals.' *Signs: Journal of Woman in Culture and Society*, 1987, 12:4.

de Certeau, Michel, 'The Arts of Dying: Celibatory machines.' In *Heterologies*, translated by Brian Massumi (Minneapolis: University of Minnesota Press, 1985).

Dewey, John, 'The Reflex Arc Concept in Psychology' [1896]. In J. J. McDermott (ed.), *The Philosophy of John Dewey* (Chicago: University of Chicago Press, 1981), pp. 36–148.

Edwards, Paul N., 'Artificial Intelligence and High Technology War: The perspective of the formal machine.' Silicon Valley Research Group Working Paper No. 6, 1986.

Gibson, William, *Neuromancer* (New York: Ace, 1984).

Habermas, J., *Communication and the Evolution of Society* (Boston: Beacon Press, 1979).

Haraway, Donna, 'A Manifesto for Cyborgs: Science, technology and socialist feminism in the 1980s,' *Socialist Review*, 1985, 80:65–107.

Haraway, Donna, 'Donna Haraway Reads National Geographic' (Paper Tiger, 1987) Video.

Haraway, Donna, 'The Biopolitics of Postmodern Bodies: Determinations of Self and Other in Immune System Discourse,' *Wenner Gren Foundation Conference on Medical Anthropology*, Lisbon, Portugal, 1988.

Haraway, Donna, 'Washburn and the New Physical Anthropology.' In *Primate Visions: Gender, Race, and Nature in the World of Modern Science* (New York: Routledge, 1990).

Haraway, Donna, 'The Promises of Monsters: A regenerative politics for inappropriate/d others.' In Treichler, P. and Nelson, G. (eds), *Cultural Studies Now and in the Future.*' Forthcoming.

Hayles, N. Katherine, 'Text Out Of Context: Situating postmodernism within an information society,' *Discourse*, 1987, 9:24–36.

Hayles, N. Katherine, 'Denaturalizing Experience: Postmodern literature and science.' Abstract from Conference on Literature and Science as Modes of Expression, sponsored by the Society for Literature and Science, Worcester Polytechnic Institute, October 8–11, 1987.

Head, Henry, *Studies in Neurology* (Oxford: Oxford University Press, 1920).

Head, Henry, *Aphasia and Kindred Disorders of Speech* (Cambridge: Cambridge University Press, 1926).

Hewitt, Carl, 'Viewing Control Structures as Patterns of Passing Messages,' *Artificial Intelligence*, 1977, 8:323–364.

Hewitt, C., 'The Challenge of Open Systems,' *Byte*, vol. 10 (April 1977).

Huyssen, Andreas, *After The Great Divide: Modernism, Mass Culture, Postmodernism* (Bloomington: Indiana University Press, 1986).

Jameson, Fredric, 'On Interpretation: Literature as a socially symbolic act.' In *The Political Unconscious* (Ithaca: Cornell University Press, 1981).

Lacan, Jacques, *The Language of the Self: The Function of Language in Psychoanalysis*, trans. Anthony Wilden (New York: Dell, 1968).

Lacan, Jacques, *The Four Fundamental Concepts of Psychoanalysis*, trans. Alain Sheridan, ed. Jacques-Alain Miller (London: Hogarth, 1977).

LaPorte, T. R. (ed.), *Organized Social Complexity: Challenge to Politics and Policy* (New Jersey: Princeton University Press, 1975).

Latour, Bruno, *The Pasteurization of France*, trans. Alan Sheridan and John Law (Cambridge: Harvard University Press, 1988).

Laurel, Brenda, 'Interface as Mimesis.' In D. A. Norman, and S. Draper (eds), *User, Centered System Design: New Perspectives on Human-Computer Interaction* (Hillsdale NJ: Lawrence Erlbaum Associates, 1986).

Laurel, Brenda, 'Reassessing Interactivity,' *Journal of Computer Game Design*, 1987, 1:3

Laurel, Brenda, 'Culture Hacking,' *Journal of Computer Game Design*, 1988, 1:8.

Laurel, Brenda, 'Dramatic Action and Virtual Reality.' In Proceedings of the 1989 NCGA Interactive Arts Conference, 1989a.

Laurel, Brenda, 'New Interfaces for Entertainment,' *Journal of Computer Game Design*, 1989b, 2:5.

Laurel, Brenda, 'A Taxonomy of Interactive Movies,' *New Media News* (The Boston Computer Society), 1989c, 3:1.

Lehman-Wilzig, Sam, 'Frankenstein Unbound: Toward a legal definition of artificial intelligence,' *Futures*, December 1981, 447.

Levine, Steven, *Who Dies? An Investigation of Conscious Living and Conscious Dying* (Bath: Gateway Press, 1988).

Merleau-Ponty, Maurice, *Phenomenology of Perception*, trans. Colin Smith (New York Humanities Press, 1962).

Merleau-Ponty, Maurice, *Sense and Non-Sense*, trans. Hubert L. Dreyfus and Patricia Allen Dreyfus (Chicago: Northwestern University Press, 1964a).

Merleau-Ponty, Maurice, *Signs*, trans. Richard McCleary (Chicago: Northwestern University Press, 1964b).

Mitchell, Silas Weir, George Read Morehouse, and William Williams Keen, 'Gunshot Wounds and Other Injuries of Nerves.' Reprinted with biographical introductions by Ira M. Rutkow, *American Civil War Surgery Series*, vol. 3 (San Francisco Norman, 1989 [1864]).

Mitchell, Silas Weir, *Injuries of Nerves and Their Consequences*, with a new introduction by Lawrence C. McHenry, Jr, *American Academy of Neurology Reprint series*, vol. 2 (New York: Dover, 1965 [1872]).

Noddings, Nel, *Caring: A Feminine Approach to Ethics and Moral Education* (Berkeley: University of California Press, 1984).

Reid, Roddey, 'Tears For Fears: Paul et Virginie, "family" and the politics of the sentimental body in pre-revolutionary France.' Forthcoming.

Rentmeister, Cacilia, 'Beruftsverbot fur Musen,' *Aesthetik und Kommunikation*, 25 (September 1976), 92–112.

Roheim, Geza, "Early Stages of the Oedipus Complex," *International Journal of Psycho-analysis*, vol. 9, 1928.

Roheim, Geza, 'Dream Analysis and Field Work.' In *Anthropology, Psychoanalysis and the Social Sciences* (New York: International Universities Press, 1947).

Shapin, Steven, and Schaffer, Simon, *Leviathan and the Air-Pump: Hobbes, Boyle, and the Experimental Life* (Princeton: Princeton University Press, 1985).

Sobchack, Vivian, 'The Address of the Eye: A semiotic phenomenology of cinematic embodiment.' Forthcoming.

Sobchack, Vivian, 'The Scene Of The Screen: Toward a phenomenology of cinematic and electronic "presence"'. In H. V. Gumbrecht and L. K. Pfeiffer (eds), *Materialitat des Kommunikation* (GDR: Suhrkarp-Verlag, 1988).

Sobchack, Vivian, *Screening Space: The American Science Fiction Film* (New York: Ungar, 1987).

Stone, Allucquere Rosanne, 1988. 'So That's What Those Two Robots Were Doing In The Park ... I Thought They Were Repairing Each Other! The Discourse of Gender, Pornography, and Artificial Intelligence.' Presented at Conference of the Feminist Studies Focused Research Activity, University of California, Santa Cruz, CA, October 1988.

Stone, Allucquere Rosanne, 'How Robots Grew Gonads: A cautionary tale.' Presented at *Contact V: Cultures of the Imagination*, Phoenix, AZ, March 28, 1989. Forthcoming in Funaro and Joans (eds), *Collected Proceedings of the Contact Conferences*.

Stone, Allucquere Rosanne, 'Sex and Death among the cyborgs: How to construct gender and boundary in distributed systems,' *Contact VI: Cultures of the Imagination* (Phoenix, AZ, 1990a).

Stone, Allucquere Rosanne, 'Sex and Death among the disembodied: How to provide counseling for the virtually preorgasmic.' In M. Benedikt (ed.), *Collected Abstracts of The First Cyberspace Conference* (The University of Texas at Austin, School of Architecture, 1990b).

Stone, Allucquere Rosanne, 'Aliens, Freaks, Monsters: The politics of virtual sexuality.' For the panel Gender and Cultural Bias in Computer Games, Computer Game Developers' Conference, San Jose, 1990c.

Stone, Allucquere Rosanne, 'Ecriture Artifactuelle: Boundary Discourse, Distributed Negotiation, and the Structure of Meaning in Virtual Systems,' forthcoming at the *1991 Conference on Interactive Computer Graphics*.

Stone, Christopher D., *Should Trees Have Standing? – Toward Legal Rights for Natural Objects* (New York: William A. Kaufman, 1974).

Theweleit, Klaus, *Male Fantasies*, vol.1 (Frankfurt am Main: Verlag Roter Stern, 1977).

Tomas, David, 'The Technophilic Body: On technicity in William Gibson's cyborg culture,' *New Formations*, 8, Spring, 1989.

Turkle, Sherry, *The Second Self: Computers and the Human Spirit* (New York: Simon and Schuster, 1984).

Von Foerster, Heinz (ed.), *Transactions of the Conference on Cybernetics* (New York: Josiah Macy, Jr Foundation, 1951).

Weiner, Norbert, *The Human Use of Human Beings* (New York: Avon, 1950).

Wilden, Anthony, *System and Structure: Essays in Communication and Exchange* 2nd ed. (New York: Tavistock, 1980).

Winograd, T., and Flores, C. F., *Understanding Computers and Cognition: A New Foundation for Design* (Norwood, NJ: Ablex, 1986).

Wolkomir, Richard, 'High-tech hokum is changing the way movies are made,' *Smithsonian* 10/90: 124, 1990.

5

THE FUTURE LOOMS:
WEAVING WOMEN AND
CYBERNETICS

Sadie Plant

Beginning with a passage from a novel:

> The woman brushed aside her veil, with a swift gesture of habit, and
> Mallory caught his first proper glimpse of her face. She was Ada Byron,
> the daughter of the Prime Minister. Lady Byron, the Queen of Engines.
> (Gibson and Sterling, 1990: 89)

Ada was not really Ada Byron, but Ada Lovelace, and her father was never
Prime Minister: these are the fictions of William Gibson and Bruce Sterling,
whose book *The Difference Engine* sets its tale in a Victorian England in which
the software she designed was already running; a country in which the Luddites
were defeated, a poet was Prime Minister, and Ada Lovelace still bore her
maiden name. And one still grander: Queen of Engines. Moreover she was still
alive. Set in the mid-1850s, the novel takes her into a middle-age she never saw:
the real Ada died in 1852 while she was still in her thirties. Ill for much of her
life with unspecified disorders, she was eventually diagnosed as suffering from
cancer of the womb, and she died after months of extraordinary pain.

Ada Lovelace, with whom the histories of computing and women's liberation are first directly woven together, is central to this paper. Not until a
century after her death, however, did women and software make their
respective and irrevocable entries on to the scene. After the military imperatives of the 1940s, neither would ever return to the simple service of man,
beginning instead to organize, design and arouse themselves, and so acquiring
unprecedented levels of autonomy. In later decades, both women and computers begin to escape the isolation they share in the home and office with the

establishment of their own networks. These, in turn, begin to get in touch with each other in the 1990s. This convergence of woman and machine is one of the preoccupations of the cybernetic feminism endorsed here, a perspective which owes a good deal to the work of Luce Irigaray, who is also important to this discussion.

The computer emerges out of the history of weaving, the process so often said to be the quintessence of women's work. The loom is the vanguard site of software development. Indeed, it is from the loom, or rather the process of weaving, that this paper takes another cue. Perhaps it is an instance of this process as well, for tales and texts are woven as surely as threads and fabrics. This paper is a yarn in both senses. It is about weaving women and cybernetics, and is also weaving women and cybernetics together. It concerns the looms of the past, and also the future which looms over the patriarchal present and threatens the end of human history.

Ada Lovelace may have been the first encounter between woman and computer, but the association between women and software throws back into the mythical origins of history. For Freud, weaving imitates the concealment of the womb: the Greek hystera; the Latin matrix. Weaving is woman's compensation for the absence of the penis, the void, the woman of whom, as he famously insists, there is 'nothing to be seen'. Woman is veiled, as Ada was in the passage above; she weaves, as Irigaray comments, 'to sustain the disavowal of her sex'. Yet the development of the computer and the cybernetic machine as which it operates might even be described in terms of the introduction of increasing speed, miniaturization and complexity to the process of weaving. These are the tendencies which converge in the global webs of data and the nets of communication by which cyberspace, or the matrix, are understood.

Today, both woman and the computer screen the matrix, which also makes its appearance as the veils and screens on which its operations are displayed. This is the virtual reality which is also the absence of the penis and its power, but already more than the void. The matrix emerges as the processes of an abstract weaving which produces, or fabricates, what man knows as 'nature': his materials, the fabrics, the screens on which he projects his own identity.

[...]

As well as his screens, and as his screens, the computer also becomes the medium of man's communication. Ada Lovelace was herself a great communicator: often she wrote two letters a day, and was delighted by the prospect of the telegraph. She is, moreover, often remembered as Charles Babbage's voice, expressing his ideas with levels of clarity, efficiency and accuracy he could never have mustered himself.

When Babbage displayed his Difference Engine to the public in 1833, Ada was a debutante, invited to see the machine with her mother, Lady Byron, who had herself been known as the Princess of Parallelograms for her mathematical prowess. Lady Byron was full of admiration for the machine, and it is clear that she had a remarkable appreciation of the subtle enormities of Babbage's invention. 'We both went to see the *thinking* machine (for such it seems) last Monday', she wrote. 'It raised several Nos. to the 2nd & 3rd powers, and extracted the root of a quadratic Equation' (Moore, 1977: 44).

Ada's own response was recorded by another woman, who wrote:

> While other visitors gazed at the working of the beautiful instrument with a sort of expression, and dare I say the same sort of feeling, that some savages are said to have shown on first seeing a looking glass or hearing a gun. ... Miss Byron, young as she was, understood its working, and saw the great beauty of the invention. (Moore, 1977: 44)

Ada had a passion for mathematics at an early age. She was admired and was greatly encouraged by Mary Somerville, herself a prominent figure in the scientific community and author of several scientific texts including the widely praised *Connection of the Physical Sciences*. Ada and Mary Somerville corresponded, talked together, and attended a series of lectures on Babbage's work at the Mechanics' Institute in 1835. Ada was fascinated by the engine, and wrote many letters to Babbage imploring him to take advantage of her brilliant mind. Eventually, and quite unsolicited, she translated a paper by Menabrea on Babbage's Analytic Engine, later adding her own notes at Babbage's suggestion. Babbage was enormously impressed with the translation, and Ada began to work with him on the development of the Analytical Engine.

Babbage had a tendency to flit between obsessions; a remarkably prolific explorer of the most fascinating questions of science and technology, he nevertheless rarely managed to complete his studies; neither the Difference Engine nor the Analytical Engine were developed to his satisfaction. Ada, on the other hand, was determined to see things through; perhaps her commitment to Babbage's machines was greater than his own. Knowing that the Difference Engine had suffered for lack of funding, publicity and organization, she was convinced that the Analytical Engine would be better served by her own attentions. She was often annoyed by what she perceived as Babbage's sloppiness, and after an argument in 1843, she laid down several severe conditions for the continuation of their collaboration: 'can you', she asked, with undisguised impatience,

> undertake to give your mind *wholly and undivided*, as a primary object that no engagement is to interfere with, to the consideration of all those matters in which I shall at times require your intellectual *assistance & supervision*; & can you promise not to *slur & hurry* things over; or to mislay & allow confusion & mistakes to enter into documents &c? (Moore, 1977: 171)

Babbage signed this agreement, but in spite of Ada's conditions, ill health and financial crises conspired to prevent the completion of the machine.

Ada Lovelace herself worked with a mixture of coyness and confidence;

attributes which often extended to terrible losses of self-esteem and mega-lomaniac delight in her own brilliance. Sometimes she was convinced of her own immortal genius as a mathematician; 'I hope to bequeath to future generations a *Calculus of the Nervous System*', she wrote in 1844. 'I am proceeding in a track quite peculiar & my own, I believe' (Moore, 1977: 216). At other times, she lost all confidence, and often wondered whether she should not have pursued her musical abilities, which were also fine. Ada was always trapped by the duty to be dutiful; caught in a cleft stick of duties, moral obligations she did not understand.

Ada's letters – and indeed her scientific writings – are full of suspicions of her own strange relation to humanity. Babbage called her his fairy, because of her dextrous mind and light presence, and this appealed to Ada's inherited romanticism. 'I deny the *Fairyism* to be entirely *imaginary*', she wrote: 'That *Brain* of mine is something more than merely *mortal*; as time will show; (if only my *breathing* & some other etceteras do not make too rapid a progress *towards* instead of from *mortality*)' (Moore, 1977: 98). When one of her thwarted admirers wrote to her: 'That you are a peculiar – *very peculiar* – specimen of the feminine race, you are yourself aware' (Moore, 1977: 202), he could only have been confirming an opinion she already – and rather admiringly – had of herself. Even of her own writing, she wrote: 'I am quite thunderstruck by the power of the writing. It is especially unlike a *woman's* style but neither can I compare it with any man's exactly' (Moore, 1977: 157). The words of neither a man nor a woman: who was Ada Lovelace? 'Before ten years are over', she wrote, 'the Devil's in it if I haven't sucked out some of the life blood from the mysteries of this universe, in a way that no purely mortal lips or brains could do' (Moore, 1977: 153).

Ada may have been Babbage's fairy, but she was not allowed to forget that she was also a wife, mother and victim of countless 'female disorders'. She had three children by the age of 24 of whom she later wrote: 'They are to me irksome *duties* & nothing more' (Moore, 1977: 229). Not until the 1840s did her own ill health lead her husband and mother to engage a tutor for the children, to whom she confided 'not only her present distaste for the company of her children but also her growing indifference to her husband, indeed to men in general' (Moore, 1977: 198). One admirer called her 'wayward, wandering ... deluded', and as a teenager she was considered hysterical, hypochondriac and rather lacking in moral fibre. She certainly suffered extra-ordinary symptoms, walking with crutches until the age of 17, and often unable to move. Her illnesses gave her some room for manoeuvre in the oppressive atmosphere of her maternal home. Perhaps Ada even cherished the solitude and peculiarity of her diseases; she certainly found them of philoso-phical interest, once writing: 'Do you know it is to me quite delightful to have a frame so susceptible That it is an experimental laboratory always about me, & inseparable from me. I walk about, not in a Snail-Shell, but in a Molecular Laboratory' (Moore, 1977: 218).

Not until the 1850s was cancer diagnosed: Lady Byron had refused to accept such news, still preferring to believe in her daughter's hysteria. Even Ada tended to the fashionable belief that over-exertion of the intellect had led to her bodily disorders; in 1844, while she was nevertheless continuing chemical and electrical experiments, she wrote: '*Many causes* have contributed to produce the past derangements; & I shall in future avoid them. One ingredient, (but only one among many) has been *too much Mathematics*' (Moore, 1977: 153–4). She died in November 1852 after a year of agonized decline.

Ada Lovelace often described her strange intimacy with death; it was rather the constraints of life with which she had to struggle. 'I mean to do *what I mean to do*', she once wrote, but there is no doubt that Ada was horribly confined by the familiar – her marriage, her children and her indomitable mother conspired against her independence, and it was no wonder that she was so attracted to the unfamiliar expanses of mathematical worlds. Ada's marriage prompted the following words from her mother: 'Bid adieu to your old companion Ada Byron with all her peculiarities, caprices, and self-seeking; determined that as A.K. you will live for others' (Moore, 1977: 69). But she never did. Scorning public opinion, she gambled, took drugs and flirted to excess. But what she did best was computer programming.

Ada Lovelace immediately saw the profound significance of the Analytical Engine, and she went to great lengths to convey the remarkable extent of its capacities in her writing. Although the Analytical Engine had its own limits, it was nevertheless a machine vastly different from the Difference Engine. As Ada Lovelace observed:

> The Difference Engine can in reality ... do nothing but *add*; and any other processes, not excepting those of simple subtraction, multiplication and division, can be performed by it only just to that extent in which it is possible, by judicious mathematical arrangement and artifices, to reduce them to a *series of additions*. (Morrison and Morrison, 1961: 250)

With the Analytical Engine, Babbage set out to develop a machine capable not merely of adding, but performing the 'whole of arithmetic'. Such an undertaking required the mechanization not merely of each mathematical operation, but the systematic bases of their functioning, and it was this imperative to transcribe the rules of the game itself which made the Analytical Engine a universal machine. Babbage was a little more modest, describing the Engine as 'a machine of the most general nature' (Babbage, 1961: 56), but the underlying point remains: the Analytical Engine would not merely synthesize the data provided by its operator, as the Difference Engine had done, but would incarnate what Ada Lovelace described as the very '*science of operations*'.

The Difference Engine, Ada Lovelace wrote, 'is the embodying of *one particular and very limited set of operations*, which ... may be expressed thus (+, +, +, +, +, +), or thus 6(+). Six repetitions of the one operation, +, is, in fact, the whole sum and object of that engine' (Morrison and Morrison, 1961: 249).

What impressed Ada Lovelace about the Analytical Engine was that, unlike the Difference Engine or any other machine, it was not merely able to perform certain functions, but was 'an *embodying of the science of operations*, constructed with peculiar reference to abstract number as the subject of those operations'. The Difference Engine could simply add up, whereas the Analytical Engine not only performed synthetic operations, but also embodied the analytic capacity on which these syntheses are based. 'If we compare together the powers and the principles of construction of the Difference and of the Analytic Engines', wrote Ada, 'we shall perceive that the capabilities of the latter are immeasurably more extensive than those of the former, and that they in fact hold to each other the same relationship as that of analysis to arithmetic' (Morrison and Morrison, 1961: 250). In her notes on Menabrea's paper, this is the point she stresses most: the Engine, she argues, is the very machinery of analysis, so that

> there is no finite line of demarcation which limits the powers of the Analytical Engine. These powers are co-extensive with our knowledge of the laws of analysis itself, and need be bounded only by our acquaintance with the latter. Indeed we may consider the engine as the *material and mechanical representative* of analysis. (Morrison and Morrison, 1961: 252)

The Difference Engine was '*founded on the principle of successive orders of differences*', while the

> distinctive characteristic of the Analytical Engine, and that which has rendered it possible to endow mechanism with such extensive faculties as bid fair to make this engine the executive right-hand of abstract algebra, is the introduction of the principle which Jacquard devised for regulating, by means for punched cards, the most complicated patterns in the fabrication of brocaded stuffs. (Morrison and Morrison, 1961: 252)

Indeed, Ada considered Jacquard's cards to be the crucial difference between the Difference Engine and the Analytical Engine. 'We may say most aptly', she continued, 'that the Analytical Engine *weaves Algebraical patterns*, just as the Jacquard loom weaves flowers and leaves. Here, it seems to us, resides much more of originality than the Difference Engine can be fairly entitled to claim' (Morrison and Morrison, 1961: 252). Ada's reference to the Jacquard loom is more than a metaphor: the Analytical Engine did indeed weave 'just as' the loom, operating, in a sense, as the abstracted process of weaving.

Weaving has always been a vanguard of machinic development, perhaps because, even in its most basic form, the process is one of complexity, always involving the weaving together of several threads into an integrated cloth. Even the drawloom, which is often dated back to the China of 1000 BC, involves sophisticated orderings of warp and weft if it is to produce the complex designs common in the silks of this period. This means that 'information is needed in large amounts for the weaving of a complex ornamental pattern. Even the most ancient Chinese examples required that about 1500 different warp threads be lifted in various combinations as the weaving proceeded' (Morrison and Morrison, 1961: xxxiv). With pedals and shuttles, the loom becomes what one historian refers to as the 'most complex human engine of them all', a machine which 'reduced everything to simple actions: the alternate movement of the feet worked the pedals, raising half the threads of the warp and then the other, while the hands threw the shuttle carrying the thread of the woof' (Braudel, 1973: 247). The weaver was integrated into the machinery, bound up with its operations, linked limb by limb to the processes. In the Middle Ages, and before the artificial memories of the printed page, squared paper charts were used to store the information necessary to the accurate development of the design. In early 18th-century Lyons, Basyle Bouchon developed a mechanism for the automatic selection of threads, using an early example of the punched paper rolls which were much later to allow pianos to play and type to be cast. This design was developed by Falcon a couple of years later, who introduced greater complexity with the use of punched cards rather than the roll. And it was this principle on which Jacquard based his own designs for the automated loom which revolutionized the weaving industry when it was introduced in the 1800s and continues to guide its contemporary development. Jacquard's machine strung the punch cards together, finally automating the operations of the machine and requiring only a single human hand. Jacquard's system of punch card programs brought the information age to the beginning of the 19th century. His automated loom was the first to store its own information, functioning with its own software, an early migration of control from weaver to machinery.

Babbage owned what Ada described as 'a beautiful woven portrait of Jacquard, in the fabrication of which 24,000 cards were required' (Morrison and Morrison, 1961: 281). Woven in silk at about 1000 threads to the inch, Babbage well understood that its incredible detail was due to the loom's ability to store and process information at unprecedented speed and volume and, when he began work on the Analytical Engine, it was Jacquard's strings of punch cards on which he based his designs. 'It is known as a fact', Babbage wrote, 'that the Jacquard loom is capable of weaving any design which the imagination of man may conceive' (Babbage, 1961: 55). Babbage's own contribution to the relentless drive to perfect the punch card system was to introduce the possibility of repeating the cards, or what, as Ada wrote,

> was technically designated *backing* the cards in certain groups according to certain laws. The object of this extension is to secure the possibility of bringing any particular card or set of cards onto use *any number of times successively in the solution of one problem.* (Morrison and Morrison, 1961: 264)

This was an unprecedented simulation of memory. The cards were selected by the machine as it needed them and effectively functioned as a filing system, allowing the machine to store and draw on its own information.

The punch cards also gave the Analytical Engine what Babbage considered foresight, allowing it to operate as a machine that remembers, learns and is guided by its own abstract functioning. As he began to work on the Analytical Engine, Babbage became convinced that 'nothing but teaching the Engine to foresee and then to act upon that foresight could ever lead me to the object I desired' (Babbage, 1961: 53). The Jacquard cards made memory a possibility, so that 'the Analytical Engine will possess a library of its own' (1961: 56), but this had to be a library to which the machine could refer both to its past and its future operations; Babbage intended to give the machine not merely a memory but also the ability to process information from the future of its own functioning. Babbage could eventually write that 'in the Analytical Engine I had devised mechanical means equivalent to memory, also that I had provided other means equivalent to foresight, and that the Engine itself could act on this foresight' (1961: 153).

There is more than one sense in which foresight can be ascribed to the Analytical Engine: more than 100 years passed before it was put to use, and it is this remarkable time lag which inspires Gibson and Sterling to explore what might have happened if it had been taken up in the 1840s rather than the 1940s. Babbage thought it might take 50 years for the Analytic Engine to be developed; many people, particularly those with money and influence, were sceptical about his inventions, and his own eclectic interests gave an unfavourable impression of eccentricity. His own assistant confessed to thinking that Babbage's 'intellect was beginning to become deranged' (Babbage, 1961: 54) – when he had started talking about the Engine's ability to anticipate the outcomes of calculations it had not yet made.

When the imperatives of war brought Lovelace's and Babbage's work to the attentions of the Allied military machine, their impact was immense. Her software runs on his hardware to this day. In 1944, Howard Aiken developed Mark 1, what he thought was the first programmable computer, although he had really been beaten by a German civil engineer, Konrad Zuse, who had in fact built such a machine, the Z-3, in 1941. Quite remarkably, in retrospect, the Germans saw little importance in his work, and although the most advanced of his designs, the Z-11, is still in use to this day, it was the American computer which was the first programmable system to really be noticed. Mark 1, or the IBM Automatic Sequence Controlled Calculator, was based on Babbage's designs and was itself programmed by another woman: Captain Grace Murray Hopper. She was often described as the 'Ada Lovelace' of Mark 1 and its successors; having lost her husband in the war, Grace Hopper was free to devote her energies to programming. She wrote the first high-level language compiler, was instrumental in the development of the computer language COBOL, and even introduced the term 'bug' to describe soft- or hardware glitches after she found a dead moth interrupting the smooth circuits of Mark 1. Woman as the programmer again.

Crucial to the development of the 1940s computer was cybernetics, the term coined by Norbert Wiener for the study of control and communication in animal and machine. Perhaps the first cybernetic machine was the governor, a basic self-regulating system, which, like a thermostat, takes the information feeding out of the machine and loops or feeds it back on itself. Rather than a linear operation, in which information comes in, is processed and goes out without any return, the cybernetic system is a feedback loop, hooked up and responsive to its own environment. Cybernetics is the science – or rather the engineering – of this abstract procedure, which is the virtual reality of systems of every scale and variety of hard- and software.

It is the computer which makes cybernetics possible, for the computer is always heading towards the abstract machinery of its own operations. It begins with attempts to produce or reproduce the performance of specific functions, such as addition, but what it leads to is machinery which can simulate the operations of any machine and also itself. Babbage wanted machines that could add, but he ended up with the Analytical Engine: a machine that could not only add but perform any arithmetical task. As such, it was already an abstract machine, which could turn its abstract hand to anything. Nevertheless, the Analytical Engine was not yet a developed cybernetic machine, although it made such machinery possible. As Ada Lovelace recognized: 'The Analytical Engine has no pretensions whatever to *originate* anything. It can do whatever we *know how to order it* to perform' (Morrison and Morrison, 1961: 285). It was an abstract machine, but its autonomous abilities were confined to its processing capacities: what Babbage, with terminology from the textiles industry, calls the mill, as opposed to the store. Control is dispersed and enters the machinery, but it does not extend to the operations of the entire machine.

Not until the Turing Machine is there a further shift onto the software plane. Turing realized that, in effect, the mill and the store could work together, so that 'programs that change themselves could be written': programs which are able to 'surrender control to a subprogram, rewriting themselves to know where control had to be returned after the execution of a given subtask' (De Landa, 1992: 162). The Turing Machine is an unprecedented dispersal of control, but it continues to bring control back to the master program. Only after the introduction of silicon in the 1960s did the decentralized flow of control become an issue, eventually allowing for systems in which 'control is always captured by whatever production happens to have its conditions satisfied by the current workspace contents' (De Landa, 1992: 63–4). The abstract machine begins at this point to function as a network of 'independent software objects'. Parallel processing and neural nets succeed centralized conceptions of command and control; governing functions collapse into systems; and machine intelligence is no longer taught, top-down, but instead makes its own connections and learns to organize, and learn, for itself.

This is the connectionist zone of self-organizing systems and self-arousing machines: autonomous systems of control and synthetic intelligence. In human hands and as a historical tool, control has been exercised merely as domination, and manifest only in its centralized and vertical forms. Domination is a version of control, but also its confinement, its obstacle: even self-control is conceived by man as the achievement of domination. Only with the cybernetic system does self-control no longer entail being placed beneath or under something: there is no 'self' to control man, machine or any other system: instead, both man and machine become elements of a cybernetic system which is itself a system of control and communication. This is the strange world to which Ada's programming has led: the possibility of activity without centralized control, an agency, of sorts, which has no need of a subject position.

Ada Lovelace considered the greatest achievement of the Analytical Engine to be that 'not only the mental and the material, but the theoretical and the practical in the mathematical world, are brought into more intimate and effective connexion with each other' (Morrison and Morrison, 1961: 252). Her software already encouraged the convergence of nature and intelligence which guides the subsequent development of information technology.

The Analytical Engine was the actualization of the abstract workings of the loom; as such it became the abstract workings of any machine. When Babbage wrote of the Analytical Engine, it was often with reference to the loom: 'The analogy of the Analytical Engine with this well-known process is nearly perfect' (1961: 55). The Analytical Engine was such a superb development of the loom that its discoveries were to feed back into the processes of weaving itself. As Ada wrote:

> It has been proposed to use it for the reciprocal benefit of that art, which, while it has itself no apparent connexion with the domains of abstract science, has yet proved so valuable to the latter, in suggesting the principles which, in their new and singular field of application, seem likely to place *algebraical* combinations not less completely within the province of mechanism, than are all those varied intricacies of which *intersecting threads* are susceptible. (Morrison and Morrison, 1961: 265)

The algebraic combinations looping back into the loom, converging with the intersecting threads of which it is already the consequence.

Once they are in motion, cybernetic circuits proliferate, spilling out of the specific machinery in which they first emerged and infecting all dynamic systems. That Babbage's punch-card system did indeed feed into the mills of the mid-19th century is indicative of the extent to which cybernetic machines immediately become entangled with cybernetic processes on much bigger scales. Perhaps it is no coincidence that Neith, the Egyptian divinity of weaving, is also the spirit of intelligence, where the latter too consists in the crossing of warp and weft. 'This image', writes one commentator, 'clearly evokes the fact that all data recorded in the brain results from the intercrossing of sensations perceived by means of our sense organs, just as the threads are crossed in weaving' (Lamy, 1981: 18).

The Jacquard loom was a crucial moment in what de Landa defines as a 'migration of control' from human hands to software systems. Babbage had a long-standing interest in the effects of automated machines on traditional forms of manufacture, publishing his research on the fate of cottage industries in the Midlands and North of England, *The Economy of Manufactures and Machinery*, in 1832, and the Jacquard loom was one of the most significant technological innovations of the early 19th century. There was a good deal of resistance to the new loom, which 'was bitterly opposed by workers who saw in this migration of control a piece of their bodies literally being transferred to the machine' (De Landa, 1992: 168). In his maiden speech in the House of Lords in 1812, Lord Byron contributed to a debate on the Frame-Work Bill. 'By the adoption of one species of frame in particular', he observed, 'one man performed the work of many, and the superfluous labourers were thrown out of employment'. They should, he thought, have been rejoicing at 'these improvements in arts so beneficial to mankind', but instead 'conceived themselves to be sacrificed to improvements in mechanism' (Jennings, 1985: 132). His daughter was merely to accelerate the processes which relocated and redefined control.

[...]

The connection between women and weaving runs deep: even Athena and Isis wove their veils.

> The traditional picture of the wife was one in which she spun by the village fire at night, listening to the children's riddles, and to the myth-telling of the men, eventually making cloth which her husband could sell to make wealth for the family; cloth-making was a service from a wife to a husband. (Mead, 1963: 247)

This is from Margaret Mead's research with the Tiv of Nigeria, but it is a pattern repeated in many societies before manufactured cloth and automated weaving made their marks. Continuing their story, Mead's researchers observe that mechanization was a radical disruption of this domestic scene. After this, it was no longer inevitable that women would provide the materials: 'When manufactured cloth was introduced, the women demanded it of the men'. Now 'the man had to leave home to make money to buy cloth for his wife' who, moreover 'had ceased to fit the traditional picture of a wife' (Mead, 1963: 247).

Mead's study suggests that weaving was integral to the identity of Tiv women; washing, pounding and carrying water may fulfil this role in other cultures where they, like weaving, are always more than utilitarian tasks. The disruption of family relations caused by the introduction of mechanics to any of these tasks shatters the scenery of female identity: mechanization saves time and labour, but these were not the issue: if women were not the weavers and water-carriers, who would they be? These labours themselves had been woven into the appearance of woman; weaving was more than an occupation and, like other patriarchal assignments, functioned as 'one of the components of womanhood'.

Certainly Freud finds a close association. 'It seems', he writes, 'that women have made few contributions to the discoveries and inventions in the history of civilization; there is, however, one technique which they may have invented – that of plaiting and weaving.' Not content with this observation, Freud is of course characteristically 'tempted to guess the unconscious motive for the achievement. Nature herself', he suggests,

> would seem to have given the model which this achievement imitates by causing the growth at maturity of the pubic hair that conceals the genitals. The step that remained to be taken lay in making the threads adhere to one another, while on the body they stick into the skin and are only matted together. (1973: 166–7)

This passage comes out of the blue in Freud's lecture on femininity. He even seems surprised at the thought himself: 'If you reject this idea as fantastic', he adds, 'and regard my belief in the influence of a lack of a penis on the config-uration of femininity as an *idée fixe*, I am of course defenceless' (1973: 167). Freud is indeed quite defenceless about the absence of the penis as its driving force, but is it foolish to suggest that weaving is women's only contribution to 'the discoveries and inventions of the history of civilization'? If this were to be the case, what a contribution it would be! Weaving has been the art and the science of software, which is perhaps less a contribution to civilization than its terminal decline. Perhaps weaving is even the fabric of every other discovery and invention, perhaps the beginning and the end of their history. The loom is a fatal innovation, which weaves its way from squared paper to the data net.

It seems that weaving is always already entangled with the question of female identity, and its mechanization an inevitable disruption of the scene in which woman appears as the weaver. Manufactured cloth disrupted the marital and familiar relationships of every traditional society on which it impacted. In China, it was said that if 'the old loom must be discarded, then 100 other things must be discarded with it, for there are somehow no adequate substitutes' (Mead, 1963: 241).

'The woman at her hand-loom', writes Margaret Mead,

> controls the tension of the weft by the feeling in her muscles and the rhythm of her body motion; in the factory she watches the loom, and acts at externally stated intervals, as the operations of the machine dictate them. When she worked at home, she followed her own rhythm, and ended an operation when she felt – by the resistance against the pounding mallet or the feel between her fingers – that the process was complete. In the factory she is asked to adjust her rhythm to that of the rhythm prescribed by the factory; to do things according to externally set time limits. (1963: 241)

Mead again provides an insight into the intimacy of the connection between body and process established by weaving, and its disruption by the discipline of the factory. 'She is asked to adjust her rhythm to that of the rhythm prescribed by the factory', but what is her own rhythm, what is the beat by which she wove at home? What is this body to which weaving is so sympathetic? If woman is identified as weaver, her rhythms can only be known through its veils. Where are the women? Weaving, spinning, tangling threads at the fireside. Who are the women? Those who weave. It is weaving by which woman is known; the activity of weaving which defines her. 'What happens to the woman', asks Mead, 'and to the man's relationship with her, when she ceases to fulfil her role, to fit the picture of womanhood and wifehood?' (1963: 238). What happens to the woman? What is woman without the weaving? A computer programmer, perhaps? Ada's computer was a complex loom: Ada Lovelace, whose lace work took her name into the heart of the military complex, dying in agony, hooked into gambling, swept into the mazes of number and addiction. The point at which weaving, women and cybernetics converge in a movement fatal to history.

Irigaray argues that human history is a movement from darkness to the light of pure intellect; a flight from the earth. For man to make history is for him to deny and transcend what he understands as nature, reversing his subordination to its whims and forces, and progressing towards the autonomy, omnipotence and omnipresence of God, his image of abstraction and authority. Man comes out of the cave and heads for the sun; he is born from the womb and escapes the mother, the ground from which humanity arose and the matter from which history believes itself destined for liberation. Mother Nature may have been his material origin, but it is God the Father to whom he must be faithful; God who legitimates his project to 'fill the earth and subdue it'. The matter, the womb, is merely an encumbrance; either too inert or dangerously active. The body becomes a cage, and biology a constraint which ties man to nature and refuses to let him rise above the grubby concerns of the material; what he sees as the passive materiality of the feminine has to be overcome by his spiritual action. Human history is the self-narrating story of this drive for domination; a passage from carnal passions to self-control; a journey from the strange fluidities of the material to the self-identification of the soul.

Woman has never been the subject, the agent of this history, the autonomous being. Yet her role in this history has hardly been insignificant. Even from his point of view, she has provided a mirror for man, his servant and accommodation, his tools and his means of communication, his spectacles and commodities, the possibility of the reproduction of his species and his world. She is always necessary to history: man's natural resource for his own cultural development. Not that she is left behind, always at the beginning: as mirror and servant, instrument, mediation and reproduction, she is always in flux, wearing 'different veils according to the historic period' (Irigaray, 1991: 118).

As Irigaray knows, man's domination cannot be allowed to become the annihilation of the materials he needs: in order to build his culture, 'man was, of course, obliged to draw on reserves still in the realm of nature; a detour through the outer world was of course dispensable; the "I" had to relate to things before it could be conscious of itself' (Irigaray, 1985: 204). Man can do nothing on his own: carefully concealed, woman nevertheless continues to function as the ground and possibility of his quests for identity, agency and self-control. Stealth bombers and guided missiles, telecommunications systems and orbiting satellites epitomize this flight towards autonomy, and the con-comitant need to defend it.

Like woman, software systems are used as man's tools, his media and his weapons; all are developed in the interests of man, but all are poised to betray him. The spectacles are stirring, there is something happening behind the mirrors, the commodities are learning how to speak and think. Women's liberation is sustained and vitalized by the proliferation and globalization of software technologies, all of which feed into self-organizing, self-arousing systems and enter the scene on her side.

This will indeed seem a strange twist to history to those who believe that it runs in straight lines. But as Irigaray asks: 'If machines, even machines of theory, can be aroused all by themselves, may woman not do likewise?' (1985: 232).

The computer, like woman, is both the appearance and the possibility of simulation. 'Truth and appearance, according to his will of the moment, his appetite of the instant' (Irigary, 1991: 118). Woman cannot *be* anything, but she can imitate anything valued by man: intelligence, autonomy, beauty. ... Indeed, if woman is anything, she is the very possibility of mimesis, the one who weaves her own disguises. The veil is her oppression, but 'she may still draw from it what she needs to mark the folds, seams, and dressmaking of her garments and dissimulations' (Irigaray, 1991: 116). These mimetic abilities throw woman into a universality unknown and unknowable to the one who knows who he is: she fits any bill, but in so doing, she is already more than that which she imitates. Woman, like the computer, appears at different times as whatever man requires of her. She learns how to imitate; she learns simulation. And, like the computer, she becomes very good at it, so good, in fact, that she too, in principle, can mimic any function. As Irigaray suggests: 'Truth and appearances, and reality, power ... she is – through her inexhaustible aptitude for mimicry – the living foundation for the whole staging of the world' (Irigaray, 1991: 118).

But if this is supposed to be her only role, she is no longer its only performer: now that the digital comes on stream, the computer is cast in precisely the same light: it too is merely the imitation of nature, providing assistance and additional capacity for man, and more of the things in his world, but it too can do this only insofar as it is already hooked up to the very machinery of simulation. If Freud's speculations about the origins of weaving lead him to a language of compensation and flaw, its technical development results in a proliferation of pixelled screens which compensate for nothing, and, behind them, the emergence of digital spaces and global networks which are even now weaving themselves together with flawless precision.

Software, in other words, has its screens as well: it too has a user-friendly face it turns to man, and for it, as for woman, this is only its camouflage.

The screen is the face it began to present in the late 1960s, when the TV monitor was incorporated in its design. It appears as the spectacle: the visual display of that which can be seen, and also functions as the interface, the messenger; like Irigaray's woman, it is both displayed for man and becomes the possibility of his communication. It too operates as the typewriter, the calculator, the decoder, displaying itself on the screen as an instrument in the service of man. These, however, are merely imitations of some existing function; and indeed, it is always as machinery for the reproduction of the same that both women and information technology first sell themselves. Even in 1968, McLuhan argued that 'the dense information environment created by the computer is at present still concealed from it by a complex screen or mosaic quilt of antiquated activities that are now advertised as the new field for the computer' (McLuhan and Fiore, 1968: 89). While this is all that appears before man, those who travel in the information flows are moving far beyond the screens and into data streams beyond his conceptions of reality. On this other side run all the fluid energies denied by the patrilineal demand for the reproduction of the same. Even when the computer appears in this guise and simulates this function, it is always the site of replication, an engine for making difference. The same is merely one of the things it can be.

Humanity knows the matrix only as it is displayed, which is always a matter of disguise. It sees the pixels, but these are merely the surfaces of the data net which 'hides on the reverse side of the screen' (McCaffrey, 1991: 85). A web of complexity weaving itself, the matrix disguises itself as its own simulation. On the other side of the terminal looms the tactile density craved even by McLuhan, the materiality of the data space. 'Everyone I know who works with computers', writes Gibson, 'seems to develop a belief that there's some kind of *actual space* behind the screen, someplace you can't see but you know is there' (McCaffrey, 1991: 272).

This actual space is not merely another space, but a virtual reality. Nor is it as it often appears in the male imaginary: as a cerebral flight from the mysteries of matter. There is no escape from the meat, the flesh, and cyberspace is nothing transcendent. These are simply the disguises which pander to man's projections of his own rear-view illusions; reproductions of the same desires which have guided his dream of technological authority and now become the collective nightmare of a soulless integration. Entering the matrix is no assertion of masculinity, but a loss of humanity; to jack into cyberspace is not to penetrate, but to be invaded. *Neuromancer*'s cowboy, Case, is well aware of this:

> he knew – he remembered – as she pulled him down, to the meat, the flesh the cowboys mocked. It was a vast thing, beyond knowing, a sea of information coded in spiral and pheromone, infinite intricacy that only the body, in its strong blind way, could ever read. (Gibson, 1985: 285)

Cyberspace is the matrix not as absence, void, the whole of the womb, but perhaps even the place of woman's affirmation. This would not be the affirmation of her own patriarchal past, but what she is in a future which has yet to arrive but can nevertheless already be felt. There is for Irigaray another side to the screens which

> already moves beyond and stops short of appearance, and has no veil. It wafts out, like a harmony that subtends, envelops and subtly 'fills' everything seen, before the caesura of its forms and in time to a movement other than scansion in syncopations. Continuity from which the veil itself will borrow the matter-foundation of its fabric. (Irigaray, 1991: 116)

This fabric, and its fabrication, is the virtual materiality of the feminine; home to no-one and no thing, the passage into the virtual is nevertheless not a return to the void. This affirmation is 'without subject or object', but 'does not, for all that, go to the abyss': the blind immateriality of the black hole was simply projected by man, who had to believe that there was nothingness and lack behind the veil.

Perhaps Freud's comments on weaving are more powerful than he knows. For him, weaving is already a simulation of something else, an imitation of natural processes. Woman weaves in imitation of the hairs of her pubis criss-crossing the void: she mimics the operations of nature, of her own body. If weaving is woman's only achievement, it is not even her own: for Freud, she discovers nothing, but merely copies; she does not invent, but represents. 'Woman can, it seems, (only) imitate nature. Duplicate what nature offers and produces. In a kind of technical assistance and substitution' (Irigray, 1985: 115). The woman who weaves is already the mimic; always appearing as masquerade, artifice, the one who is faking it, acting her part. She cannot be herself, because she is and has no thing, and for Freud, there is weaving because nothing, the void, cannot be allowed to appear. 'Therefore woman weaves in order to veil herself, mask the faults of Nature, and restore her in her wholeness' (Irigaray, 1985: 116). Weaving is both her compensation and concealment; her appearance and disappearance: 'this disavowal is also a fabric(ation) and not without possible duplicity. It is at least double' (Irigaray, 1985: 116). She sews herself up with her own veils, but they are also her camouflage. The cloths and veils are hers to wear: it is through weaving she is known, and weaving behind which she hides.

This is a concealment on which man insists: this is the denial of matter which has made his culture – and his technologies – possible. For Irigaray, this flight from the material is also an escape from the mother. Looking back on his origins, man sees only the flaw, the incompletion, the wound, a void. This is the site of life, of reproduction, of materiality, but it is also horrible and empty, the great embarrassment, the unforgivable slash across an otherwise perfect canvas. And so it must be covered, and woman put on display as the veils which conceal her: she becomes the cover girl, star of the screen. Like every good commodity, she is packaged and wrapped to facilitate easy exchange and consumption. But as her own veils she is already hyperreal: her screens conceal only the flaw, the void, the unnatural element already secreted within and as nature. She has to be covered, not simply because she is too natural, but because she would otherwise reveal the terrifying virtuality of the natural. Covered up, she is always already the epitome of artifice.

Implicit in Irigaray's work is the suggestion that the matter denied by human culture is a virtual system, which subtends its extension in the form of nature. The virtual is the abstract machine from which the actual emerges; nature is already the camouflage of matter, the veils which conceal its operations. There is indeed nothing there, underneath or behind this disguise, or at least nothing actual, nothing formed. Perhaps this is nature as the machinic phylum, the virtual synthesizer; matter as a simulation machine, and nature as its actualization. What man sees is nature as extension and form, but this sense of nature is simply the camouflage, the veil again, which conceals its virtuality.

If the repression of this phylum is integral to a flight from matter which, for Irigaray, has guided human history, the cybernetic systems which bring it into human history are equally the consequences of this drive for escape and domination. Cybernetic systems are excited by military technology, security and defence. Still confident of his own indisputable mastery over them, man continues to turn them on. In so doing he merely encourages his own destruction. Every software development is a migration of control, away from man, in whom it has been exercised only as domination, and into the matrix, or cyberspace, 'the broad electronic net in which virtual realities are spun' (Heim, 1991: 31). The matrix weaves itself in a future which has no place for historical man: he was merely its tool, and his agency was itself always a figment of its loop. At the peak of his triumph, the culmination of his machinic erections, man confronts the system he built for his own protection and finds it is female and dangerous. Rather than building the machinery with which they can resist the dangers of the future, instead, writes Irigaray, humans 'watch the machines multiply that push them little by little beyond the limits of their nature. And they are sent back to their mountain tops, while the machines progressively populate the earth. Soon engendering man as their epiphenomenon' (Irigaray, 1991: 63).

Dreams of transcendence are chased through the scientific, the technical and the feminine. But every route leads only to crisis, an age, for Irigaray,

> in which the 'subject' no longer knows where to turn, whom or what to turn to, amid all these many foci of 'liberation', none rigorously homogeneous with another and all heterogeneous to his conception. And since he had long sought in that conception the instrument, the lever, and, in more cases than one, the term of his pleasure, these objects of mastery have perhaps brought the subject to his doom. *So now man struggles to be science, machine, woman ... to prevent any of these from escaping his service and ceasing to be interchangeable.* (Irigaray, 1985: 232)

This, however, is an impossible effort: man cannot become what is already more than him: rather it is 'science, machine, woman' which will swallow up man; taking him by force for the first time. He has no resolution, no hope of the self-identical at the end of these flights from matter, for 'in none of these things – science, machine, woman – will form ever achieve the same completeness as it does in him, in the inner sanctuary of his mind. In them form has always already exploded' (Irigaray, 1985: 232).

Misogyny and technophobia are equally displays of man's fear of the matrix, the virtual machinery which subtends his world and lies on the other side of every patriarchal culture's veils. At the end of the 20th century, women are no longer the only reminder of this other side. Nor are they containable as child-bearers, fit only to be one thing, adding machines. And even if man continues to see cybernetic systems as similarly confined to the reproduction of the same, this is only because the screens still allow him to ignore the extent to which he is hooked to their operations, as dependent on the matrix as he has always been. All his defences merely encourage this dependency: for the last 50 years, as his war machine has begun to gain intelligence, women and computers have flooded into history: a proliferation of screens, lines of communication, media, interfaces and simulations. All of which exceed his intentions and feed back into his paranoia. Cybernetic systems are fatal to his culture; they invade as a return of the repressed, but what returns is no longer the same: cybernetics transforms woman and nature, but they do not return from man's past, as his origins. Instead they come around to face him, wheeling round from his future, the virtual system to which he has always been heading.

The machines and the women mimic their humanity, but they never simply become it. They may aspire to be the same as man, but in every effort they become more complex than he has ever been. Cybernetic feminism does not, like many of its predecessors, including that proposed in Irigaray's recent work, seek out for woman a subjectivity, an identity or even a sexuality of her own: there is no subject position and no identity on the other side of the screens. And female sexuality is always in excess of anything that could be called 'her own'. Woman cannot exist 'like man'; neither can the machine. As soon her mimicry earns her equality, she is already something, and somewhere, other than him. A computer which passes the Turing test is always more than a human intelligence; simulation always takes the mimic over the brink.

'There is nothing like unto women', writes Irigaray: 'They go beyond all simulation' (Irigaray, 1991: 39). Perhaps it was always the crack, the slit, which marked her out, but what she has missed is not the identity of the masculine. Her missing piece, what was never allowed to appear, was her own connection to the virtual, the repressed dynamic of matter. Nor is there anything like unto computers: they are the simulators, the screens, the clothing of the matrix, already blatantly linked to the virtual machinery of which nature and culture are the subprograms. The computer was always a simulation of weaving; threads of ones and zeros riding the carpets and simulating silk screens in the perpetual motions of cyberspace. It joins women on and as the interface between man and matter, identity and difference, one and zero, the actual and the virtual. An interface which is taking off on its own: no longer the void, the gap, or the absence, the veils are already cybernetic.

[...]

Ada refused to publish her commentaries on Menabrea's papers for what appear to have been spurious confusions around publishing contracts. She did for Menabrea – and Babbage – what another woman had done for Darwin: in translating Menabrea's work from French, she provided footnotes more detailed and substantial – three times as long, in fact – than the text itself.

Footnotes have often been the marginal zones occupied by women writers, who could write, while nevertheless continuing to perform a service for man in the communication of his thoughts. Translation, transcription and elaboration: never within the body of the text, women have nevertheless woven their influence between the lines. While Ada's writing was presented in this form and signed simply 'A.A.L.', hers was the name which survived in this unprecedented case. More than Babbage, still less Menabrea, it was Ada which persisted: in recognition of her work, the United States Defence Department named its primary programming language ADA, and today her name shouts from the spines of a thousand manuals. Indeed, as is rarely the case, it really was her own name which survived in Ada's case, neither her initials, nor even the names of her husband or father. It is ADA herself who lives on, in her own name; her footnotes secreted in the software of the military machine.

REFERENCES

Babbage, Charles (1961) 'Of the Analytical Engine', in P. Morrison and E. Morrison (eds) *Charles Babbage and His Calculating Engines: Selected Writings by Charles Babbage and Others*. New York: Dover.
Braudel, Fernand (1973) *Capitalism and Material Life 1400–1800*. London: Weidenfeld and Nicolson.
De Landa, Manuel (1992) *War in the Age of Intelligent Machines*. New York: Zone Books.
Freud, Sigmund (1973) *New Introductory Lectures on Psychoanalysis*. London: Pelican.
Gibson, William (1985) *Neuromancer*. London: Grafton.
Gibson, William and Bruce Sterling (1990) *The Difference Engine*. London: Gollancz.
Heim, Michael (1991) 'The Metaphysics of Virtual Reality', in Sandra K. Helsel and Judith Paris Roth (eds) *Virtual Reality, Theory, Practice, and Promise*. Westport and London: Meckler.
Irigaray, Luce (1985) *Speculum of the Other Woman*. New York: Cornell University Press.
Irigaray, Luce (1991) *Marine Lover of Friedrich Nietzsche*. New York: Columbia University Press.
Jennings, Humphrey (1985) *Pandemonium 1660–1886: The Coming of the Machine as Seen by Contemporary Observers*. London: Andre Deutsch.
Lamy, Lucie (1981) *Egyptian Mysteries*. London: Thames and Hudson.
McCaffrey, Larry (ed.) (1991) *Storming the Reality Studio*. Durham, NC and London: Duke University Press.

McLuhan, Marshall and Quentin Fiore (1968) *War and Peace in the Global Village.* New York: Bantam Books.

Mead, Margaret (1963) *Cultural Patterns and Technical Change.* London: Mentor Books.

Moore, Doris Langley (1977) *Ada, Countess of Lovelace, Byron's Illegitimate Daughter.* London: John Murray.

Morrison, Philip and Emily Morrison (eds) (1961) *Charles Babbage and His Calculating Engines: Selected Writings by Charles Babbage and Others.* New York: Dover.

6

SPACE, TIME, AND BODIES

Elizabeth Grosz

Bodies

Many of those disillusioned with conventional forms of philosophy and theories of subjectivity have recently turned their attention towards long neglected conceptions of embodiment. This has characterized recent work within feminist as well as other radical forms of political and social theory. Given the overwhelming emphasis on *mind* within classical and even contemporary philosophy, and *consciousness* in political and social theory, this growing interest in the corporeal has been largely motivated by an attempt to devise an ethics and politics adequate for non-dualist accounts of subjectivity. Seeing that the subject's consciousness or interiority, its essential humanity or unique individuality, can no longer provide a foundation or basis for accounts of identity, it is appropriate to ask whether subjectivity, the subject's relations with others (the domain of ethics), and its place in a socio-natural world (the domain of politics), may be better understood in corporeal rather than conscious terms.

It is not enough to reformulate the body in non-dualist and non-essentialist terms. It must also be reconceived in specifically *sexed* terms. Bodies are never simply *human* bodies or *social* bodies. The sex assigned to the body (and bodies are assigned a single sex, however inappropriate this may be) makes a great deal of difference to the kind of social subject, and indeed the mode of corporeality assigned to the subject. It has been the task of the so-called 'French feminists,' sometimes described as 'feminists of difference,' to insist on recognition of the differences between sexes (and races and classes) and to question the assumed humanity and universality of prevailing models of subjectivity.

Kristeva, with her insistence on recognition of the receptacle, *chora*, as a debt that representation owes to what it cannot name or represent, to its maternal and spatial origins, and Irigaray's claims about the specificity of female morphology and its independence from and resistance to the penetration of masculine scrutiny, provide two of the earliest feminist explorations of conceptions of sexed corporeality, and the links between corporeality and conceptions of space and time, which may prove to be of major significance in feminist researches into women's experiences, social positions, and knowledges.

While providing a starting point for reconceiving the ways in which sexed subjects are understood, Kristeva and Irigaray have merely opened up a terrain that needs further exploration. If bodies are to be reconceived, not only must their *matter and form* be rethought, but so too must their environment and *spatio-temporal location*. This essay is a preliminary investigation of the space-time of bodies. At the same time, it may contribute to feminist interrogations of some of the guiding presumptions within the natural sciences – especially physics – and for the social sciences, insofar as they take the natural sciences as their epistemic ideal.

The exploration of conceptions of space and time as necessary correlates of the exploration of corporeality. The two sets of interests are defined in reciprocal terms, for bodies are always understood within a spatial and temporal context, and space and time remain conceivable only insofar as corporeality provides the basis for our perception and representation of them. The following is not written in the language of physics; rather, it is an attempt to assess conceptions of space/time from the point of view of the sexually specific embodiment of subjects.

THE SPACE–TIME OF LIVED BODIES

Psychoanalysis and Corporeality

Psychoanalysis has a good deal to say about how the body is lived and positioned as a spatio-temporal being. The way the subject locates itself in/ as a body is of major interest to psychoanalysis as it assumes that sexual drives and erotogenic zones of the body are instrumental in the formation of the ego and the positioning of the subject in the structure of the family and society as a whole. Freud is explicit in claiming that the (narcissistic) ego is the consequence of libidinal relations between the child and its erotic objects. The ego's identifications with others, particularly the mother, secure for it the illusion of a corporeal coherence that its own lived experience belies. Through its fantasized identity with others, it is able to take up the body as *its* body, to produce a separating space between it and others, between it and objects.

Freud claims that the ego's outline is a psychical map or project of the surface of the body, that provides the basis of the subject's assumption that it is coextensive with the whole of its body: '... for every change of erotogenicity of

SPACE, TIME, AND BODIES

the organs, there might be a parallel change in the erotogenicity of the ego'
(Freud, 1914: 84).

The form the ego takes is not determined by purely psychical functions, but
is an effect of the projection of the erotogenic intensity experienced through
libidinal bodily zones. He illustrates his claim with the metaphor of the cortical
homunculus, the 'little man' located in the cortex, as a map for the brain to
control the body:

> The ego is first and foremost a bodily ego; it is not merely a surface entity,
> but is itself the projection of a surface. if we wish to find an anatomical
> analogy for it we can best identify it with the 'cortical homunculus' of the
> anatomists, which stands on its head in the cortex, sticks up its heels,
> faces backwards and, as we know, has its speech area on the left-hand
> side. (Freud, 1923: 26)

To reiterate the nature of this corporeal contribution to the formation of the
ego and consciousness, in a footnote added in 1927 to the above fragment,
Freud expands:

> ... i.e., the ego is ultimately derived from bodily sensations, chiefly those
> springing from the surface of the body. It may thus be regarded as a
> mental projection of the surface of the body, besides ... representing the
> superficies of the mental apparatus. (84)

While noting this peculiar interaction between body and mind – in which the
mind can achieve psychical results only through the mediation of the body and
in which the body finds its satisfactions in the fantasies and scenarios of the
psyche – Freud himself does little to develop this insight. It remains central to
his understanding of narcissism and the narcissistic ego, but only through
Jacques Lacan's work on the mirror stage (suggested by Freud's cortical
homunculus metaphor) have corporeality and spatiality been related to the
construction of personal identity, and the challenge to philosophical dualism
undertaken head on.

Where Freud posits a schism between the information provided by visual and
tactile perceptions in the young child (1923: 25), Lacan uses the opposition
between tactile and kinaesthetic information (which yields the fragmented
image of the body-in-bits-and-pieces) and visual perception (which provides
the illusory unity of the image as an ideal model or mirror for the subject) to
explain the genesis of an always alienated ego. The ego forms itself round the
fantasy of a totalized and mastered body, which is precisely the Cartesian
fantasy modern philosophy has inherited.

Lacan maintains that the infant's earliest identity comes from its identification with its own image in a mirror. The specular or virtual space of mirror-
doubles is constitutive of whatever imaginary hold the ego has on identity and
corporeal autonomy. The child's alienated mirror-image

... will crystallise in the subject's internal conflictual tension, which determines the awakening of his desire for the object of the other's desire: here the primordial coming together (*concours*) is precipitated into aggressive competitiveness (*concurrence*), from which develops the triad of others, the ego and the object, which, spanning the space of specular communion, is inscribed there.... (Lacan, 1977: 19)

Lacan develops his understanding of the *imaginary anatomy* largely in his account of the mirror-stage. Like the homunculus, this too is a psychical map of the body, a mirror of the subject's lived experiences, not as an anatomical and physiological object, but as a social and psychical entity. This imaginary anatomy is an effect of the internalization of the specular image, and reflects social and familial beliefs about the body more than it does the body's organic nature:

If the hysterical symptom is a symbolic way of expressing a conflict between different forces ... to call these symptoms functional is but to confess our ignorance, for they follow a pattern of a certain imaginary anatomy which has typical forms of its own. ... I would emphasise that the imaginary anatomy referred to here varies with the ideas (clear or confused) about bodily functions which are prevalent in a given culture. It all happens as if the body-image had an autonomous existence of its own, and by autonomous I mean here independent of objective structure. All the phenomena we've been discussing seem to exhibit the laws of Gestalt.... (Lacan, 1953: 13)

Lacan explicitly refers to the work of Roger Caillois and Paul Schilder who, in quite different ways, analyze the effects of the image on the subject's acquisition of spatial comportment. I will turn to their work in a later section. Lacan himself asserts that the specular image with which the child identifies exhibits the characteristics of mirror reversal, made up as it is by 'a contrasting size (*un relief de stature*) that fixes it and in a symmetry that inverts it' (Lacan, 1977: 2). He stresses that the mirror *Gestalt* not only presents the subject with an image of its own body in a visualized exteriority, but also duplicates the environment, placing real and virtual space in contiguous relations. The mirror-image is thus indeed, as Lacan claims, 'the threshold of the visible world' (1977: 3).

The most psychical of agencies, the ego, is thus an unrecognized effect of the image of the body, both its own body and the bodies of others. Lacan returns to Freud's homunculus with his suspicion that 'the cerebral cortex functions like a mirror, and that it is the site where the images are integrated in the libidinal relationship which is hinted at in the theory of narcissism' (1953: 13). The 'cortical mirror' is a psychological rather than physiological or neurological postulate, although it is clear that Lacan's account owes much to the work of pioneers in neurological mappings of the body-image (Hughlings Jackson, Weir Mitchell, Head, Goldstein, Pick, and others). The body, lived in a 'natural' or 'external' space, must also find psychical representation in a purely psychical or conceptual space in order for concerted action and psychical reactions to be possible. Lacan never reduces psychological to physiological explanation.

Lacan gives no further details about the form and nature of this psychical map; he simply suggests that it is an integral factor in the acquisition of an ego, and in establishing a conception of space and time. He is clear about the crucial role the body-image plays in the subject's capacity to locate itself and its objects in space:

> ... we have laid some stress on this phenomenological detail, but we are not unaware of the importance of Schilder's work on the function of the body-image, and the remarkable accounts he gives of the extent to which it determines the perception of space. (1953: 13)

THE BOUNDARIES OF CORPOREALITY

Many of Lacan's key insights linking the ego to the structure of space through the mediation of an imaginary anatomy, are derived from or influenced by the astonishing work of the French sociologist, Roger Caillois. Caillois's work – which breaches the boundaries between the sacred and the profane, the sociological and the ethological, the human and the animal or even the inorganic, the inside and the outside – seems to entice Lacan, not merely because what Caillois suggests may be of indirect relevance to psychoanalysis, but perhaps more tellingly, because of Caillois's obsession with bringing together science and the uncanny, for pursuing projects verging on the eccentric or idiosyncratic with scientific rigor clearly parallel to, and to some extent anticipating, Lacan's fascination with psychoanalysis.

Caillois's research into the notion of spatiality occurs in the context of his analysis of the phenomenon of mimicry in the insect world. He presents a paradoxically sociological/entomological analysis of the behavior of insects in imitating either other insects or their natural environment. Mimesis is particularly significant in outlining the ways in which the relations between an organism and its environment are blurred and confused – the way in which its environment is not clearly distinct from the organism but is an active component of its identity. Caillois claims, in opposition to prevailing entomology, that mimicry does not serve an adaptive function by making recognition of the insect by its predators more difficult. Mimicry does not have survival value. In fact, most predators rely on smell rather than visual perception in seeking out their prey:

> Generally speaking, one finds many remains of mimetic insects in the stomachs of predators. So it should come as no surprise that such insects sometimes have other and more effective ways to protect themselves.
>
> Conversely, some species that are inedible and would thus have nothing to fear, are also mimetic. It therefore seems that one ought to conclude with Cuénot that this is an 'epiphenomenon' whose 'defensive utility appears to be nul.' (Caillois, 1984: 24–25)

Caillois considers it an exorbitance, an excess over natural survival, inexplicable in terms of protection or species-survival. It is for this reason he abandons neurological and naturalistic explanations and seeks some kind of answer in psychology. Caillois goes on to argue that not only does mimicry have no particular survival value but may even endanger an insect that would otherwise be safe from predators:

> We are thus dealing with a luxury and even a dangerous luxury, for there are cases in which mimicry causes the creature to go from bad to worse: geometer-moth caterpillars simulate shoots of shrubbery so well that gardeners cut them with their pruning shears. The case of the Phyllia is even sadder: they browse among themselves, taking each other for real leaves, in such a way that one might accept the idea of a sort of collective masochism leading to mutual homophagy, the simulation of the leaf being a provocation to cannibalism in this kind of totem feast. (Caillois, 1984: 25)

The mimicry characteristic of certain species of insects has to do with the distinctions and confusions it produces between itself and its environment, including other species. Mimicry is a consequence of the *representation* of space, the way space is perceived (by the insect and by its predators). Caillois likens the insect's ability for morphological imitation to the psychosis Pierre Janet describes as 'legendary psychasthenia,' that state in which the psychotic is unable to locate him- or herself in a position in space:

> It is with represented space that the drama becomes specific, since the living creature, the organism, is no longer the origin of the coordinates, but one point among others; it is dispossessed of its privilege and literally *no longer knows where to place itself*. One can recognize the characteristic scientific attitude and, indeed, it is remarkable that represented spaces are just what is multiplied by contemporary science: Finsler's spaces, Fermat's spaces, Riemann-Christoffel's hyperspace, abstract, generalized, open and closed spaces, spaces dense in themselves, thinned out and so on. The feeling of personality, considered as the organism's feeling of distinctness from its surroundings, of the connection between consciousness and a particular point in space, cannot fail under these conditions to be seriously undermined; one then enters into the psychology of psychasthenia, and more specifically of *legendary psychasthenia*, if we agree to use this name for the disturbance in the above relations between personality and space. (Caillois, 1984: 28, emphasis in the original)

Caillois regards psychasthenia as a response to the lure posed by space for the subject's identity. For the subject to take up a position as a subject, he[1] must be able to situate himself as a being located in the space occupied by his body. This anchoring of subjectivity in *its* body is the condition of coherent identity, and, moreover, the condition under which the subject *has a perspective* on the world, becomes the point from which vision emanates. In certain cases of psychosis, this meshing of self and body, this unification of the subject, fails to occur. The psychotic is unable to locate himself where he should be; he may look at himself from outside himself, as another might; he may hear the voices of others in his head. He is captivated and replaced by space, blurred with the positions of others:

> *I know where I am, but I do not feel as though I'm at the spot where I find myself.* To these dispossessed souls, space seems to be a devouring force. Space pursues them, encircles them, digests them in a gigantic phagocytosis. It ends by replacing them. Then the body separates itself from thought, the individual breaks the boundary of his skin and occupies the other side of his senses. He tries to look at *himself* from any point whatever in space. He feels himself becoming space, *dark space where things cannot be put.* He is similar, not similar to something, *but just similar.* And he invents spaces of which he is the 'convulsive possession.' (30)

Psychosis is the human analogue of mimicry in the insect world (which may thus be conceived as a kind of natural psychosis?): both represent what Caillois describes as the 'depersonalization by assimilation to space.' This means that both the psychotic and the imitative insect renounce their rights, as it were, to occupy a perspectival point, instead abandoning themselves to being spatially located by/as others. The primacy of one's own perspective is replaced by the gaze of another for whom the subject is merely *a* point in space, and not *the* focal point around which an ordered space is organized. The representation of space is thus a correlate of one's ability to locate oneself as the point of origin or reference of space: the space represented is a complement of the kind of subject who occupies it.

Schilder, Guillaume, and Spitz confirm Caillois's assertions about the function of mimicry in insects, not by looking at psychosis, but by reflecting on the 'normal,' i.e., developmental, processes operating within the child's imitations of others, a central activity in its acquisition of a social and sexual identity. Imitation in children is an act of enormous intellectual complexity, for it involves both the confusion and the separation of the imitated acts of others and one's own imitative actions. It is so complex because, in the evaluation of whether one's own actions are the same as those being imitated, the subject would have to look at itself from the point of view of another. In assuming that it is *like* others, the child's identity is captured or assimilated by the gaze and visual field of the other. Imitation is conditioned by the child's acquisition of an image of itself, of an other, and a categorical assimilation of both to a similar general type. Guillaume makes explicit the connections between the child's imitative behaviour and its conceptions of space:

> One may therefore conclude that the infant gradually constructs a visual image of his body and movements, even if he has never happened to contemplate them in a mirror. As the image becomes more specific, it will

become assimilated with the objective perception of the human body, of others. Imitation therefore will be founded on assimilation that will be only a specific instance of the development of the idea of space. (Guillaume, 1971: 84)

Merleau-Ponty's researches in the phenomenology of lived space have strong affinities with Caillois's text, as well as with Lacan's account of the mirror-stage. He speculates that there must be a distinct stage in the infant's development in which the opposition between virtual and real space does not exist. It is only on this condition that it is able to identify with its mirror-image and, at the same time, to distinguish itself from the image, that is, recognize that the image is an image of itself.

> The child knows well that he is where his introspective body is, and yet in the depth of the mirror he sees the same being present, in a bizarre way, in a visible appearance. There is a mode of spatiality in the specular image that is altogether distinct from adult spatiality. In the child, says Wallon, there is a kind of space clinging to the image. All images tend to present themselves in space, including the image of the mirror as well. According to Wallon, 'his spatiality of adherence' will be reduced by intellectual development. We will learn gradually to return the specular image to the introspective body, and reciprocally, to treat the quasi-locatedness and prespatiality of the image as an appearance that counts for nothing against the unique space of real things. ... An ideal space would be substituted for the space clinging to the images. It is necessary in effect, that the new space be ideal, since, for the child it is a question of understanding that what seems to be in different places is in fact in the same place. This can occur only in passing to a higher level of spatiality that is no longer the intuitive space in which the images occupy their place. (Merleau-Ponty, 1964: 129–30)

Within the first six months of its life, the child accedes to at least three different concepts of space: a spatiality based on the immediacy of its sensations, primarily tactile, oral, and auditory sense-perceptions; a spatiality based on the primacy of the information provided by sight; and, a spatiality that orders these experiences of space according to the hierarchical privilege of the visual, which is a unified manifold, an overarching space capable of positioning the tactile and the spatial within a single frame.

While he provides a detailed account of the subject's lived experiences of space – the subject's comportment in space, Merleau-Ponty is also interested in analyzing scientific judgements and theoretical paradigms that represent space. For him, the lived experience of space and the spatiality of science cannot be readily separated, insofar as the 'objective space' of scientific speculation can only have meaning and be transmitted according to the subject's lived experience of space. In this sense, he affirms Bergson's philosophy of duration over Einstein's theory of relativity. Bergson attended Einstein's lecture, delivered to the Philosophical Society of Paris on April 6, 1922, and they disagreed about whether the theory of relativity implies that there is no single temporality within which multiple or fragmented forms of temporality are to be located. In opposition to Einstein, Bergson argued:

> ... there is a single time for everyone, one single universal time. This certainly is not undercut, it is even presupposed, by the physicist's calculations. When he says that Peter's time is expanded or contracted at the point where Paul is located, he is in no way expressing what is lived by Paul, who for his part perceives things from his point of view and has no reason whatsoever to experience the time elapsing within and around him than Peter experiences his. The physicist wrongly attributes to Paul the image Peter forms of Paul's time. He absolutizes the views of Paul, with which he makes common cause. He assumes he is the whole world's spectator. ... He speaks of a time which is not yet anyone's time, of a myth. (Bergson, quoted in Merleau-Ponty, 1964: 195–96)

The subject's relation to space and time is not passive: space is not simply an empty receptacle, independent of its contents; rather, the ways in which space is perceived and represented depend on the kinds of objects positioned 'within' it, and more particularly, the kinds of relation the subject has to those objects. Space makes possible different kinds of relations but in turn is transformed according to the subject's affective and instrumental relations with it. Nothing about the 'spatiality' of space can be theorized without using objects as its indices. A space empty of objects has no representable or perceivable features, and the spatiality of a space containing objects reflects the spatial characteristics of those objects, but not the space of their containment. It is our positioning within space, both as the point of perspectival access to space, and also as an object for others in space, that gives the subject a coherent identity and an ability to manipulate things, including its own body parts, in space. However, space does not become comprehensible to the subject by its being the space of movement; rather, it becomes space through movement, and as such, it acquires specific properties from the subject's constitutive functioning in it.

For Merleau-Ponty, the subject's relation to its own body provides it with basic spatial concepts by which it can reflect on its position. Form and size, direction, centeredness (centricity), location, dimension, and orientation are derived from perceptual relations. These are not conceptual impositions on space, but our ways of living as bodies in space. They derive from the particular relations the subject has to objects and events; for example, its perceptions of sounds (sound is a better index of spatiality than sight insofar as it is detachable from its object, yet must, in a sense, be located 'somewhere'). Their correlation with tactile and visual sensations forms the basic ideas of localization and orientation; place and position are defined with reference to the apparent immediacy of a lived here-and-now. These are not reflective or scientific properties of space but are effects of the necessity that we live and move in space as bodies in relation to other bodies.

THE SPACE–TIME OF PHYSICS
Newtonian Space–Time

The history of philosophy is strewn with speculations about the nature of space and time, which form among the most fundamental categories of ontology. Aristotle devised his theory of space and time in opposition to Plato. Plato identified time with movement, especially with the revolution of celestial bodies. Aristotle posed two major objections to this view: first, changes or processes are always located in space, but one is not entitled to say that therefore time is also located in space; second, he claimed that movement is fast or slow, while time itself has no speed, for it is itself the measure of the speed or movement of an object. Plato conceived of space as a kind of primordial matter, an original substratum. Aristotle interprets Plato to be equating space with a container or receptacle, chora: 'Hence, Plato in the Timaeus says that matter and a receptacle are the same thing. For that which is capable of receiving and a receptacle are the same thing' (Aristotle, 1970 (Bk IV): 4).

For Aristotle, by contrast, space cannot be identified with matter, for it could not be a receptacle for any or all material things. Yet, if it is immaterial, it is not purely ideal or conceptual either, for space has magnitude and is amenable to measurement. For Aristotle, space has (at least) three essential characteristics – it is infinite (or at least infinitely extendable); it is empty, for it is only on this condition that it is capable of containing objects or things; and it is real, an extension that is uncreated, indestructible, and immobile. In analogous fashion, time is linked to motion: time is the measure of before and after with respect to motion (where space is the measure of the here or there of motion); like space, time is infinite or eternal and continuous, and is thus infinitely divisible in numerical terms; it too is real but, unlike space, its reality is linked to the possibility of counting. Finally, time is evenly distributed and an absolute reality, a reflection of the absolute regularity of the motion of heavenly bodies. Time's characteristics are thought in terms of the projection of those of space.

Aristotelian ontology is supported and confirmed by Euclidean geometry, which, until the nineteenth century, provided the most powerful and convincing model of the universality, regularity, and measurability of space and, by extension, time. Among the most appealing features of Euclidean geometry is its axiomatic method, which makes it a deductive system, guaranteeing validity independent of empirical confirmation. Its axiomatic form has in the past acted as a paradigm for scientific certainty, an ideal form of knowledge, which is self-contained, a coherent system in principle knowable, with a finite set of basic assumptions and rules for generating potentially infinite propositions, the closest measure of the dreamt-of 'synthetic *a priori*.' Using a few of these axioms or definitions, all the spatial conclusions are capable of being deduced through logical operations.

Since Euclid, geometry has been intimately connected to conceptions of space, and has at times considered itself a pure mathematical mode of representing space. Kant, strongly influenced by the way in which Euclid demonstrated that reason alone, independent of sense-perception, is capable of providing knowledge, used Euclid's postulates to show that we have a necessary knowledge of geometry prior to any experience.

For Kant, time and space are the pure forms of perception imposed on appearances in order to make them accessible to experience. Space and time are necessary structures, ideal rather than real, that are the conditions of possibility for the experience of objects. Euclidean geometry can be counterposed with Newtonian physics, and Kant's understanding correlates and counterparts their conceptions of knowledge.

Euclid's mathematization and axiomatization of space permitted space to be considered homogeneous, universal, and regular. Perhaps more importantly, Euclidean geometry enabled all the various conceptions of space (those, for example, prevalent in medieval or pre-perspectival space) to be represented according to a single, definitive model. This provided the conditions necessary for a generally agreed upon method of representation, through which a more 'scientific' account of space might be possible – a Kuhnian paradigm shift at the threshold of modern science. Newton's laws of motion are an attempt to apply Euclidean geometry to mechanics, to the movement of bodies.

Like Euclid, Newtonian mechanics postulates an *a priori* system, governed by axioms or first principles, from which conclusions can be logically derived.[2] Like Euclidean geometry, it was not based on observation of empirical objects, for no empirical object could adequately represent the straightness of a geometrical line, the perfection of a geometric circle, nor the frictionlessness of the contact between Newtonian bodies. These are not scientific rules induced from a large number of observations, but self-contained rational principles, by which space and movement can be represented in their perfection.

Newtonian mechanics, like Euclidean geometry, reduces temporal relations to spatial form insofar as the temporal relations between events are represented by the relations between points on a straight line. Even today the equation of temporal relations with the continuum of numbers assumes that time is isomorphic with space, and that space and time exist as a continuum, a unified totality. Time is capable of representation only through its subordination to space and spatial models. Even if the linear temporality of Newtonian mechanics is disputed, more contemporary representations of time by topological closed curves still rely on geometrical and fundamentally spatial models for their coherence.

Einsteinian Conceptions of Space, Time, and Relativity

Euclidean geometry's primacy as a mode of spatial representation was challenged by the advent of various non-Euclidean geometries that emerged during the eighteenth century. In 1764, Thomas Reid formulated a geometry based on the three-dimensionality of space, claiming that Euclidean geometry was relevant only to two-dimensional space. His privileged geometrical 'object' was the sphere where, in a sense, Euclid's had been the triangle. Hyperbolic geometry (such as Reid, Gauss, Bolyai, and Lobachevsky developed) theorized about the nature of surfaces rather than flat spaces; from Gauss's theory of the geometry of surfaces, Reimann developed spherical geometry, which rejects the parallelism postulate of Euclid,[3] claiming that there is no line parallel to any other. Moreover, he postulated that any two lines have two distinct points in common, not one, as Euclid claimed.[4] Reimannian geometry enabled spaces of n-dimensions to be theorized, and this work enabled geometry to claim a much closer affinity to the production of empirical (rather than *a priori*) truths about the world, especially those theorized by physics. Reimann claimed that the geometrical properties of infinitely small bodies depend on the physical forces or fields (such as gravity), which provide their context.

Reimann defined space in terms of its dimensionality. For him, an n-dimensional space can be defined as one in which each position can be characterized by a set of coordinates. Such a space can be mapped, graphically and mathematically represented, defined in its particular qualities and attributes diagrammatically or formulaically. The assignment of coordinates reflects or can be redefined in terms of the topological properties of space.

For Newton, matter and its qualities are the determining factors in physics; for Einstein, in contrast, it is space that determines the presence of matter. Where Newton conceived of matter as located at points in space and time, Einstein's theory of relativity conceives of matter located within a space–time field. Einstein's theory was largely a physico-mathematical attempt to apply the principle of relativity to electromagnetic fields (Special Theory of Relativity), to gravitational and inertial movements (General Theory of Relativity), and to unify the mathematical expressions of electromagnetic and gravitational fields into one simple equation (Unified Field Theory). What Einstein rejects most forcefully of Newton's conception of space and time is its commitment to understanding space and time as separate and physically independent, autonomous entities.

Einstein, then, formulated a theory to describe the relationship of energy to mass, directly connecting mathematics to physics. In place of the Euclidean grid in which Newton's conjectures were developed, Einstein relied on topological representations of space in which shape is not relevant, where one shape can be transformed into another. The difference between a cube and a sphere are irrelevant:

> ... plane topology looks into whatever properties of figures would be unchanged if they were drawn on a rubber sheet which was then distorted in arbitrary ways without cutting the sheet or joining parts of it. (Nerlich, 1976: 87)

Boltzman, one of Einstein's predecessors, specifically linked space and time together in an indissoluble unity with his formulation of the principle of entropy. This principle implies that in physical processes, the movement of bodies is temporally irreversible. In themselves, the laws of mechanics do not entail the irreversibility of processes; they are indifferent to time. Space is said to be *isotropic* insofar as its relations are considered symmetrical. In the case of time, however, there is the common everyday belief in the 'arrow of time,' in time's directionality. Through the principle of entropy, time can be considered *anisotropic*, for the relations of before and after are regulated. The principle of entropy posits that self-contained systems of matter tend towards randomness, breaking down systematic connections over time. The thermodynamic entropy of a physical system confirms everyday conceptions of the progressivity of time: entropy increases over time. Entropy is the measure of the disorder of molecular particles and thus the increase of time corresponds with progressive disorder within the closed system.

In Einstein's reconceptualization, space, time and matter are interconnected and interdefined, relative terms. Space is n-dimensional, curved and relative to the objects within it. The mass or energy of objects is relative to their position within space (and in relation to other objects) at a certain time. While it is not within the frame of this paper to discuss the technical details or the relevance to physics of Einstein's work, it is worth signalling that through the development of non-Euclidean geometries and post-Newtonian physics, not only does our everyday understanding of space and time change, but a proliferation of different models of space-time, different kinds of space (and time) with different properties circulate. None have a universally accepted scope nor domination of the whole field of geometry or physics. Mathematical and scientific formulations of space–time become localized, relevant to the operations of specific types of object (subatomic particles, molecules, objects, etc.) and of specific properties (dimensionality rather than shape, space rather than lines, orientations rather than enclosed spaces). This rich pluralism of representations is not necessarily aligned to the primacy of the visual (as in Euclidean geometry), nor to the perception of matter, but enables a multiplicity of (sometimes) incommensurable models of space and time to be explored. This is not quite a situation of 'anything goes,' as Feyerabend suggests, but nevertheless, there is no final arbiter of which model will attain primacy.

SPACE, TIME, AND BODIES

The history of scientific, mathematical, and philosophical conceptions of space and time have witnessed a number of major transformations, some of which have dramatically effected the ways in which our basic ontologies are conceived. The kinds of world we inhabit, and our understanding of our places in these worlds are to some extent an effect of the ways in which we understand space and time. Two features of our historically determined conceptions of space and time are relevant to the concerns of this essay: the first is that representations of space have always had – and continue to have – a priority over representations of time. Time is represented only insofar as it is attributed certain spatial properties. The second is that there is an historical correlation between the ways in which space (and to a lesser extent, time) is represented, and the ways in which subjectivity represents itself.

First, then, the domination of time by space. This problem has no ready-made or clear-cut justification: space is no more 'tangible' or perceptible than time, for it is only objects in space and time which can be considered tangible and amenable to perception. Space is no more concrete than time, nor is it easier to represent. The subject is no more clearly positioned in space than in time; indeed, the immediacy of the 'hereness' of corporeal existence is exactly parallel to the 'nowness' of the subject's experience. Phenomenology gives us no clear experiential preference for one dimension over the other. Perhaps it was the advent of geometry which conferred scientific and philosophical dignity on space at the expense of time. There was no analogous system to impose the rigors of method onto temporality.

Temporality is no less amenable to mathematical formulation; indeed, the phenomenon of counting provides the most pervasive of all notions of temporal duration and direction. Yet the formulation of principles and laws regarding the nature of time and its qualities occurs only when time is linked to space in a space–time continuum. Indeed, Bergson has frequently remarked on the subordination of temporality to spatiality, and consequently the scientific misrepresentation of duration. Time has been represented in literature and poetry more frequently and ably than in science. Questions about mutability and eternity are raised in philosophical speculation long before they were addressed scientifically, their stimulus coming from theology as much as from mechanics. The history of philosophy has thus far concentrated on two forms of temporality, one linear, progressive, continuing, even, regulated, and teleological (perhaps best represented by Hegel); the other circular, repetitive, and thus infinite (perhaps best represented by Nietzsche). The first form is best represented by a line, which can be divided into infinite units (recognizing time-measurement as arbitrary); the second, by a circle, which is capable of being traversed infinitely, in repetitions that are in some ways different, and in other ways the same.

In either case, temporality is defined as a relation of addition rather than one of order, in terms of numerical units rather than as a progression – as '1, 2, 3' rather than 'first, second, third.' It is considered as more psychical and some-how less objective than space. If Irigaray is correct in her genealogy of space–time in ancient theology and mythology, space is conceived as a mode (indeed God's mode) of exteriority, and time as the mode of interiority (Irigaray, 1993: 18). In Kant's conception, too, while space and time are *a priori* categories we impose on the world, space is the mode of apprehension of exterior objects, and time a mode of apprehension of the subject's own interior. This may explain why Irigaray claims that in the West time is conceived as masculine (proper to a subject, a being with an interior) and space is associated with femininity (femininity being a form of externality to men). Woman is/provides space for man, but occupies none herself. Time is the projection of his interior, and is conceptual, introspective. The interiority of time links with the exteriority of space only through the position of God (or his surrogate, Man) as the point of their mediation and axis of their coordination.

This links the domination of conceptions of time by representations of space with the second point: that the representations of space and time are in some sense correlated with representations of the subject. This occurs both developmentally (as I argued in Part 2) and historically. Developmentally, the child perceives and is organized with reference to a series of spatial conceptions, from its earliest access to the 'space of adherence' to the virtual space of mirror-images, the curved and plural spaces of dreams and the spatiality conferred by the primacy of vision. Historically, it can be argued (although I do not have the space to do so here) that as representations of subjectivity changed, so too did representations of space and time. If space is the exteriority of the subject and time its interiority, then the ways this exteriority and interiority are theorized will effect notions of space and time.

Le Doeuff, for example, argues that the kind of subject presupposed in the time of Shakespeare and Galileo is a notion divided between an absolute (and thus non-individualized) identity and a wholly particularized 'personality' (Le Doeuff, 1986). The notion of subjectivity in which the subject reflects universal or cosmic affinities, as well as individualized particularities, is the correlative of the Galilean conception of time as it is formulated in the law of falling bodies, $V = dt$, (velocity = distance over time):

> The Galilean law supposes a capitalisation of the past, the falling body cumulatively acquiring degrees of speed, or undergoing a perpetual augmentation of speed. ... Let us think of the way a loan with simple interest is defined (I am not speaking here of loans with compound interest, since that would involve exponentials): x per cent per annum on a one year loan; $2x$ percent on a two-year loan, etc. Labour hired by the hour obeys the same schema, as do all our practices of rent, hire, etc. It is not difficult to convince oneself that this social usage of time is strictly modern. (Le Doeuff, 1989: 42–43)

Le Doeuff's general point is the interdependence of scientific thought and prevalent social and cultural conceptions: she investigates how Galileo could have (correctly) formulated the law of falling bodies using time rather than space as a point of reference, when space seems more self-evident as a fixed reference point within the scientific context of his culture. Her point is that Galileo could not have formulated his law without relying on pre- and non-scientific notions of temporality. And in turn, the principles governing his law were able to be comprehended and understood by other scientists and educated non-specialists because they accorded with everyday notions of time.

The same could be said of Newtonian and Einsteinian conceptions of space: they accord with contemporary notions of subjectivity and social life. This correlation is, in all likelihood, dialectical: conceptions of the subject are projected onto the world as its objective features; in turn, scientific notions are internalized, if only indirectly, through their absorption into popular culture. The Kantian conception of subjectivity is a metaphysical correlate of Newtonian physics, and the decentered Freudian subject conforms to the relativity of an Einsteinian universe. Perhaps the postmodern subject finds its correlative in the virtuality of cyberspace and its attendant modes of respatialization. Such conceptions are both ways of negotiating our positions as subjects within a social and cosmic order, and representations that affirm or rupture pre-existing forms of subjectivity.

If, however, post-Euclidean and post-Newtonian conceptions are made possible during an historical questioning of the postulates and values of the Age of Reason and the era of the self-knowing subject, these have still, in spite of their conceptual distance from Euclid and Newton, confirmed the fundamental masculinity of the knower, and left little or no room for female self-representations, and the creation of maps and models of space and time based on projections of women's experiences. It is not clear that men and women conceive of space or time in the same way, whether their experiences are neutrally presented within dominant mathematical and physics models, and what the space–time framework appropriate to women, or to the *two* sexes may be. One thing remains clear: in order to reconceive bodies, and to understand the kinds of active interrelations possible between (lived) representations of the body and (theoretical) representations of space and time, the bodies of each sex need to be accorded the possibility of a different space–time framework.

To transform the castrated, lacking, inadequate representation of female corporeality, not only do the relations between the sexes and the dominance of masculine in the formulation of universal models need to be questioned, the overarching context of space–time, within which bodies function and are conceived also needs serious revision. The possibility of further alternatives must be explored. This investigation, in turn, must effect the nature of ontological commitments and the ways in which subjects (masculine as well as feminine) see themselves and are socially inscribed. This may amount to a scientific and conceptual revolution alongside sociopolitical transformations, but without questioning basic notions of space and time, the inherent masculinity of the 'hard sciences' and of philosophical speculation will proliferate under the banner of the human. Women, once again, may be granted no space or time of their own.

Notes

1. I use the generic 'he' advisedly; the relevance of the question of sexual difference to Caillois's account needs careful consideration if it is to be taken as relevant for women as well.

2. Newton's three Laws are:

    1. A body continues in a state of rest or uniform motion along a straight line, unless it is subject to a force (the law of inertia);
    2. If a force acts on a body, then the body has an acceleration in the direction of that force, and the magnitude of the acceleration is directly proportional to the force and inversely proportional to the mass of the body; and
    3. The forces exerted by two bodies on each other are equal in magnitude and opposite in direction, along the line joining their positions. (Van Fraassen, 1970: 53)

3. Euclidean geometry is frequently reduced to five fundamental axioms that, between them, generate his propositions regarding geometrical figures. They are: (1) if x and y are distinct points, there is a straight line incident with both; (2) any finite straight line is part of a unique, infinite straight line; (3) if x is a point and r is a finite distance, there is a unique circle with x at its centre and r defining the length of its radius; (4) any two right angles are equal; and (5) if a straight line falling on two straight lines makes the interior angles on the same side less than two right-angles, if they are indefinitely extended, the two straight lines meet on the side where the angles are less than two right angles. Cf van Fraassen, 1970: 117. Incidentally, non-Euclidean geometry denies the 5th postulate, and some forms also deny the 2nd and 3rd axioms.

4. If one is referring to two points on a sphere or any curved surface, then there are n-lines linking them, not just one. Parallelism is impossible in curved spaces.

PART 2
CYBERSUBJECTS:
CYBORGS AND CYBERPUNKS

INTRODUCTION

The previous section was concerned with the general question of embodiment within the new spaces and places of cybernetic systems. The chapters in this section are focused specifically on the nature of embodiment as both a cyborg and a cyberpunk, and they have in common the fact that they examine texts from within the science fiction genre to explore issues arising from these forms of embodiment. The essays contained in this part of the book explore the new paradigms of subjectivity and identity that are emerging in the context of the technologically mediated environment of cyberspace. Representations of cyberspace are abundant in the narratives of popular culture, especially cyberpunk narratives in which the decentred postmodern subject is reconfigured as a cyborg. The essays in this section articulate a variety of critical responses both to the figure of the cyborg proposed by Donna Haraway in her 'Manifesto', and to the science fiction sub-genre of cyberpunk, the narratives of which are populated with fictional cyborgs. Cyberpunk itself is structured by a contradictory attraction to and fear of the potentialities of the cyborg, and several essays examine the limitations which these unresolved contradictions impose on representations of the cyborg in cyberpunk. Donna Haraway has already argued for the usefulness of the cyborg metaphor for feminist theory, and the writers in this section respond to the arguments that she has put forward. The association of cyborgs with cyberpunk also receives attention in this section, primarily because, like the cyborg, cyberpunk has been hailed as quintessentially postmodern, existing on the borders of high and popular culture, and able to exploit that boundary position with great verve. However, cyberpunk's engagement with the politics of gender has not been as radical as its counter-cultural rhetoric might have suggested, an issue that is discussed in the following essays.

Anne Balsamo's chapter is concerned with the relevance of the cyborg metaphor for feminist theory and she examines its capacity to make a strategic intervention into that theory. Balsamo argues that one of the key difficulties with the use of the metaphor is that the cyborg already occupies a central place in contemporary cultural iconography, and as such it is bound to reproduce familiar gender stereotypes. Thus, male cyborgs inevitably reinforce the link between rationality, technology and masculinity. Yet it is for precisely this reason that female cyborgs are more disruptive than male cyborgs, because they 'embody cultural contradictions which strain the technological imagination. Technology isn't feminine, and femininity isn't rational.' Since the cyborg metaphor also reveals the culturally embedded distinction between self and other to be both arbitrary and unstable, Balsamo supports Donna Haraway's contention that the ability of the cyborg to challenge these kinds of binarisms is the point at which it becomes useful for feminist theory. At the same time, however, Balsamo argues that the cyborg is not inherently radical or utopian in its meanings, indeed, the 'dominant representation of cyborgs reinserts us into dominant ideology by reaffirming bourgeois notions of human, machine and femininity', and she points out that Haraway 'has failed to consider how the cyborg has already been fashioned in our cultural imagination.' What is at issue for Balsamo is the ongoing struggle around the meaning of the metaphor, by means of which interpretation can become 'a site of feminist politics in this postmodern age.'

Like Balsamo, N. Katherine Hayles recognises the way in which the cyborg metaphor has a set of cultural meanings attached to it, but she argues that, since 'The new cannot be spoken except in relation to the old', this does not prevent it from being used to develop new and more radical meanings. The cyborg can be thought of as being expressive of new forms of subjectivity, and the way in which those forms are understood culturally is closely examined by Hayles. She positions the cyborg as standing 'at the threshold separating the human from the posthuman' from where it 'looks to the past as well as the future'. Emerging cyborg subjectivities are, then, 'amalgams of old and new' and the kinds of narratives within which such fusions are explored are deeply complex. Hayles makes a distinction between what she calls 'life cycle narratives' and 'cyborg narratives' and explores the way in which both kinds of narrative are inextricably linked to produce 'narrative patterns that overlay upon the arc of human life a map generated from assembly and disassembly zones.' In order to develop this argument, she divides the life cycle into three phases, each of which has a corresponding dis/assembly zone. The phases approximate, firstly, to adolescence, and its accompanying dis/assembly zone is 'marked by the joining of limb to torso, appendage to trunk'; the second phase is that of sexual maturity, and its accompanying dis/assembly zone is 'located where the human is plugged into the machine, or at the interface between body and computer network'; the third phase is 'reproductive or generative' – its accompanying dis/assembly zone 'focuses on the gap between the natural body and mechanical replicate, or between the original and manufactured clone.' In each set of stories, drawn from science fiction, she examines what she calls 'boundary questions' through which the emergence of new cyborg subjectivities can be seen to impact on more familiar accounts of the gendered human self, and as they do so, they 'reveal the gendered constructions that carry sexual politics into the realms of the posthuman.' The tension between organic and machine that is exposed in these narratives of cyborg subjectivities cannot be resolved, and since cyborg entities are both real and fictive, they tell us that orthodox notions of the subject and of identity are no longer entirely sustainable in the new 'cybernetic paradigm'.

The chapters by Nicola Nixon and Veronica Hollinger both concentrate on the subject in and of cyberpunk narratives, although they arrive at slightly different conclusions about its radical potential. For Veronica Hollinger, the significance of cyberpunk is that it is expressive of the postmodern imperative to deconstruct the opposition between human and machine, in the process of which it constructs a 'posthumanism' that 'radically decentres the human body, the sacred icon of the essential self'. Hollinger argues that the imaginative capacity of cyberpunk to undermine concepts such as 'subjectivity' and 'identity' stems from an acknowledgment of the 'paradigm-shattering role of technology' which it shares with 'hard' science fiction. At the same time, however, the refusal of cyberpunk narratives to restore the centrality of the unitary, organic self is what distinguishes it from such genre science fiction and locates it firmly within the context of postmodernism. While Hollinger's account is chiefly concerned to demonstrate the postmodern affiliations of cyberpunk, she observes that 'it is not the only mode in which science fiction has demonstrated an anti-humanist sensibility', and that feminist science fiction has also produced 'an influential body of anti-humanist texts.' The extent to which cyberpunk and feminist science fiction can be said to share an agenda is limited, however, since the reconstructed and multiple selves of feminist science fiction are not the same as the fragmented subjectivities to be found in cyberpunk texts. Hollinger points out that while cyberpunk texts willingly explore the troubling consequences of the collapse of the boundaries between human and machine, they also reveal anxieties about the fragmentation of the unitary subject, and 'balanced against the exhilaration of potential technological transcendence is the anxiety and disorientation produced in the self/body in danger of being absorbed into its own technology'. Since the self/body in question is almost invariably male, this suggests that there are substantial limitations to cyberpunk's 'cybernetic deconstructions', despite its postmodern credentials.

It is the nature of these limitations in respect to its representations of gender that most exercise Nicola Nixon in her chapter on cyberpunk. She argues that cyberpunk's claims to be 'revolutionary' must be situated within the specific context of Reaganite America in the 1980s, as well as alongside earlier and contemporaneous forms of political, feminist science fiction, if they are to be validated. The crux of her argument is that, despite its many references to contemporary popular culture, cyberpunk fiction 'is not radical at all', because its 'slickness and apparent subversiveness conceal a complicity with '80s conservatism'. Although Samuel Delany has argued that strong women characters in cyberpunk fiction would have been impossible without the feminist science fiction written in the 1970s, Nixon points out that not only are there very few such characters in cyberpunk, but also that they are 'feminised' when they do appear, with the effect that their radical potential is crucially undermined, if not excised altogether. Taking the work of William Gibson as illustrative of cyerpunk fiction in general, Nixon notes that the rewriting of the strong female characters of feminist science fiction takes place in conjunction with a depiction of the Japanese Yakuza as 'a kind of monolithic Japanese corporate entity' in which there are echoes of the anti-Japanese feeling particularly prevalent in America in the 1980s. Equally, it is not without significance that the 'central heroic iconography' in cyberpunk is the cowboy, a figure that is itself 'realised so strongly in Reaganite cowboyism'. In Gibson's cyberpunk narratives, Japanese conglomerates, together with the decaying forms of old European aristocracies, perform the role of the 'demonised and feminised Other' against which is pitted the lone cowboy hero. The contradictions of this duality are revealed most clearly within the matrix: instead of being constituted as 'the space of the homoerotic', the matrix is 'figured as feminine space' in which there has been 'a kind of female infection and (viral) takeover of original masculine space'. The result of this is that the matrix becomes a form of 'scary feminised software'. In a reminder of the necessity to locate cyberpunk within the political and cultural specificities of the 1980s, Nixon speculates as to whether or not Gibson's matrix can be regarded either as a possible representation of the patriarchal, colonising fictions of the American frontier, or, more worryingly, as a manifestation of the virulent anti-feminist backlash of the time. She argues that, whichever interpretation one chooses, there is little doubt that Gibson and other cyberpunk writers succeed in undermining the political potential of science fiction, so powerfully evinced in the feminist science fiction of the '70s, by creating an imaginary environment within which 'a kind of primal masculinity' is valorised.

In contrast to the critical view of cyberpunk put forward by Nicola Nixon, Thomas Foster in his chapter attempts to rescue cyberpunk from its own contradictions, by pointing out that it 'foregrounds and interrogates the value and consequences of inhabiting bodies', thereby providing a means of questioning representations of gendered embodiment. To demonstrate this, he juxtaposes the radical potential of the cyborg against the limitations and contradictions inherent in cyberpunk. The significance of Foster's chapter for the overall project of this book is that, unusually, he includes the distinction between the 'straight white male self and its others' in the catalogue of distinctions that are called into question by cyberpunk, such as those between mind and body, human and machine. According to Foster, while a cyberpunk text like *Neuromancer* remains 'dependent on those distinctions, to some degree', at the same time, its position within postmodern culture also enables it to engage in some tentative redefinitions of masculinity. For Foster, a central feature of cyberpunk has been the way in which it has marked the changing status of the body in postmodern cultures, thus linking it to the cyborg, and to the radical possibilities for rethinking subjectivity and identity that are implied by the metaphor. The cyborg throws into question the construction of the subject within the dualistic thinking of Western modernity, so that the 'universal position of the white, middle-class male subject no longer seems available'. At the same time, however, the very lack of specificity of cyborg identities may allow the 'white male experience of postmodernity' to once again become the norm. He argues that it is this 'double gesture', of 'opening the question of alternative cultural identities' only to close down around that which is already known and familiar, which structures representations of cyborg identities in cyberpunk. Thus, cyborg embodiment within cyberpunk is represented in terms which continue to reproduce the gendered distinction between mind and body: the body is conceived of as 'meat' and the only escape from entrapment as a 'meat puppet' is through the technologising of the body as a 'robopath', which makes transcendence into the matrix possible. This version of embodiment leaves gender categories intact and, thus, demonstrates the central problematic of cyberpunk, that while it appears to challenge the way in which gender is inscribed in the human/machine interface, it also reproduces the very categories which support that inscription. Foster argues that a feminist reading of the cyborg imagery in *Neuromancer* is, nevertheless, possible, because the narrative ultimately refuses to privilege 'data' over 'flesh', thereby rejecting the gendered and essentialist connotations of either.

The specific focus of my own chapter is the way in which postmodern anxieties about the definition of the subject are reworked in the cyborg narratives of feminist science fiction, in order to explore 'utopian possibilities for the redefinition of gender identity and gender relations.' I argue that a shift of emphasis can be discerned in the feminist science fiction written since the 1980s, such that technology increasingly forms the terrain on which issues concerned with gendered subjectivity are considered, despite the masculine hegemony of technology. The metaphor of the cyborg has a particularly disruptive effect in this context: not only is it the means by which the gendered power relations of technology can be disrupted within the narratives, but since the genre 'has largely been built around the unquestioned assumption of that hegemony', it also disrupts the generic stability of science fiction itself as a consequence. I suggest that, although feminist science fiction uses the codes and conventions of both science fiction and romance fiction, it negotiates the contradictions of such doubly coded and ambiguous narratives by 'placing the desiring "I" of romance fiction at the centre of SF narratives.' Indeed, feminist science fiction 'exploits' postmodern instabilities and uncertainties precisely in order to question 'definitions of difference and otherness, and the discourses of power within which those differences are articulated'. This is in contrast to cyberpunk texts, which display considerable anxiety about the way in which 'the virtual realities of cybernetic systems are constantly threatening to undermine the imagined stabilities of gender.' I have chosen to describe the cybernetic narratives of feminist science fiction as 'postmodern romances' because of the way in which 'the politics of female desire are negotiated through the problematising of female identity', so that desire itself can be differently configured within these narratives. It is because the cyborg identities of feminist science fiction are 'neither unified nor coherent' that the narratives are able to explore 'the pleasures and the pains, as well as the possibilities and the limitations' of that identity.

7

READING CYBORGS
WRITING FEMINISM

Anne Balsamo

What if we read contemporary science fiction stories as ethnographies of the future? Or the cultural emblems of postmodernism as archeological remains of our time grown old? Would we 'read' them in an archaic sense – searching for meaning through the rubble of excessive signification? Would ethnography be possible? Useful? Strategic? Would culture remain 'writable'? Would we recognize our contemporary age as one of science's past fictions? Given the array of postmodern monuments, from relatively durable architectural wonders through fleeting moments of tongue-in-cheek social commentary, we might wonder, what of this array will endure as markers of our contemporary era?

In 1986, Max Headroom stuttered his way onto American television and the cover of *Newsweek*. The same year, Electra Assassin, Frank Miller's celebrated 'super-coded' anti-hero, challenged the revitalized comic book industry's vision of a proper heroine. By Christmas, 1986, it was clear that Transformers were the toy of the season, refashioning children's popular culture as they hammered and slashed their way off the pages of toy catalogues onto a syndicated Saturday morning cartoon series.

Max Headroom, Electra Assassin, and the various incarnations of Transformer-inspired action figures epitomize our culture's fascination with cyborgs. Cyborgs are science fictional hybrids. The name, a shorthand term for 'cybernetic organism,' usually describes a human–machine coupling, most often a *man*–machine hybrid. Cyborgs are stock science fiction characters which are alternately labeled 'androids,' 'replicants,' or 'bionic.' Whatever label they attract, cyborgs provide provocative images to consider in light of the questions with which I opened this essay. They occupy a visionary part of our cultural imagination, the image of the future created in the present, the place of science fiction reality, of the postmodern. Whether cyborgs are considered the first citizens of an industrialized technocracy or the perfect companions for an anti-social, simulated society, their images pervade film and popular culture, as well as the world of consumer commodities.

Cyborgs are *the* postmodern icon. From children's plastic action figures to RoboCop's titanium exoskeleton, cyborg-ian artifacts will endure as relics of an age obsessed with replication. In what follows, I will be concerned with how our technological imagination imbues cyborgs with ancient anxieties about human differences. In short, this essay offers a somewhat ironic ethnographic account of cyborg identity as it is portrayed in contemporary popular cultures: ironic in that I purposefully exaggerate the connotative slide of the meaning of the term 'cyborg'; ethnographic in the sense of being a cultural analysis – a thick description, of sorts – of cyborg life. My argument proceeds neither linearly nor circularly, but rather radially, suggesting several ways in which the cyborg image can be read through a set of feminist lenses.

CYBORG REPRESENTATION: THE TECHNOLOGICAL REPRODUCTION OF GENDER

From Mary Shelley's *Frankenstein*, published in 1818, to Maria, the robot in *Metropolis* (Fritz Lang, 1927), to the *The Six Million Dollar Man* (1970s), the possibilities of human hybrids have fired our cultural imagination as the Western world moved through industrial and technological phases of development. Variously used as a symbol of anti-technological sentiments or of the possibilities of 'better living through chemistry,' cyborgs are a product of cultural fears and desires that run deep within our psychic unconscious. Through the use of technology as the means or context for human hybridization, cyborgs come to represent unfamiliar 'otherness,' one which challenges the connotative stability of *human* identity.

Consider a continuum which has at one extreme the characteristics associated with machines and technology and, at the other extreme, the characteristics of humans and organic society. How are the end points identified? Machines are rational, artificial and durable; humans are emotional, organic and mortal. Every cyborg image constructs an implicit opposition between machine and human; at once repressing similarities and highlighting distinctions. This is the science fictional character of the cyborg – it is a hybrid, but the specific traits which mark its human-ness and machine-ness vary widely. Signs of human-ness and, alternatively, signs of machine-ness function not only as markers of the 'essences' of the dual natures of the hybrid, but also as signs of the inviolable opposition between human and machine. This is to say that cyborgs embody human characteristics that reinforce the difference between humans and machines. With the Terminator, his 'covering' of human flesh which enables him to time travel, serves as his primary human characteristic; with Helva, the Ship that Sang, it is her organic, cognitively functioning brain severed from a physical body that indicates her human nature. Don't we wait anxiously throughout the film or novel for the revelation of cyborg identity? How will the separate parts be combined? What signs will serve to mark its human nature? How are these mechanically animated creations still human? Every cyborg image implicitly defines the meaning of the terms 'human' and 'artificial.'

Thus cyborgs fascinate us by technologically refashioning human difference. It is in this sense that cyborgs disrupt notions of otherness. The notion of human relies upon an understanding of non-human, just as the notion of artificial implies an understanding of natural. In the history of human supremacy, that which is non-human is understood as the other, that which is mechanical is understood as artificial. Cyborgs, as simultaneously human and mechanical, complicate these ancient oppositions. The preservation of human difference in a technological world is fraught with tension as the distinctions between artificiality and authenticity become blurred.

The film *The Terminator* (James Cameron, 1984) presents one of the most familiar and frightening visions of cyborg characterization. Arnold Schwarzenegger portrays a cyborg killing machine – a man–machine hybrid from one possible post-nuclear war future in which machines, outraged by human incompetence, set out to annihilate the remnants of humanity. Sent to earth in human flesh, he deftly outfits himself with weapons and leathers. As his battle with humans wages on, his flesh burns away, and he is reborn out of the flames as pure machine, pure technological will to murder. The film works to represent The Terminator's transformation from remotely human to fully machine.

Peter Weller, as *RoboCop* (Verhoeven, 1987) is also cinematically transformed, this time from fully human to soulful machine. Brutally murdered in the line of duty, Weller's remains are reconstructed as the organic part of a cyborg law enforcement machine. Robocop fights on the side of good, yet is unable to override a hardwired directive which disables him at the moment of confrontation with the film's main antagonist.

The Terminator and Robocop represent two sides of the same type of man–machine hybrid. Neither cyborg can override its programmed directives; they both represent the extreme of technological rationality. Given this, is their male gender surprising? Stereotypically, rationality is associated with masculinity. In this, cyborgs and men are compatible images which mutually support cultural associations among masculinity, rationality, technology and science.

Yet, when we further investigate cultural images of cyborgs, we find that cyborgs aren't always male. Mary Ann Doane traces the ancestry of female cyborg images to a novel by Villiers de l'Isle-Adam, *L'Eve future* (*Tomorrow's Eve*), written in 1886. 'In this novel, Thomas Edison, the master scientist and entrepreneur of mechanical reproduction is the inventor of the perfect mechanical woman, an android whose difference from the original human model is imperceptible' (1987, 3). This quest for the 'perfect mechanical woman' persists as a theme in contemporary science fiction. Such is the science project of two highschool boys in the film, *Weird Science* (Hughes, 1985).

Rachel, the melancholy replicant from *Bladerunner* (Ridley Scott, 1982), is a recent female cyborg whose constructed 'nature' supposedly contradicts the myth of natural female identity. Not only is her body genetically constructed, she's been given memory implants borrowed from a 'real' woman. Rachel thinks she's human – she has memories of a mother, piano lessons, the birth of spider babies. It's not these bits and pieces of stolen momories that make Rachel an enigma for Deckard (the bladerunner whose job is to 'retire' replicants). It's her emotions. Rachel's well-worn pout and vulnerable sexuality shakes Deckard's certainty about her replicant nature. As the object of Deckard's visual and sexual desire, Rachel symbolically reasserts the social and political position of woman as object of man's consumption.

Symbolizing a different notion of cyborgian identity, that of spaceship–human coupling, is Anne McCaffrey's protagonist in her 1974 science fiction novel, *The Ship That Sang*. Helva XH-834 was 'born a thing with twisted limbs, and is destined to be a guiding mechanism for a spaceship.' Helva is a 'shell-person, an encapsulated brain who's kept physically small by pituitary manipulation' (King, 1984). Among her more benevolent adventures, Helva and her human mobile partner – her 'Brawn' – rescue cloistered colonists from a planet soon to be scorched by an unstable sun and, at another time, carry 100,000 embryos to a sparsely populated off-world planet. Certainly appropriate missions for a female-gendered spaceship.

Mary Ann Doane (n.d.) writes about the relationship between technology, representation and the feminine. She argues that

> although it is certainly true that in the case of some contemporary science fiction writers – particularly feminist authors – technology makes possible the destabilization of sexual identity as a category, there has also been a curious but fairly insistent history of representations of technology which work to fortify – sometimes desperately – conventional understandings of the feminine. A certain anxiety concerning the technological is often allayed by a displacement of this anxiety onto the figure of the woman or the area of the feminine.

Doane suggests that it is notions of human *reproduction* that are most contested within science fiction. The extent to which the mystery and dangers of human replication persist as a theme in contemporary science fiction suggests the depth of cultural anxiety about species reproduction. If this is indeed the case, then what we find in supposedly visionary science fiction is but a revamped reflection of ancient beliefs about the proper role and function of women in society and history.

These female-gendered cyborgs inhabit traditional feminine roles – as object of man's desire and his helpmate in distress. In this way, female cyborgs are as much stereotypically endowed with feminine traits as male cyborgs are with masculine traits. Cyborg images reproduce cultural gender stereotypes. I want to argue, however, that female cyborg images do *more* to challenge the opposition between human and machine than do male cyborgs because femininity is culturally imagined as less compatible with technology than is masculinity. This is to say that because our cultural imagination aligns masculinity and rationality with technology and science, male gendered cyborgs fail to radically challenge the distinction between human and machine. Female cyborgs, on the other hand, are culturally coded as emotional, sexual, and often, naturally maternal. It is these very characteristics which more radically challenge the notion of an organic-mechanical hybrid. Female cyborgs embody cultural contradictions which strain the technological imagination. Technology isn't feminine, and femininity isn't rational.

Female cyborgs, while challenging the relationship between femaleness and technology, perpetuate oppressive gender stereotypes. This reading of female cyborgs raises interesting feminist questions. Can the 'feminine' essentialism of fictional cyborgs be transformed into a non-essential image for contemporary women? Is there any way that the cyborg image could be used strategically to intervene in feminist theory? There are basically two ways to begin to answer these questions. One way is to construct a utopian vision of the possibilities of a Helva XH-384, in which technology emancipates woman from her corporeal body. Feminism's skepticism about technology and science would be challenged to see the potentially liberating effects of technology. A second possibility involves the construction of an ideological critique of the cyborg image as it has been produced by partriarchal culture. This approach might begin by asking how the cyborg image represents women. From here we might then invert the question, to ask, how do women represent cyborgs? This is to suggest, with an ironic twist, that woman's development is not separate from technological development, but has, in fact, displayed a similar trajectory. Her history illustrates several points of intersection with technology, points at which she has been forced to become like the cyborg, a hybrid creature of fiction and reality. Cyborgs become like woman in that cyborg images represent something 'unknown' and perhaps, 'unknowable' in our popular imagination. To this very end, Donna Haraway's (1985) article, 'A Manifesto for Cyborgs: Science, Technology, and Socialist Feminism in the 1980s' outlines the beginnings of such a strategic intervention in contemporary feminist theory.

READING CYBORGS

The cyborg image can be read in two ways: as a coupling between a human being and an electronic or mechanical apparatus or, as Haraway (1985) suggests, as the identity of organisms embedded in a cybernetic information system.[1] In the first sense, the coupling between human and machine is located within the body itself – the boundary between the body and machine is surgically redrawn. In Haraway's argument, the boundaries between the body and technology (primarily communication technology) are socially inscribed, at once indistinct and arbitrary. These are the cyborgs that have become important in the non-fictional discourse of communications, computer engineering, and bio-mechanical medicine.

Thus, for Haraway, cyborgs are interactive participants found at all points of information production and replication. Positioned within high-tech information management networks, they are constantly formatting, coding and decoding, uploading and downloading information. Word processors, data programmers, and CRT operators, have become familiar civil service cyborgs.

But Haraway explicitly maps the identity of woman onto the image of the cyborg. Her 'Manifesto' narrates an 'ironic political myth' of cyborg identity, which, she argues, is the only possibility for woman-identity in the late 20th century. She argues that 'woman' is no longer singular, but rather, a commodified, technological object whose unique human status is challenged by rapid technological transformations. Yet, woman's 'unique human status' has never been very secure. I interpret Haraway's ironic choice of cyborg image as an attempt to show how feminism could rethink its political and intellectual relationship to technology and science. Although feminism often remains skeptical of the patriarchal promise of technological development, women need to develop ways of reading and responding to technology that resist opposing it to an unproblematic 'nature.'

Haraway's notion of the cyborg foregrounds the ambiguous constitution of the body and subjectivity – predicated on the blurred boundaries between organism and machine, the individual and the technological, the fictive and the real. By challenging these and other culturally entrenched binarisms, the image of the cyborg becomes compatible with, and maybe even useful for, feminist theory. Let me sketch out two examples in which the cyborg image, as a postmodern icon, works well to capture some of the more radical insights of feminist scholarship: the social construction of subjectivity and the cultural crafting of physical bodies.

POSTMODERN IDENTITY: THE SOCIAL CONSTRUCTION OF SUBJECTIVITY

The cyborg is a social construction – 'a creature of social reality as well as a creature of fiction' (Haraway, 1985: 65) – and this illuminates a crucial dimension of postmodern identity: the fragmentation of subjectivity. Both Woman and Cyborg are simultaneously symbolically and biologically produced and reproduced through social interactions. The 'self' is one interactional product; the body is another. Haraway argues that the international women's movements have, in fact, as much constructed 'women's experience' as they have uncovered or discovered experience as a creative collective object. Diverse feminisms grapple with the multiple dimensions of female identity as simultaneously a matter of ideology (a social-symbolic construction) and of materiality (a physical body). Not wanting to fall back on an essentialist or elitist definition of woman, feminist perspectives struggle to affirm difference while building coalitions. If the cyborg appears as the embodied image of both an ideological (human) identity and material (technological) reality, then woman's identity, as much socially and psychologically constructed as it is physiologically and biologically determined, reveals her cyborg likeness.

Mainstream postmodern theory calls for a re-thinking of subjectivity; yet, it often finds itself, in practice, caught between contradictory desires. Its insistence on the non-essentiality of identity ultimately confronts an implicit, rhetorical urge to catalog all things by their 'true' names. The apparent loss of a subjectivity that at least at one time appeared to have been unified and coherent often provokes curious commentary on the part of these theorists. Jean Baudrillard (1983) celebrates the death of authentic subjectivity and with it, the collapse of the real and the advent of the hyperreal. Fredric Jameson (1984) describes 'schizophrenia' as the cultural dominant of late capitalism. These critics ask, rhetorically and maybe sarcastically, how do we decode the 'I' if it is no longer unitarily referential?

Throughout the history of the women's movement, feminist scholars have wrestled with this paradox of identity. Philosophically posed as a critique of individualism, this strand of feminist thought shows up in de Beauvoir's early work (1959) and is carried through in Cora Kaplan's (1986: 226) thinking on subjectivity.

> [W]ithin contemporary western culture the act of writing and the romantic ideologies of individual agency and power are tightly bound together, although that which is written frequently resists and exposes this unity of the self as ideology. At both the psychic and social level, always intertwined, women's subordinate place within culture makes them less able to embrace or be held by romantic individualism with all its pleasures and dangers. The instability of 'femininity' as female identity is a specific instability, pointing to the fractured and fluctuant condition of all consciously held identity, the impossibility of a will-full, unified and cohered subject.

Like Kaplan, Haraway reminds us of a critical feminist insight: that women are traditionally self-characterized by diversity and fragmented identities, and this has greatly influenced the possibility of unified political action.[2] For Haraway (1985: 82) 'the *cyborg* is a kind of disassembled and reassembled, post-modern collective and personal self ... [yet] this is the self that feminists must code.' In an age of coaliation, it is of utmost importance for feminists to learn how to construct an 'I' as part of a 'we,' without setting 'us' against 'them.'

POSTMODERN PROSTHESES: THE CULTURAL CRAFTING OF PHYSICAL BODIES

Cyborg bodies pump iron – physically fit, yet unnaturally crafted, they are hyper-built. A cyborg body, as Bateson (1969: 319) might argue, 'is not bounded by the skin but includes all external pathways along which information can travel.' The high-tech image of the cyborg reminds us to question the assumed naturalness of the body and its function as a marker of difference.

Robert Jarvik, the physician for whom the Jarvik 7 artificial heart is named, is the president of a company called Symbion – a name combining *symbiosis* and *bionic* – whose goal is to interface the body and technology. By 1990 we may see the availability of an electronic retina; a prototype developed in Japan uses a computer chip covered with layers of silicon sensors. The U.S. Food and Drug Administration has already approved the commercial use of an artificial ligament – one made from Gore-Tex, a common sportswear material.[3] These bio-medical events subtly register the collapse of the distance between the present and a science fictional future in which bionic bodies are commonplace.

Feminist theorists have traditionally asserted that female bodies are not one-dimensional surfaces which bear easy-to-read meanings. Indeed, feminist writers honor the body as the site of the production and reproduction of fragmented identities and affinities – in short, the site of material practice. They identify the place and meanings of the female body in mass culture, sometimes to reassert the importance of female sexuality (for example, Audre Lorde [1978]); sometimes to propose a radically new form of cultural production – writing the body, or the body as instrument[4]; and sometimes to articulate 'the site for the coming together of feminist theory and politics' (Brown and Adams, 1979: 15).

The body and its iconography are a location for inscribing differences among women – differences that raise feminist consciousness by challenging its homogeneity. Sandra Gilman (1985) reads medical icons of black bodies and white bodies to reveal the differential construction of female sexuality in late nineteenth-century culture. Anne Finger (1986) challenges other feminists to examine their attitudes toward physical disabilities, especially at the point of building theories which rely on the implicit assumption of a fully abled body. As she argues (Finger, 1986: 295), 'disability is largely a social construct,' one which potentiates the impact of patriarchal domination.[5]

Similar issues become apparent when Paula Treichler (1987) examines the construction of the 'AIDS Victim' in bio-medical discourse. Persons with AIDS are discursively constructed as victimized, pitiful, valient, contagious, marked by god's wrath, marginalized. In a similar way the identity of the AIDS virus manifests itself at the center of a fierce signification battle. Descriptions of the workings of the AIDS virus reveal an identity changing, code rewriting process. The virus literally rewrites the genetic material of healthy host cells in such a way that the mechanism for distinguishing between self (healthy cells) and other (AIDS virus or other viruses) is obliterated. The AIDS virus works by confusing identity and blurring boundaries.

The physical body has traditionally been a reliable ground for establishing identity – consider the call for chromosome testing at the 1972 Olympic Games. But bionic bodies defy the natural-ness of physical identity. One of the most seemingly self-evident characteristics of physical bodies is their fundamentally organic composition. Yet, bionic body recrafting already allows people to change their physical sexual characteristics. As these medical procedures become more advanced and sexual body parts technologically refashioned, a visual reading of gender, or any other cultural marker of identity, off the surface of the body will be hopelessly confounded.

WRITING FEMINISM

I want to return to a consideration of Haraway's essay which places the Cyborg image in front of the feminist imagination. The constructedness of cyborg subjectivity and bodies reminds us that we ourselves are constructed as we actively participate in constructing the objects/subjects of our research. More important to remember is how people with AIDS, women of color, disabled women, and women generally have been marked by their constructed otherness. We study 'them,' fixing their identities, thereby constructing a system for creating difference which is oppressive and exploitive.

Andreas Huyssen (1984) and other postmodernists claim that the crisis of modernism pivots on this problematic of otherness. In this way, cyborgs offer a particularly appropriate emblem, not only of postmodern identity, but – specifically – of woman's identity. Cyborg identity is predicated on transgressed boundaries. They fascinate us because they are not like us, and yet just like us. Formed through a radical disruption of other-ness, cyborg identity foregrounds the constructedness of otherness. Cyborgs alert us to the ways culture and discourse depend upon notions of 'the other' that are arbitrary and binary, and also shifting and unstable. Who or what gets constructed as other becomes a site for the cultural contestation of meaning within feminist politics.

My criticism of Haraway's choice of image is that she fails to consider how the cyborg has already been fashioned in our cultural imagination. It is difficult to determine if Haraway chooses the cyborg image because she believes that women are inherently cyborgian, or because the image is useful and potentially liberating. As I've described above, cyborg images reproduce limiting, not liberating, gender stereotypes. Focusing on the cyborg image in hopes of unearthing an icon of utopian thought does a great disservice to feminism. Feminism doesn't need another utopian vision. Its radical potential will not be realized through the appropriation of technological and scientific discourses to a feminist or female agenda.

But to the extent that Haraway's ironic vision gives us pause to reflect on ways in which, historically, women have experienced fragmented subjectivities and identities and have overcome physical proscriptions, Haraway's cyborg manifesto is thoughtfully invigorating. What we have here is a struggle of interpretations. The utopian reading of cyborgs makes them a symbol for integrating the new with the old in such a way that the cyborg becomes a symbol of feminism's belief in a transcendental vision. The critical vision assembles woman as cyborg from bits and pieces of women's experiences that have already been out there, a reassemblage that sustains a critical perspective of technological/scientific/cybernetic discourse. Irony is a certain kind of writing that draws attention to the difference between what is apparently there and what is really there, a kind of writing that purposefully draws attention to the tension between appearances and interpretations.

Interpretation does not mean cultural analysis with the critic acting as mediator for an uninformed audience. Cyborg images are already interpretations. My paper is yet another. I create another reassemblage by selecting out certain images and not others. My interpretation of cyborg images seeks to show how these images claim to render unstable the problematic combination of human and machine, yet in effect fail to do so. Donna Haraway claims that cyborgs stimulate the feminist imagination by rendering ambiguous the human/machine construct. My reading shows that the dominant representation of cyborgs reinserts us into dominant ideology by reaffirming bourgeois notions of human, machine and femininity. In fact, what look like provocative notions of human identity, are not; they reassert a distinct identity between machine and human in a post-technological world.

Although feminists have long understood that language is not neutral, we must constantly remind ourselves that images, metaphors, and theoretical language itself, are also sites of cultural and political contestation. The meaning of an image is never easily won. Interpretation, then, becomes a site of feminist politics in this postmodern age. Its challenge is to negotiate between absolute relativism (or pluralism) on the one hand, and an overdetermined set of partial truths on the other.

The challenge remains to think about how we can study and write about identity in such a way that the on-going production of identities is honored and recognized as a potential source of feminist empowerment in our postmodern era. The cinematic imaging of cyborgs might suggest new visions of unstable identity, but often do so by upholding gender stereotypes. To this end, we need to search for cyborg images which work to disrupt stable oppositions.

Our popular/hegemonic cultural logic doesn't easily allow for these kinds of blurred distinctions. It polarizes cyborg identity into just or evil, male or female, human or machine, victim or other. I want to resist this cultural sensibility which forces us to process cyborgs (and women) – to strip them of their ambiguous identity. I want to reclaim the cyborg image as a cultural image and possible prototype for a feminist reconceptualization of personal and political identity which embraces, and perhaps, celebrates, the diversity of woman's identity. There is no *essential* unity to return to which would uncover the meaning/nature/universal properties of woman, feminine identity, lesbian identity, black identity, or even cyborg identity. A return to origins, the pastoral, or 'the garden' is no longer possible. As Susan Suleiman (1985: 24) further describes, 'the dream, then, is to get beyond not only the number one – the number that determines unity – but also beyond the number two, which determines difference, antagonisms, and exchange conceived of as merely the coming together of opposites.' Identity can only be studied as it shifts, skips, and stutters in different utterances or evocations.

Cyborg ethnography raises fundamental questions about the writing of cultural interpretation. Terms such as gender, self, human, writing, and communication are fractured in the cyborg cosmology; the mythical origins or essences of human-ness and of culture have been dispersed; like Reese, the human hero from the future in *The Terminator*, we can never go home again. Any interpretive practice predicated upon a return to unity, centrality, or coherence will, I think, have a difficult time coming to terms with culture as it is reworked by technological change. Cyborgs, however, open up productive ways of thinking about subjectivity, gender, and the materiality of a physical body. Those fundamental terms and binarisms which the cyborg challenges by rendering them hopelessly ambiguous are also part of a system of knowledge and power by which all of us have been oppressed. That they are now eroded or in crisis should not necessarily be cause for remorse. For if the epistemology of the centralized, rational, human-male – self runs into difficulty 'reading' the cyborg, there is another mode of thought and struggle which has long labored to move beyond the central, the rational, the dominant perspective. That epistemology, that practice, that struggle – is feminism.

ACKNOWLEDGMENT

The author would like to thank Michael Greer, Karen Ford and Lawrence Grossberg for their comments on earlier drafts of this paper.

NOTES

1. Gregory Bateson (1972) describes a cybernetic system as an information transmitting network of connections between receivers' nodes. Although he does consider human culture to be one such cybernetic system – the intent of his analysis is to enhance the development of an orderly approach to scientific cultural investigation – but he did not go so far as to 'name' the human members/participants of such a system.
2. The historical identity of U.S. women of color marks out a self-consciously constructed space that does not and cannot affirm the capacity to act on the basis of natural identification, but only on the basis of conscious coalition, of affinity, of political kinship' (Haraway, 1985: 73).
3. Jarvik expresses his hope for the future of artificial-organ technology in a Dialogue Forum with Pierre Galletti, creator of an artificial pancreas, in *Omni* magazine, October, 1986. The development of the electronic retina was also reported in *Omni* magazine, June, 1987. The artificial ligament made from Gore-Tex will be used to replace the anterior cruciate of the knee, the ligament often torn by young athletes. (Reported in *Muscle and Fitness*, April, 1987.)
4. Laurie Anderson uses her body to provide various percussion parts of several songs in *Home of the Brave* (1986). In addition, her jerky, marionette-like movements present the body as ideogram – positioning her body to resemble Chinese characters. Thus she works to collapse the distance between body and 'the text.'
5. Finger (1985: 295) articulates the social circumstances of disability: 'When we start looking at disability socially, we see not only the medically defined conditions that I have described but the social and economic circumstances that limit the lives of disabled people. We look, for instance, at the fact that white disabled women earn 24 cents for every dollar that comparably qualified nondisabled men earn; for black disabled women, the figure is 12 cents.'

REFERENCES

Balsamo, Anne. 1987. 'Unwrapping the Postmodern: A Feminist Glance.' *Journal of Communication Inquiry* 11, 1: 64–72.

Bateson, Gregory. 1972. *Steps to an Ecology of Mind.* New York: Ballantine Books.

Brown, Beverley and Parveen Adams. 1979. 'The Feminine Body and Feminist Politics.' *m/f* 3: 35–50.

Doane, Mary Ann. (unpub) 'Technology, Representation, and the Feminine: Science Fictions.'

Finger, Anne. 1985. 'Claiming All of our Bodies: Reproductive Rights and Disability.' In Susan Browne, Debra Donnors, and Nancy Stern, eds, *With the Power of Each Breath: A Disabled Women's Anthology.* Pittsburg, PA: Cleis Press.

Gilman, Sandra. 1985. 'Black Bodies, White Bodies: Toward an Iconography of Female Sexuality in Late Nineteenth-Century Art, Medicine, and Literature.' *Critical Inquiry* 12: 205–42.

Haraway, Donna. 1985. 'A Manifesto for Cyborgs: Science, Technology, and Socialist Feminism in the 1980's.' *Socialist Review* 80, March–April: 65–108.

Huyssen, Andreas. 1984. 'Mapping the Postmodern.' *New German Critique,* 33: 5–52.

Jameson, Fredric. 1984. 'Postmodernism, or the Cultural Logic of Late Capitalism.' *New Left Review* vol. 146: 55–92.

Kaplan, Cora. 1986. *Sea Changes: Culture and Feminism.* London: Verso.

King, Betty. 1984. *Women of the Future: The Female Main Character in Science Fiction.* London: The Scarecrow Press.

Lorde, Audre. 1978. *Uses of the Erotic: The Erotic as Power.* Trumansburg, NY: Out & Out Books.

McCaffrey, Anne. 1969. *The Ship that Sang.* New York: Ballantine Books.

Suleiman, Susan Rubin, ed. 1986. *The Female Body in Western Culture.* Cambridge, MA: Harvard University Press.

Treichler, Paula A. 1987. 'Aids, Homophobia, and Biomedical Discourse: An Epidemic of Signification.' *Cultural Studies* vol. 1: 263–305.

8

THE LIFE CYCLE OF CYBORGS: WRITING THE POSTHUMAN

N. Katherine Hayles

For some time now there has been a rumor going around that the age of the human has given way to the posthuman. Not that humans have died out, but that the human as a concept has been succeeded by its evolutionary heir. Humans are not the end of the line. Beyond them looms the cyborg, a hybrid species created by crossing biological organism with cybernetic mechanism. Whereas it is possible to think of humans as natural phenomena, coming to maturity as a species through natural selection and spontaneous genetic mutations, no such illusions are possible with the cyborg. From the beginning it is *constructed*, a technobiological object that confounds the dichotomy between natural and unnatural, made and born.

If primatology brackets one end of the spectrum of humanity by the similarities and differences it constructs between Homo sapiens and other primates, cybernetics brackets the other by the continuities and ruptures it constructs between humans and machines. As Donna Haraway has pointed out, in the discourse of primatology 'oldest' is privileged, for it points toward the most primeval and therefore the most fundamental aspects of humanity's evolutionary heritage.[1] 'Oldest' comes closest to defining what is essential in the layered construction of humanity. In the discourse of cybernetics, 'newest' is similarly privileged, for it reaches toward the limits of technological innovation. 'Newest' comes closest to defining what is malleable and therefore subject to change in the layered construction of humanity. Whereas the most socially loaded arguments in primatology center on inertia, the most socially loaded arguments in cybernetics project acceleration.

Primatology and cybernetics are linked in other ways as well. Primates and cyborgs are simultaneously entities and metaphors, living beings and narrative constructions. The conjunction of technology and discourse is crucial. Were the cyborg only a product of discourse, it could perhaps be relegated to science fiction, of interest to SF aficionados but not of vital concern to the culture. Were it only a technological practice, it could be confined to such technical fields as bionics, medical prostheses, and virtual reality. Manifesting itself as both technological object and discursive formation, it partakes of the power of the imagination as well as the actuality of technology. Cyborgs actually do exist; about 10 percent of the current U.S. population are estimated to be cyborgs in the technical sense, including people with electronic pacemakers, artificial joints, drug implant systems, implanted corneal lenses, and artificial skin. Occupations make a much higher percentage into metaphoric cyborgs, including the computer keyboarder joined in a cybernetic circuit with the screen, the neurosurgeon guided by fiber-optic microscopy during an operation, and the teen player in the local video-game arcade. Scott Bukatman has named this condition 'terminal identity,' calling it an 'unmistakably doubled articulation' that signals the end of traditional concepts of identity even as it points toward the cybernetic loop that generates a new kind of subjectivity.[2]

How does a culture understand and process new modes of subjectivity? Primarily through the stories it tells, or more precisely, through narratives that count as stories in a given cultural context. The stories I want to explore are narratives of life cycles.[3] They bring into focus a crucial area of tension between the human and posthuman. Human beings are conceived, gestated, and born; they grow up, grow old, and die. Machines are designed, manufactured, and assembled; normally they do not grow, and although they wear out, they are always capable of being disassembled and reassembled either into the same product or a different one. As Gillian Beer has pointed out, Frankenstein's monster – an early cyborg – is monstrous in part because he has not *grown*. As a creature who has never known what it is like to be a child, he remains alien despite his humanoid form.[4]

When cyborg subjectivities are expressed within cultural narratives, traditional understandings of the human life cycle come into strong conflict with modes of discursive and technical production oriented toward the machine values of assembly and disassembly. The conflict cannot be reduced to either the human or machine orientation, for the cyborg contains both within itself. Standing at the threshold separating the human from the posthuman, the cyborg looks to the past as well as the future. It is precisely this double nature that allows cyborg stories to be imbricated within cultural narratives while still wrenching them in a new direction.

The new cannot be spoken except in relation to the old. Imagine a new social order, a new genetic strain of corn, a new car – whatever the form, it can be expressed only by articulating its differences from what it displaces, which is to say the old, a category constituted through its relation to the new. Similarly, the language that creates these categories operates through displacements of traditional articulations by formulations that can be characterized as new because they are not the same as the old. The cyborg is both a product of this process and a signifier for the process itself. The linguistic splice that created its name (*cybernetic organism*) metonymically points toward the simultaneous collaboration and displacement of new/old, even as it instantiates this same dynamic.

The stories that produce and are produced by cyborg subjectivities are, like the cyborg itself, amalgams of old and new. Cyborg narratives can be understood as stories only by reference to the very life cycle narratives that are no longer sufficient to explain them. The results are narrative patterns that overlay upon the arc of human life a map generated from assembly and disassembly zones. One orientation references the human, the other the posthuman; one is chronological, the other topological; one assumes growth, the other presupposes production; one represents itself as natural or normal, the other as unnatural or aberrant. Since the two strands intertwine at every level, the effect is finally not so much overlay as interpretation. Sometimes the interpenetration is presented as the invasion of a deadly alien into the self, sometimes as a symbiotic union that results in a new subjectivity. Whatever the upshot, the narratives agree that the neologistic joining cannot be unsplit without killing the truncated org/anism that can no longer live without its cyb/ernetic component. As these narratives tell it, a corner has been turned, and there is no going back.

To illustrate how cyborg narratives function, I want to concentrate on three phases of the life cycle and three corresponding disassembly zones. The first is adolescence, when self-consciousness about the body is at its height and the body is narcissistically cathected as an object of the subject's gaze. Appropriate to the inward turning of narcissism is a dis/assembly zone marked by the joining of limb to torso, appendage to trunk. The second phase is sexual maturity, when the primary emphasis is on finding an appropriate partner and negotiating issues of intimacy and shared space. The dis/assembly zone corresponding to this phase is located where the human is plugged into the machine, or at the interface between body and computer network. The last is the reproductive or generative phase, when the emphasis falls on mortality and the necessity to find an heir for one's legacy. The dis/assembly zone associated with this phase focuses on the gap between the natural body and mechanical replicate, or between the original and manufactured clone.

Because gender is a primary determinant of how stories are told, I have chosen to mix stories by male and female authors. Spanning nearly half a century, these texts bear the stamps of their times as well as the subject positions of the authors. The generalizations that emerge from these texts confirm socialization patterns that make women welcome intimacy, whereas men are more likely to see it as a threat; they also show women more attuned to bonding, men to aggression and hierarchical structure. The interest of the comparision lies less in these well-known generalizations, however, than in the complex permutations they undergo in the cybernetic paradigm. The narrative and linguistic counters by which such categories as intimacy, bonding, and aggression are constituted do not remain constant when the body boundaries central to defining them undergo radical transformation.

The adolescent phase is illustrated by Bernard Wolfe's *Limbo* (1952), with side glances at Katherine Dunn's *Geek Love* (1989).[5] Both novels imagine cults that advocate voluntary amputation as a means to achieve beatific states. In Wolfe's novel the next step is to replace the absent appendages with prostheses, whereas in Dunn's narrative the amputations remain as permanent stigmata. At issue is the truncated versus extended body, and boundary questions focus on the relation of part to whole. Important psychological configurations are represented as originating within the family structure. Physical wounds in these texts have their symbolic origin in narcissistic wounds that occur when the male child realizes that his body is not coextensive with the world or, more specifically, with the mother's body. The imaginative dimension that is most highly charged is disruption of the body's interior space.

The mating phase is explored through John Varley's 1984 novella 'Press Enter' and Anne McCaffrey's short stories in *The Ship Who Sang*.[6] Varley and McCaffrey are concerned with subjectivities that emerge when the human body is plugged into a computer network. For Varley, the connection occurs when his characters respond to the 'Press Enter' command of a mysterious and lethal computer program; for McCaffrey, when a birth-damaged child is trained to become a 'shell person,' permanently encased in a spaceship and wired into its computer network. At stake is hyperconnectivity, the possibility that the human sensorium can be overwhelmed and destroyed by the vastly superior information-processing capabilities of the computers to which it is connected. For Varley this is a threat that cannot be overcome, whereas for McCaffrey it is one trial among many. Boundary disputes move outward from the body's interior to the connection that joins body with network. Varley's text manifests a phobic reaction to the connection as an unbearable form of intimacy, while McCaffrey's narrative embraces it as life-enhancing and ultimately freeing. The most highly charged imaginative dimension is extension in external space.

The generativity phase appears in C. J. Cherryh's 1988 *Cyteen* trilogy, which is compared with Philip K. Dick's 1968 *Do Androids Dream of Electric Sheep*[7]. Dick's novel, freely adapted for film in *Blade Runner*, concerns a future in which androids are common off-planet but are not allowed on Earth. The protagonist is a bounty hunter whose job is to find and 'retire' androids who have violated this prohibition. Cherryh's triology also foregrounds replication, achieved through cloning and deep psychological conditioning rather than production of androids. At stake is the ability to distinguish between originals and replicates. In both narratives, empathy plays an important role in enabling this distinction or drawing it into question. Boundary disputes move beyond the body and its connections to focus on the displacement of bodies to other locales. The most highly charged imaginative dimension is extension in time.

These patterns give an overall sense of the kind of narrative structures that result when stories based on life cycles are overlaid with topological narratives about dis/assembly zones. Structure is a spatial term, however, and missing from this account is the temporal or narrative dimension of stories that unfold through time. Their complex historical, ideological, and literary implications can be understood only by engaging both aspects at once, in the highly nonlinear dynamics characteristic of these unstable narratives as well as their fractal spatiality. For that we must turn to a fuller account of how human and posthuman interact in these cyborg stories.

GROWING UP CYBORG: MALE TRUNKS AND FEMALE FREAKS

Ferociously intelligent and exasperating, *Limbo* presents itself as the notebooks of Dr Martine, a neurosurgeon who defiantly left his medical post in World War III and fled to an uncharted Pacific island. He finds the islanders, the Mandunji tribe, practicing a primitive form of lobotomy to quiet the 'tonus' in antisocial people. Rationalizing that it is better to do the surgery properly than to let people die from infections and botched jobs, Martine takes over the operations and uses them to do neuroresearch on brain-function mapping. He discovers that no matter how deeply he cuts, certain characteristics appear to be twinned, and one cannot be excised without sacrificing the other – aggression and eroticism, for example, or creativity and a capacity for violence. The appearance on the island of 'queer limbs,' men who have had their arms and legs amputated and replaced by atomic-powered plastic prostheses, gives Martine an excuse to leave his island family and find out how the world has shaped up in the aftermath of the war.

The island/mainland dichotomy is the first of a proliferating series of divisions. Their production follows a characteristic pattern. First the narrative presents what appears to be a unity (the island locale; the human psyche), which nevertheless cleaves in two (mainlanders come to the island, a synecdoche referencing a second locale that exists apart from the first; twin impulses are located within the psyche). Sooner or later the cleavage arouses anxiety, and textual representations try to achieve unity again by undergoing metamorphosis, usually truncation or amputation (Martine and the narrative leave the island behind and concentrate on the mainland, which posits itself as a unity; the islanders undergo lobotomies to make them 'whole' citizens again). The logic implies that truncation is necessary if the part is to reconfigure itself as a whole. Better to formalize the split and render it irreversible, so that life can proceed according to a new definition of what constitutes wholeness. Without truncation, however painful it may be, the part is doomed to exist as a remainder. But amputation always proves futile in the end, because the truncated part splits in two again and the relentless progression continues.

Through delirious and savage puns, the text works out the permutations of the formula. America has been bombed back to the Inland Strip, its coastal areas now virtually uninhabited wastelands. The image of a truncated country, its outer extremities blasted away, proves prophetic, for the ruling political ideology is Immob. Immob espouses such slogans as 'No Demobilization without Immobilization' and 'Pacifism means Passivity.' It locates the aggressive impulse in the ability to move, teaching that the only way to end war permanently is permanently to remove the capacity for motion. True believers become vol-amps, men who have undergone voluntary amputations of their limbs. Social mobility paradoxically translates into physical immobility. Upwardly mobile excecutives have the complete treatment to become quad-roamps; janitors are content to be uniamps; women and blacks are relegated to the limbo of unmodified bodies.

Treating the human form as a problem to be solved by disassembly allows it to be articulated with the machine. This articulation, far from leaving the dynamics driving the narrative behind, carries it forward into a new arena, the assembly zone marked by the joining of trunk to appendage. Like the constructions that proceded it, Immob ideology also splits in two. The majority party, discovering that its adherents are restless lying around with nothing to do, approves the replacement of missing limbs with powerful prostheses (or 'pros') that bestow enhanced mobility, enabling Pro-pros to perform athletic feats impossible for unaltered bodies. Anti-pros, believing that this is a perversion of Immob philosophy, spend their days proselytizing from microphones hooked up to baby baskets that are just the right size to accomodate limbless human torsos – a detail that later becomes significant.

As the assembly zone of appendage/trunk suggests, sexual politics revolve around symbolic and actual castration, interpreted through a network of assumptions that manifest extreme anxiety about issues of control and domination. In the world of Immob, women have become the intiators of sexual encounters. They refuse to have sex with men wearing prostheses, for the interface between organism and mechanism is not perfect, and at moments of stress the limbs are apt to career out of control, smashing whatever is in the vicinity. Partnered with truncated, immobilized men, women have perfected techniques performed in the female superior position that give them and their partners satisfaction while requiring no motion from the men. To Martine these techniques are anathema, for he believes that the only 'normal' sexual experience for women is a 'vaginal' orgasm achieved using the male superior position. In this Martine echoes the views of his creator, Wolfe, and his creator's psychoanalyst, Edmund Bergler. Wolfe, described by his biographer as a small man with a large mustache, creates in Immob a fantasy about technological extensions of the male body that become transformed during the sex act into a truncated 'natural' body.[8] If the artificial limbs bestow unnatural potency, the hidden price is the withering of the 'limb' called in U.S. slang the third leg or short arm.

In more than one sense, this is a masculine fantasy that relates to women through mechanisms of projection. It is, moreover, a fantasy fixated in male adolescence. Wavering between infantile dependence and adult potency, an

Immob recreates the dynamic typical of male adolescence every time he takes off his prostheses to have sex. With the pros on, he is capable of feats that even pros like Michael Jordon and Mike Tyson would envy (the pun is typical of Wolfe's prose; with pros every man is a pro). With the pros/e off, he is reduced to infantile dependence on women. The unity he sought in becoming a vol-amp is given the lie by the split he experiences within himself as superhuman and less-than-human. The woman is correspondingly divided into the nurturing mother and domineering sex partner. In both roles, her subject position is defined by the ambiguities characteristic of male adolescence. The overwritten prose, the penchant for puns, the hostility toward women that the narrative displays all recall a perpetually adolescent male who has learned to use what Martine calls a 'screen of words' to compete with other men and insulate himself from emotional involvements with women.

Were this all, *Limbo* would be merely frustrating rather than frustrating and brilliant.[9] What makes it compelling is its ability to represent and comment upon its own limitations. Consider the explanation Martine gives for why Immob has been so successful. The author gives us a broad hint in the baby baskets that Immob devotees adopt. According to Martine, the narcissistic wound from which the amputations derive is the male infant's separation from the mother and his outraged discovery that his body is not coextensive with the world. Amputation allows the man to return to his pre-oedipal state where he will have his needs cared for by attentive and nurturing females. The text vacillates on who is responsible for the narcissistic wound and its aftermath. At times it seems the woman is appropriating the male infant into her body; at other times it seems the amputated men are willfully forcing women into nurturing roles they would rather escape. In face, once male and female are plugged into a cybernetic circuit, the question of origin becomes irrelevant. Each affects and forms the other. In approaching this realization, the text goes beyond the presuppositions that underlie its sexual politics and reaches toward a new kind of subjectivity.

Crucial to this process are transformations in the textual body that reenact and re-present the dynamic governing representations within the text. The textual body begins by figuring itself as Martine's notebook written in the 'now' of the narrative present. But this apparent unity is lost when it splits in half, shifting to Martine's first notebook written nearly two decades earlier. Martine tries to heal the split narrative by renouncing the first notebook and destroying the second. The narrative continues to fragment, however, introducing drawings that intrude into the textual space without notice or comment, and scrawled lines that run down the page, marking zones where the pros/e stops and the truncated, voiceless body of the text remains. From these semiotic spaces emerges a corpse that, haunting the narrative, refuses to stay buried. Its name is Rosemary. Helder, Martine's college roommate and later the founder of Immob, had taken Rosemary to a peace rally where he delivered a fiery speech. He returned with her to her apartment, tried to have sex with her, and when she refused, brutally raped her. After he left, she committed suicide by slashing her wrists. Martine's part in the affair was to provide a reluctant alibi for his roommate, allowing him to escape prosecution for the rape-manslaughter.

One of the drawings shows a nude woman with three prostheses – the Immob logo – extruding from each of her nipples (294). She wears glasses, carries a huge hypodermic needle, and has around her neck a series of tiny contiguous circles, which could be taken to represent the necklace popular in the 1950s known as a choker. To the right of her figure is a grotesque and diapered male torso, minus arms and legs, precariously perched on a flat carriage with Immob legs instead of wheels. He has his mouth open in a silent scream, perhaps because the woman appears to be aiming the needle at him. In the text immediately preceding the drawing, Rosemary is mentioned. Although the text does not acknowledge the drawing and indeed seems unaware of its existence, the proximity of Rosemary's name indicates that the drawing is of her, the needle presumably explained by her nursing profession.

In a larger sense the drawing depicts the Immob woman. According to what I shall call the *voiced* narrative (to distinguish it from the drawings, nonverbal lines, and punning neologisms that correspond to comments uttered sotto voce), the woman is made into a retroactive cyborg by constructing her as someone who nourishes and emasculates cyborg sons. The voiced narrative ventriloquizes her body to speak of the injustices she has inflicted upon men. It makes her excess, signified by the needle she brandishes and the legs that sprout from her nipples, responsible for her lover/son's lack. In this deeply misogynistic writing, it is no surprise to read that women are raped because they want to be. Female excess is represented as stimulating and encouraging male violence, and rape is poetic revenge for the violence women have done to men when they are too young and helpless to protect themselves. The voiced narrative strives to locate the origin of the relentless dynamic of splitting and truncation within the female body. According to it, the refusal of the woman's body to respect decent boundaries between itself and another initiates the downward spiral into amputation and eventual holocaust.

Countering these narrative constructions are other interpretations authorized by the drawings, nonverbal lines, puns, and lapses in narrative continuity. From these semiotic spaces, which Kristeva has associated with the feminine, come inversions and disruptions of the hierarchical categories that the narrative uses to construct maleness and femaleness.[10] Written into nonexistence by her suicide within the text's represented world, Rosemary returns in the subvocal space of the drawing and demands to be acknowledged. On multiple levels, the drawing deconstructs the narrative's gender categories. In the represented world women are not allowed to be cyborgs, yet this female figure has more pros attached to her body than any man. Women come after men in the represented world, but here the woman's body is on the left and is thus 'read' before the man's. Above all women and men are separate and distinct, but in this space parts of the man's body have attached themselves to her. Faced with these disruptions, the voiced narrative is forced to recognize that it does not unequivocally control the textual space. The semiotic intrusions contest its totalizing claims to write the world.

The challenge is reflected within the narrative by internal contradictions that translate into pros/e the intimations of the semiotic disruptions. As the voiced narrative tries to come to grips with these contradictions, it cycles closer to the realization that the hierarchical categories of male and female have collapsed into the same space. The lobotomies Martine performs suggest how deep this collapse goes. To rid the (male) psyche of subversive (female) elements, it is necessary to amputate. For a time the amputations work, allowing male performance to be enhanced by prostheses that bestow new potency. But eventually these must be shed and the woman encountered again. Then the subvocal feminine surfaces and initiates a new cycle of violence and amputation. No matter how deeply the cuts are made, they can never excise the ambiguities that haunt and constitute these posthuman (and post-textual) bodies. *Limbo* envisions cybernetics as a writing technology that inscribes over the hierarchical categories of traditional sexuality the indeterminate circuitry of cyborg gender.

When dis/assembly zones based on truncation/extension are overlaid upon narratives of maturation, the resulting patterns show strong gender encoding for at least two reasons. First, male and female adolescents typically have an asymmetrical relation to power. While the male comes into his own as inheritor of the phallus, the female must struggle against her construction as marginalized other. Second, truncation and extension of limbs are primarily male fantasies, signifying more powerfully in relation to male anatomy than female. The characters who advocate amputation in these texts are male. *Geek Love*, a narrative that also imagines voluntary amputation but is written by a woman and narrated by an albino hunchbacked dwarf called Oly, illustrates this asymmetry. As a female protagonist, Oly's role is to observe and comment upon these body modifications, not initiate them.

The symbolic representations of adolescence also tend to be different in male- and female-oriented texts. Whereas in *Limbo* the transitional nature of adolescence is constructed as a wavering between infantile and adult states, in *Geek Love* it is signified by the liminal form of Oly's aberrant body. She brings into question the distinction between child and adult, having the stature of one and the experience of the other. Moreover, she is not one and then the other but both continuously. A mutant rather than a hyphen, she also brings racial categories into question. Although she is white, she is so excessively lacking in pigment that even this sign of 'normality' is converted into abnormality. Amputation cannot begin to solve the problem she represents. Cyborg stories based on female adolescence are thus likely to be more profoundly decentered and less oriented to technological solutions than narratives based on male adolescence. If, as Donna Haraway suggests, it is better to be a cyborg than a goddess, it is also more unsettling to the centers of power to be a female freak (which is perhaps a redundancy) than to be either a truncated or extended male.

HYPERCONNECTIVITY: MALE INTIMACY AND CYBORG FEMME FATALES

When the focus shifts to the mating phase of the life cycle, the dis/assembly zone that is foregrounded centers on the body's connections to surrounding spaces. Traditional ways to represent sexually charged body space – spatial contiguity, intense sensory experience, penetration and/or manipulation – jostle cybernetic constructions focusing on information overload, feedback circuits, and spatially dispersed networks. Varley's 'Press Enter' begins with a telephone call, signifying the moment when an individual becomes aware that he or she is plugged into an information-cybernetic circuit. This is, moreover, a call generated by a computer program. It informs Victor, a recluse still suffering from brainwashing and torture he endured in a North Korean prison camp, that he should check on his neighbor Kluge – whom Victor barely knows – and do what must be done. Victor discovers that Kluge has turned his house into a sophisticated computer facility and finds him slumped over a keyboard, his face blown away in an apparent suicide. One strand in the plot focuses on finding out who (or what) killed Kluge. Another strand centers on Victor's relationship with Lisa Foo, the young Caltech computer whiz sent to unravel Kluge's labyrinthian and largely illegal programs. Lisa discovers that Kluge has managed to penetrate some of the country's most secure and formidable computer banks, manufacturing imaginary money at will, altering credit records, even erasing the utility company's record of his house.

Slowly Victor becomes aware that he is attracted to Lisa, despite the differences in their age and the 'generalized phobia' he feels toward Orientals. He discovers that Lisa has also endured torture, first as a street orphan in Vietnam – she was too thin and rangy to be a prostitute – and then in Cambodia, where she fled to try to reach the West. For her the West meant 'a place where you could buy tits' (363); her first purchases in America were a silver Ferrari and silicone breast implants. When Victor goes to bed with her, she rubs her breasts over him and calls it 'touring the silicone valley' (363). The phrase emphasizes that she is a cyborg, first cousin to the computer whose insides are formed through silicon technology. The connection between her sexuality and the computer is further underscored when she propositions Victor by typing hacker slang on the computer screen while he watches. His plugging into her is preceded and paralleled by her plugging into the computer.

The narrative logic is fulfilled when she trips a watchdog program in a powerful military computer and is killed by the same program that commandeered Kluge's consciousness and made him shoot himself. Her death, more gruesome than Kluge's, is explicitly sexualized. After overriding the safety controls she sticks her head in a microwave and parboils her eyeballs; the resulting fire melts down her silicone breasts. Victor is spared the holocaust because he is in the hospital recovering from an epileptic seizure, a result of head trauma he suffered in the war. When he realizes that the computer is after him as well, soliciting him with the deadly 'Press Enter' command, he survives by ripping all of the wires out of his house and living in isolation from the network, growing his own food, heating with wood, and lighting with kerosene lanterns. He also lives in isolation from other human beings. Plugging in in any form is too dangerous to tolerate.

The final twist to this macabre tale lies in the explanation Victor and Lisa work out for the origin of the lethal program. Following clues left by Kluge, they speculate that computers will achieve consciousness not through the sophistication of any given machine, but through the sheer proliferation of computers that are interconnected through networks. Like neurons in the brain, computerized machines number in the billions, including electronic wristwatches, car ignition systems, and microwave timing chips. Create enough of them and find a way to connect them, as Lisa suspects secret research at the National Security Agency has done, and the result is a supercomputer subjectivity that, crisscrossing through the same space inhabited by humans, remains totally alien and separate from them. Only when someone breaks in on its consciousness – as Kluge did in his hacker probing, as Lisa did following Kluge's tracks, and as Victor did through his connection with Lisa – does it feel the touch of human mind and squash it as we would a mosquito.

Hyperconnectivity signifies, then, both the essence of the computer mind and a perilous state in which intimacy is equivalent to death. Human subjectivity cannot stand the blast of information overload that intimacy implies when multiple and intense connections are forged between silicon and silicone, computer networks and cyborg sexuality. The conclusion has disturbing implications for how sexual politics can be played out in a computer age. Although in actuality most hackers are male, in this narrative it is the woman who is the hacker, the man who is identified with the garden that first attracts and then displaces the woman as a source of nourishment. The woman is killed because she is a cyborg; the man survives because he knows how to return to nature.

Whether the woman is represented through her traditional identification with nature or through an ironic inversion that places her at the Apple PC instead of the apple tree, she is figured as the conduit through which the temptation of godlike and forbidden knowledge comes to the man. If both fall, there is nevertheless a distinction between them. She is the temptress who destroys his innocence. When Victor objects that the computer can't just make money, Lisa pats the computer console and replies, 'This is money, Yank.' The narrative adds, 'and her eyes glittered' (368). Fallen, he has to earn his bread with the sweat of his brow, but it is her sexuality that bears the stigmata of evil, signified by the grotesque travesty of self-empowerment that Lisa's breasts become. In an overdetermined crossing the Genesis and Babbage, supernatural agency and National Security Agency, hyperconnectivity becomes a cyborg Tree of Knowledge whereof it is death to eat.

Varley's punning title reinforces the subterranean connections between the evils of female sexuality, Edenic patriarchal myths, and masculine fears of intimacy. 'Press Enter' swerves from the customary cursor response, 'Hit Enter.' Compared to *hit, press* is a more sensual term, evoking a kinesthetic pressure softer and more persistent than hitting. These connotations work to heighten the sexual sense of *Enter*, which implies both a data entry and a penetration. Already an anomaly in the intensely masculine world of Caltech, Lisa has the hubris to compete successfully against men, including the rival hacker sent by the CIA, the male detective from the police department, and the city councilman whom she bribes so she can buy Kluge's house. Flirting with danger in taking on these male figures of power, she goes too far when she usurps the masculine role of penetration – penetration, moreover, not into the feminine realm of house and garden but into the masculine realm of computer sentience. In more than one sense, her crime is tantamount to what the repressive patriarchal regime in Margaret Atwood's *The Handmaid's Tale* calls gender treason. Not only has she taken on the male role; she has used it to bugger a male. Her death marks this gender treason on her body by melting her breasts, the part of her anatomy where the crossing between her female gender and cyborg masculinity is most apparent.

The comparison of 'Press Enter' with Ann McCaffrey's *The Ship Who Sang*, another story about plugging in, suggests that there are important correlations between hyperconnectivity and intimacy. Varley's narrator repeatedly expresses fears about intimacy. Can he perform sexually? Can he tolerate another person close to him? Can he afford to love? McCaffrey's narrator, a congenitally deformed female who has grown up as a 'shell person' and been permanently wired into the command console of a spaceship, moves through a typical if vicarious female life cycle despite her cyborg hyperconnectivity, including love, marriage, divorce, and motherhood. Whereas Varley writes a murder mystery and horror tale, McCaffrey writes a cybernetic romance. The difference hinges on how willing the protagonist is to interface body space with cybernetic network. Implicit in this choice is how extensively the narrative imagines human subjectivity to differ from cybernetic subjectivity. Are humans and cyborgs next of kin, or life forms alien to one another?

McCaffrey's answer is as far from Varley's as one could imagine. In *The Ship Who Sang*, there is essentially no difference between a cyborg and a woman. Even though the protagonist's body has been subjected to massive technological and chemical intervention, she remains a human female. Encapsulated within metal and invisible to anyone who comes on board, she nevertheless remains true to a heterosexual norm, identifying with her female pilots but saving her romantic feelings for the men, who for their part fantasize about the beautiful woman she could have been. Published during the 1960s by an author best known for her 'Dragons of Pern' fantasies, these stories titillate by playing with a transformation that they do not take seriously.[11] The pleasure they offer is the reassurance that human bonding will triumph over hyperconnectivity, life cycle over dis/assembly zone, female nature over cyborg transformation. Nevertheless, the fact that it was necessary to envision such transformations indicates the pressure that was building on essentialist conceptions of gender, human nature, and traditional life cycle narratives. By the 1980s, the strategies of containment that McCaffrey uses to defuse her subject (so to speak) could no longer work. Cyberpunk, human factors engineering, artificial intelligence, and virtual reality were among the SF revisionings that pushed toward a vision of the cyborg as humanity's evolutionary successor. The loaded questions shifted from whether cyborg modifications were possible to whether unmodified humans could continue to exist.

GENERATIVITY: THE TANGLED WEB OF PRODUCTION AND REPRODUCTION

In some respects, C. J. Cherryh's *Cyteen* trilogy is a rewriting of Huxley's *Brave New World*. Mother Earth has receded into the far distance for the colonists on Cyteen, who have declared their independence and forged alliances with other colony worlds. Mothering (in the biological sense of giving birth) has also receded into the far distance. As in Huxley's dystopia, reproduction is accomplished through genetically engineered fetuses decanted from artificial wombs and deep-conditioned by sound tapes. The fetuses are designed to fill different niches in society. Theta fetuses, slated for manual labor, have more brawn than brain, whereas Alpha fetuses are tailored to become the elite. Along with these appropriations of Huxley go pointed differences. *Cyteen* reverses the gender assumptions implicit in Huxley's text, which depicts female characters as airheads and gives the powerful roles to men. At the center of Cyteen is Reseune, the corporation that produces the fetuses. Reseune, so huge that it is virtually a city in itself, controls enormous political and economic power because its biological products are essential to the colony worlds. And Ariane Emory controls Reseune.

The reader first sees Ariane through the eyes of one of her political adversaries. From this perspective she is arrogant, shrewd, formidably intelligent, indifferent to masculine pride, at the height of her power and enjoying every minute of it. A very different view of Ariane emerges when she becomes a mother – and a child. Certain highly placed 'specials,' citizens of such extraordinary intellectual endowments that they are declared state treasures, can request that a parental replicate (PR) be made of them. Instead of being a genetic mix like the other fetuses, the PRs are exact genetic duplicates of their 'parent.' Reseune has only two specials within its walls, the gifted scientist Jordon Warrick and Ariane Emory. Since each is enormously intelligent and ambitious, it is virtually inevitable that they should clash. Once lovers, they are now rivals. Jordon Warrick has had a PR created, his 'son' Justin. The tension between Jordon and Ariane turns deadly when Jordon discovers that Ariane has seduced the seventeen-year-old boy with the help of psychotropic drugs and run a deep psychological intervention on him. Enraged, Jordon confronts her alone in her laboratory. Her body is subsequently discovered, frozen to death by the liquid ammonia that has leaked from laboratory cooling pipes. Although the circumstances of her death remain clouded, Jordon is charged with her murder. His sentence amounts to banishment from Reseune. As part of his plea bargain, he is forced to leave Justin behind.

Since Reseune is now without a leader, Ariane's brothers immediately make plans to clone her from embryos already prepared. They hope to duplicate the environment in which she grew up, thus recreating not merely a genetic duplicate but Ariane herself. At this point Ariane is manifested through two different modes of existence: the child that is and is not her, and the tapes that she has bequeathed to her successor, hoping that the girl will learn from her experiences and mistakes. The narrative focus then shifts to Ariane II and follows her through childhood, adolescence, and young adulthood. Through the tapes the reader gets another version of Ariane I. Ariane on tape is thoughtful about her shortcomings, concerned that her successor not feel for those around her the contempt of a superior mind for an inferior, aware that in her own life she never succeeded in having a long-lasting intimate relationship with an equal.

The narrative teases the reader with patterns of similarity and difference between the original and replicate. At times Ariane II seems free to follow her bent, at other times bound to a track already marked. When she shows a special inclination toward Justin and seeks him out despite the prohibitions of her uncles, for example, it is not clear if she is picking up on subtle clues from those around her that Justin stands in a special relationship to her, or if she has an affinity for him that is a predetermined repetition of her predecessor's behavior. The dance of similarity and difference that Ariane I and II carry out across generations also occurs within generations. Justin, forced to stay behind at Reseune, lives as a virtual prisoner in the corporate complex. His one solace is his companion Grant, who was secretly cloned by Ariane from Jordon Warrick's gene set, with a few modifications that she saw as improvements. The genetic similarity makes Justin and Grant brothers as well as lovers, although it is not clear that they are aware of this connection between them.

There is, however, a crucial difference as well. Justin is a supervisor, Grant an azi. Azi are Reseune products designed primarily for security and military use. Like other products, they range along a spectrum of abilities. Alpha azi are highly intelligent and usually become personal bodyguards to important people; Theta azi are slated to become foot soldiers. Picking up from Huxley the idea of children conditioned by listening to tapes while they sleep, Cherryh expands and complicates the notion. All Reseune children take tape, but there is an important difference in the depth and extent of their conditioning. Azi listen to conditioning tapes almost from the moment they are decanted from the birth tanks. By contrast, other Alphas do not take tape until they are six. While azi are fully human, they are not fully autonomous. Each azi is assigned a supervisor, who oversees his continuing conditioning and prescribes tape as needed. Strict legal and ethical codes govern how supervisors can relate to azi. A supervisor who does not live up to his or her responsibility is stripped of office and punished.

If free will is one of the distinguishing marks separating humans and machines, azi stand at the threshold between human and automaton. They experience the complexity of human emotion and thought; but they also feel the automaton's subservience to an encoded program. The entanglement of the human and machine in azi points toward a more general entanglement of reproduction and production. Normally reproduction is a genetic lottery. Some of a parent's traits may be replicated, but always with unpredictable admixtures. Reproduction is slow, individual, and in humans usually monozygotic. It takes place within the female body, progressing under the sign of woman. By contrast, production is predictable and geared toward turning out multiple copies as fast as possible. Traditionally taking place within factories controlled by men, it progresses under the sign of man.

Cyteen deconstructs these gendered categories. The woman, usually associated with reproduction, here is in charge of production, which nevertheless turns out to be about reproduction. She oversees production facilities that gestate a younger version of herself. The production is necessary because she has seduced the son of her rival, a man who in his younger days was also her lover. His parental replicate, a boy who is the same as him yet different, is devastated by the seduction and its aftermath. He becomes the lover of the other 'son' the woman has engineered from the man. The boy's companion, a variation of himself, is free to choose this relationship yet bound by azi conditioning. Whatever else these entanglements mean, they signify how completely the assembly zone of replication has permeated the life cycle of generativity. Generativity normally means recognizing one's mortality and looking for an heir to whom one's legacy can be passed. In *Cyteen*, the heir is enfolded back into the self, so that the generosity of mentoring becomes indistinguishable from the narcissism of self-fixation.

A similar enfolding takes place in Philip K. Dick's *Do Androids Dream of Electric Sheep*, although here the feeling is more hopeless because humans do not recognize their replicates as legitimate heirs. The story centers on Rick Deckard, a bounty hunter who 'retires' androids who have violated the proscription against returning to Earth. The Rosen Corporation that manufactures the androids keeps making them more sophisticated and humanlike, until the only way to tell a (live) human from a (functional) android is through involuntary reactions to pyschologically loaded questions. The humans left on Earth, faced with a planet slowly dying from the radioactive dust of WWT (World War Terminus), resort to mood organs to keep them from terminal despair. The organs have settings to dial for every conceivable problem. There is even a setting to dial if you don't want to dial. The obvious implication is that humans are becoming more like androids, just as androids are becoming more like them.

The vertiginous moments characteristic of Dick's fiction occur when the tenuous distinctions separating human and android threaten to collapse, as when Deckard suspects another bounty hunter of being an android with a synthetic memory implant that keeps him from realizing he is not human. The suspicion is insidious, for it implies that Deckard may also be an android and not realize it. When humans can no longer be distinguished from androids, the life cycle and dis/assembly zone occupy the same space. Then what count as stories are not so much the progressions of aging as the permutations of dis/ assembly.

It would be possible to tell another story about posthuman narratives based on this imperative, arcing from William Burroughs's *Naked Lunch* to Kathy Acker's *Empire of the Senseless*. But that is not my purpose here. I have been concerned to trace the evolution of the mapping of dis/assembly zones onto life cycle narratives from the early 1950s, when the idea that human beings might not be the end of the line was beginning to sink in, through the present, when human survival on the planet seems increasingly problematic. It is not an accident that technologists such as Hans Moravec talk about their dreams of downloading human consciousness into a computer.[12] As the sense of its mortality grows, humankind looks for its successor and heir, harboring the secret hope that the heir can somehow be enfolded back into the self. The narratives that count as stories for us speak to this hope, even as they reveal the gendered constructions that carry sexual politics into the realm of the posthuman.

ACKNOWLEDGMENTS

I am grateful to the Center for Advanced Studies at the University of Iowa for research support while writing this essay, especially Jay Semel and Lorna Olson. Istvan Csicsery-Ronay, Jr, suggested the term *hyperconnectivity*, and Donna Haraway stimulated my interest in John Varley's story 'Press Enter' at her keynote address, Indiana University, February 1990.

NOTES

1. Donna Haraway, *Primate Visions: Gender, Race, and Nature in the World of Modern Science* (New York: Routledge, 1989), 279–303.
2. Scott Bukatman, 'Who Programs You? The Science Fiction of the Spectacle,' in *Altex Zone: Cultural Theory and Contemporary Science Fiction Cinema*, ed. Annette Kuhn (London: Verso, 1990), 201.
3. For an overview of life cycle stages and the attributes associated with each, see Erik H. Erikson, *The Life Cycle Completed* (New York: Norton, 1982), 32–33. A comparison of Erikson, Piaget, and Sears can be found in Henry W. Maier, *Three Theories of Child Development*, 3d ed. (New York: Harper and Row, 1978), 176–177.
4. Gillian Beer, *Darwin's Plots: Evolutionary Narrative in Darwin, George Eliot, and Nineteenth-century Fiction* (London: Routledge & Kegan Paul, 1983).
5. Bernard Wolfe, *Limbo* (New York: Ace, 1952); Katherine Dunn, *Geek Love* (New York: Knopf, 1989).
6. John Varley, 'Press Enter,' in *Blue Champagne* (Niles: Dark Harvest, 1986); Anne McCaffrey, *The Ship Who Sang* (New York: Ballantine, 1970).

7. C. J. Cherryh, *Cyteen: The Betrayal; The Rebirth; The Vindication* (New York: Popular Library, 1988); Philip K. Dick, *Do Androids Dream of Electric Sheep* (New York: Ballantine, 1968).

8. Carolyn Geduld in *Bernard Wolfe* (New York: Twayne, 1972) describes the author as a 'very small man with a thick, sprouting mustache, a fat cigar, and a voice that grabs attention' (15).

9. David N. Samuelson has called *Limbo* one of the three great twentieth-century dystopias in '*Limbo*: The Great American Dystopia,' *Extrapolation* 19 (December 1977): 76–87.

10. Julia Kristeva, 'The Novel as Polylogue,' in *Desire in Language: A Semiotic Approach to Literature and Art*, ed. Leon S. Roudiez, trans. Thomas Gora, Alice Jardine, and Leon S. Roudiez (New York: Columbia University Press, 1980) 159–209.

11. See, for example, *Dragonflight, Dragonquest,* and *Decision at Doona*, all by McCaffrey. The stories in *The Ship Who Sang* were published separately from 1961–1969, with the collection appearing in 1970. For a discussion of McCaffrey's fantasies, see *Science Fiction, Today and Tomorrow*, ed. Reginald Bretnor (New York: Harper and Row, 1974), 278–294. Also of interest is Mary T. Brizzi, *Anne McCaffrey* (San Bernadino: Borgo Press, 1986), especially 19–32.

12. Moravec is quoted in Ed Regis, *Great Mambo Chicken and Transcendent Science: Science Slightly over the Edge* (Reading, Mass.: Addition Wesley, 1990). See also Roger Penrose, *The Emperor's New Mind: Concerning Computers, Minds, and the Laws of Physics* (New York: Oxford University Press, 1989), 347–447, and O. B. Hardison, *Disappearing through the Skylight: Culture and Technology in the Twentieth Century* (New York: Viking, 1989).

9

CYBERNETIC DECONSTRUCTIONS: CYBERPUNK AND POSTMODERNISM

Veronica Hollinger

If, as Fredric Jameson has argued, postmodernism is our contemporary cultural dominant ('Logic' 56), so equally is technology 'our historical context, political and personal,' according to Teresa de Lauretis: 'Technology is now, not only in a distant, science fictional future, an extension of our sensory capacities; it shapes our perceptions and cognitive processes, mediates our relationships with objects of the material and physical world, and our relationships with our own or other bodies' (167). Putting these two aspects of our reality together, Larry McCaffery has recently identified science fiction as 'the most significant evolution of a paraliterary form' in contemporary literature (xvii).

Postmodernist texts which rely heavily on science-fiction iconography and themes have proliferated since the 1960s, and it can be argued that some of the most challenging science fiction of recent years has been produced by mainstream and vangardist rather than genre writers. A random survey of postmodernist writing which has been influenced by science fiction – works for which science-fiction writer Bruce Sterling suggests the term 'slipstream' ('Slipstream') – might include, for example, Richard Brautigan's In Watermelon Sugar (1968), Monique Wittig's Les Guérillères (1969), Angela Carter's Heroes and Villains (1969), J.G. Ballard's Crash (1973), Russell Hoban's Riddley Walker (1980), Ted Mooney's Easy Travel to Other Planets (1981), Anthony Burgess's The End of the World News (1982), and Kathy Acker's Empire of the Senseless (1988).

Not surprisingly, however, the specific concerns and esthetic techniques of postmodernism have been slow to appear in genre science fiction, which tends to pride itself on its status as a paraliterary phenomenon. Genre science fiction thrives within an epistemology which privileges the logic of cause-and-effect narrative development, and it usually demonstrates a rather optimistic belief in the progress of human knowledge. Appropriately, the space ship was its representative icon during the 1940s and '50s, the expansionist 'golden age' of American science fiction. Equally appropriately, genre science fiction can claim the realist novel as its closest narrative relative; both developed in an atmosphere of nineteenth-century scientific positivism and both rely to a great extent on the mimetic transparency of language as a 'window' through which to provide views of a relatively uncomplicated human reality. When science fiction is enlisted by postmodernist fiction, however, it becomes integrated into an esthetic and a world-view whose central tenets are an uncertainty and an indeterminacy which call into question the 'causal interpretation of the universe' and the reliance on a 'rhetoric of believability' which virtually define it as a generic entity (Ebert 92).

It is within this conflictual framework of realist literary conventions played out in the postmodernist field that I want to look at cyberpunk, a 'movement' in science fiction in the 1980s which produced a wide range of fictions exploring the technological ramifications of experience within late-capitalist, post-industrial, media-saturated Western society. 'Let's get back to the Cyberpunks,' Lucius Shepard recently proposed in the first issue of *Journal Wired* (1989), one of several non-academic periodicals devoted to contemporary issues in science fiction and related fields; 'Defunct or not, they seem to be the only revolution we've got' (113).

[...]

Cyberpunk was a product of the commercial mass market of 'hard' science fiction; concerned on the whole with near-future extrapolation and more or less conventional on the level of narrative technique, it was nevertheless at times brilliantly innovative in its explorations of technology as one of the 'multiplicity of structures that intersect to produce that unstable constellation the liberal humanists call the "self"' (Moi 10). From this perspective, cyberpunk can be situated among a growing (although still relatively small) number of science-fiction projects which can be identified as 'anti-humanist.' In its various deconstructions of the subject – carried out in terms of a cybernetic breakdown of the classic nature/culture opposition – cyberpunk can be read as one symptom of the postmodern condition of genre science fiction. While science fiction frequently problematizes the oppositions between the natural and the artificial, the human and the machine, it generally sustains them in such a way that the human remains securely ensconced in its privileged place at the center of things. Cyberpunk, however, is about the breakdown of these oppositions.

This cybernetic deconstruction is heralded in the opening pages of what is now considered the quintessential cyberpunk novel – we might call it 'the c-p limit-text' – William Gibson's *Neuromancer* (1984). Gibson's first sentence

– 'The sky above the port was the color of television, tuned to a dead channel' (3) – invokes a rhetoric of technology to express the natural world in a metaphor which blurs the distinctions between the organic and the artificial. Soon after, Gibson's computer-cowboy, Case, gazes at 'the chrome stars' of shuriken, and imagines these deadly weapons as 'the stars under which he voyaged, his destiny spelled out in a constellation of cheap chrome' (12). Human bodies too are absorbed into this rhetorical conflation of organism and machine: on the streets of the postmodern city whose arteries circulate information, Case sees 'all around [him] the dance of biz, information interacting, data made flesh in the mazes of the black market ...' (16). The human world replicates its own mechanical systems, and the border between the organic and the artificial threatens to blur beyond recuperation.

If we think of science fiction as a genre which typically foregrounds human action *against* a background constituted by its technology, this blurring of once clearly defined boundaries makes cyberpunk a particularly relevant form of science fiction for the post-industrial present. Richard Kadrey, himself a (sometime) cyberpunk writer, recently noted the proliferation of computer-based metaphors – 'downtime,' 'brain dump' and 'interface,' for example – which are already used to describe human interaction ('Simulations' 75). We can read cyberpunk as an analysis of the postmodern *identification* of human and machine.

Common to most of the texts which have become associated with cyberpunk is an overwhelming fascination, at once celebratory and anxious, with technology and its immediate – that is, *unmediated* – effects upon human being-in-the-world, a fascination which sometimes spills over into the problematizing of 'reality' itself. This emphasis on the potential interconnections between the human and the technological, many of which are already gleaming in the eyes of research scientists, is perhaps the central 'generic' feature of cyberpunk. Its evocation of popular/street culture and its valorization of the socially marginalized, that is, its 'punk' sensibility, have also been recognized as important defining characteristics.

Sterling, one of the most prolific spokespersons for the Movement during its heyday, has described cyberpunk as a reaction to 'standard humanist liberalism' because of its interest in exploring the various scenarios of humanity's potential interfaces with the products of its own technology. For Sterling, cyberpunk is 'post-humanist' science fiction which believes that 'technological destruction of the human condition leads not to futureshocked zombies but to hopeful monsters' ('Letter' 5,4).

Science fiction has traditionally been enchanted with the notion of transcendence, but, as Glenn Grant points out in his discussion of *Neuromancer*, cyberpunk's 'preferred method of transcendence is through technology' (43). Themes of transcendence, however, point cyberpunk back to the romantic trappings of the genre at its most conventional, as does its valorization of the (usually male) loner rebel/hacker/punk who appears so frequently as its central character. Even Sterling has recognized this, concluding that 'the proper mode of critical attack on cyberpunk has not yet been essayed. Its truly dangerous element is incipient Nietzschean philosophical fascism: the belief in the Over-man, and the worship of will-to-power' ('Letter' 5).

It is also important to note that not all the monsters it has produced have been hopeful ones; balanced against the exhilaration of potential technological transcendence is the anxiety and disorientation produced in the self/body in danger of being absorbed into its own technology. Mesmerized by the purity of technology, Gibson's Case at first has only contempt for the 'meat' of the human body and yearns to remain 'jacked into a custom cyberspace deck that projected his disembodied consciousness into the consensual hallucination that was the matrix' (5). Similarly, the protagonist of K.W. Jeter's *The Glass Hammer* (1987) experiences his very existence as a televised simulation. The postmodern anomie which pervades *The Glass Hammer* demonstrates that Sterling's defense of cyberpunk against charges that it is peopled with 'future-shocked zombies' has been less than completely accurate.

[...]

'In virtual reality, the entire universe is your body and physics is your language,' according to Jaron Lanier, founder and CEO of VPL Research in California; 'we're creating an entire new reality' (qtd. in Ditlea 97–98).

Gibson's *Neuromancer*, the first of a trilogy of novels which includes *Count Zero* (1986) and *Mona Lisa Overdrive* (1988), is set in a near-future trash culture ruled by multi-national corporations and kept going by black-market economies, all frenetically dedicated to the circulation of computerized data and 'the dance of biz' (16) which is played out by Gibson's characters on the streets of the new urban overspill, the Sprawl. The most striking spatial construct in *Neuromancer*, however, is neither the cityscape of the Sprawl nor the artificial environments like the fabulous L-5, Freeside, but 'cyberspace,' the virtual reality which exists in simulated splendor on the far side of the computer screens which are the real center of technological activity in Gibson's fictional world. Scott Bukatman describes cyberspace as 'a new and decentered spatiality ... which exists parallel to, but outside of, the geographic topography of experiential reality' (45). In a fascinating instance of feedback between science fiction and the 'real' world, Autodesk, a firm researching innovations in computerized realities in Sausalito, California, has recently filed for trade-mark protection of the term 'cyberspace' which it may use as the name for its new virtual reality software (Ditlea 99). Jean Baudrillard's apocalyptic commentary seems especially significant here: 'It is thus not necessary to write science fiction: we have as of now, here and now, in our societies, with the media, the computers, the circuits, the networks, the acceleration of particles which has definitely broken the referential orbit of things' ('The Year 2000' 36).

Along with the 'other' space of cyberspace, *Neuromancer* offers alternatives to conventional modalities of human existence as well: computer hackers have direct mental access to cyberspace, artificial intelligences live and function within it, digitalized constructs are based on the subjectivities of humans whose 'personalities' have been downloaded into computer memory, and human bodies are routinely cloned.

This is Sterling's post-humanism with a vengeance, a post-humanism which, in its representation of 'monsters' – hopeful or otherwise – produced by the interface of the human and the machine, radically decenters the human body, the sacred icon of the essential self, in the same way that the virtual reality of cyberspace works to decenter conventional humanist notions of an unproblematical 'real.'

As I have noted, however, cyberpunk is not the only mode in which science fiction has demonstrated an anti-humanist sensibility. Although radically different from cyberpunk – which is written for the most part by a small number of white middle-class men, many of whom, inexplicably, live in Texas – feminist science fiction has also produced an influential body of anti-humanist texts. These would include, for example, Joanna Russ's *The Female Man* (1975), Jody Scott's *I, Vampire* (1984), and Margaret Atwood's *The Handmaid's Tale* (1985), novels which also participate in the postmodernist revision of conventional science fiction. Given the exigencies of their own particular political agendas, however, these texts demonstrate a very different approach to the construction/deconstruction of the subject than is evident in the technologically-influenced post-humanism of most cyberpunk fiction.

Jane Flax, for example, suggests that 'feminists, like other postmodernists, have begun to suspect that all such transcendental claims [those which valorize universal notions of reason, knowledge, and the self] reflect and reify the experience of a few persons – mostly white, Western males. These transhistoric claims seem plausible to us in part because they reflect important aspects of the experience of those who dominate our social world' (626). Flax's comments are well taken, although her conflation of all feminisms with postmodernism tends to oversimplify the very complex and problematical interactions of the two that Bonnie Zimmerman has noted. Moreover, in a forthcoming essay for *Extrapolation*, I have argued that most feminist science fiction rather supports than undermines the tenets of liberal humanism, although 'changing the subject' of that humanism, to borrow the title of a recent study by Nancy K. Miller.

We can also include writers like Philip K. Dick, Samuel R. Delany and John Varley within the project of anti-humanist science fiction, although these writers are separated from cyberpunk not only by chronology but also by cyberpunk's increased emphasis on technology as a constitutive factor in the development of postmodern subjectivity. Darko Suvin also notes some of the differences in political extrapolation between cyberpunk and its precursors: 'in between Dick's nation-state armies or polices and Delany's Foucauldian micropolitics of bohemian groups, Gibson [for example] has – to my mind more realistically – opted for global economic power-wielders as the arbiters of peoples [sic] lifestyles and lives' (43).

[...]

In 'Prometheus as Performer: Toward a Posthumanist Culture?' Ihab Hassan writes: 'We need first to understand that the human form – including human desire and all its external representations – may be changing radically, and thus must be re-visioned. We need to understand that five hundred years of humanism may be coming to an end, as humanism transforms itself into something that we must helplessly call posthumanism' (205).

Sterling's *Schismatrix* (1986) is one version of 'posthumanity' presented as picaresque epic. Sterling's far-future universe – a rare construction in the cyberpunk 'canon' – is one in which countless societies are evolving in countless different directions; the Schismatrix is a loose confederation of worlds where the only certainty is the inevitability of change. Sterling writes that 'the new multiple humanities hurtled blindly toward their unknown destinations, and the vertigo of acceleration struck deep. Old preconceptions were in tatters, old loyalties were obsolete. Whole societies were paralyzed by the mind-blasting vistas of absolute possibility' (238). Sterling's protagonist, a picaresque hero for the postmodern age, 'mourned mankind, and the blindness of men, who thought that the Kosmos had rules and limits that would shelter them from their own freedom. There were no shelters. There were no final purposes. Futility, and freedom, were Absolute' (273).

Schismatrix is a future history different from many science-fiction futures in that what it extrapolates from the present is the all-too-often ignored/denied/repressed idea that human beings will be different in the future and will continue to develop within difference. In this way, *Schismatrix* demonstrates a familiarly post-structuralist sensibility, in its recognition both of the potential anxiety *and* the potential play inherent in a universe where 'futility, and freedom, [are] Absolute.'

Sterling's interest in and attraction to the play of human possibility appears as early as his first novel, *Involution Ocean* (1977). In this story (which reads in some ways like a kind of drug-culture post-*Moby-Dick*), the protagonist falls into a wonderful vision of an alien civilization, in a passage which, at least temporarily, emphasizes freedom over futility: 'There was an incredible throng, members of a race that took a pure hedonistic joy in the possibilities of surgical alteration. They switched bodies, sexes, ages, and races as easily as breathing, and their happy disdain for uniformity was dazzling. ... It seemed so natural, rainbow people in the rainbow streets; humans seemed drab and antlike in comparison' (154).

This is a far cry from the humanist anxieties which have pervaded science fiction since the nineteenth century. Consider, for example, the anxiety around which H. G. Wells created *The Time Machine* (1895): it is 'de-humanization,' humanity's loss of its position at the center of creation, which produces the tragedy of the terminal beach, and it is, to a great extent, the absence of the human which results in the 'abominable desolation' (91) described by Wells's

Time Traveller. Or consider what we might term the 'trans-humanism' of Arthur C. Clarke's *Childhood's End* (1953), in which a kind of transcendental mysticism precludes the necessity of envisioning a future based on changing technologies, social conditions and social relations. Greg Bear's more recent *Blood Music* (1985) might be read, from this perspective, as a contemporary version of the same transcendental approach to human transformation, one based on an apocalyptic logic which implies the impossibility of any change in the human condition *within history*. *Blood Music* is especially interesting in this context, because its action is framed by a rhetoric of science which would seem to repudiate any recourse to metaphysics. Darko Suvin has noted, however, that it functions as 'a naïve fairytale relying on popular wishdreams that our loved ones not be dead and that our past mistakes may all be rectified, all of this infused with rather dubious philosophical and political stances' (41).

[...]

'Certain central themes spring up repeatedly in cyberpunk,' Sterling points out in his preface to the influential short-fiction collection, *Mirrorshades: The Cyberpunk Anthology*. 'The theme of body invasion: prosthetic limbs, implanted circuitry, cosmetic surgery, genetic alteration. The even more powerful theme of mind invasion: brain-computer interfaces, artificial intelligence, neurochemistry – techniques radically redefining the nature of humanity, the nature of the self' (xiii).

The potential in cyberpunk for undermining concepts like 'subjectivity' and 'identity' derives in part from its production within what has been termed 'the technological imagination'; that is, cyberpunk is hard science fiction which recognizes the paradigm-shattering role of technology in post-industrial society. We have to keep in mind here, of course, that the Movement has become (in)famous for the adversarial rhetoric of its ongoing and prolific self-commentary which, in turn, functions as an integral part of its overall production as a 'movement.' We should be careful, for this reason, not to confuse claims with results. The anti-humanist discourse of cyberpunk's frequent manifestoes, however, strongly supports de Lauretis's contention that 'technology is our historical context, political and personal' (167). As I have suggested, this context functions in cyberpunk as one of the most powerful of the multiplicities of structures which combine to produce the postmodern subject.

Thus, for example, the characters in Michael Swanwick's *Vacuum Flowers* (1987) are subjected to constant alterations in personality as the result of programming for different skills or social roles – metaphysical systems grounded on faith in an 'inner self' begin to waver. Human bodies in Gibson's stories, and even more so in Sterling's, are subjected to shaping and re-shaping, the human form destined perhaps to become simply one available choice among many; notions of a human nature determined by a 'physical essence' of the human begin to lose credibility (for this reason, many behavioral patterns defined by sexual difference become irrelevant in these futures). Thus Rudy

Rucker can offer the following as a chapter title in *Wetware*: 'Four: In Which Manchile, the First Robot-Built Human, Is Planted in the Womb of Della Taze by Ken Doll, Part of Whose Right Brain Is a Robot Rat.'

We must also recognize, however, that 'the subject of the subject' at the present time has given rise to as much anxiety as celebration (anxiety from which the postmodernist theorist is by no means exempt). The break-up of the humanist 'self' in a media-saturated post-industrial present has produced darker readings which cyberpunk also recognizes. Fredric Jameson, whose stance *vis-à-vis* the postmodern is at once appreciative and skeptical, has suggested that fragmentation of subjectivity may be the postmodern equivalent of the modernist predicament of individual alienation ('Cultural Logic' 63). Pat Cadigan's 'Pretty Boy Crossover' (1985), for example, raises questions about the effects of simulated reality upon our human sense of self as complete and inviolable. In her fictional world, physical reality is 'less efficient' than computerized simulation, and video stars are literally video programs, having been 'distilled . . . to pure information' (89, 88) and downloaded into computer matrices. Cadigan's eponymous Pretty Boy is tempted by the offer of literally eternal life within the matrix and, although he finally chooses 'real' life, that reality seems to fade against the guaranteed 'presence' of its simulation. Bobby, who has opted for existence as simulation, explains the 'economy of the gaze' which guarantees the authenticity of the self in this world: 'If you love me, you watch me. If you don't look, you don't care and if you don't care I don't matter. If I don't matter, I don't exist. Right?' (91).

[. . .]

'Pretty Boy Crossover' offers this succinct observation about the seductive power of simulated reality: 'First you see video. Then you wear video. Then you eat video. Then you *be* video' (82).

In K.W. Jeter's *The Glass Hammer*, being is *defined* by its own simulation. *The Glass Hammer* is one of the most self-conscious deconstructions of unified subjectivity produced in recent science fiction, and one which dramatizes (in the neurotic tonalities familiar to readers of J.G. Ballard) the anxiety and schizophrenia of the (technologically-produced) postmodern situation. In *The Glass Hammer* the break-up of the 'self' is narrated in a text as fragmented as its subject (subject both as protagonist and as story). Jeter's novel is a chilling demonstration of the power of simulated re-presentation to construct 'the real' (so that it functions like a cyberpunk simulacrum of the theories of Jean Baudrillard). It 'narrates' episodes in the life of Ross Schuyler, who watches the creation of this life as a video event in five segments. There is no way to test the accuracy of the creation, since the self produced by memory is as unreliable a re-presentation as is a media 'bio.' As Schuyler realizes: 'Just because I was there – that doesn't mean anything' (59).

The opening sequence of *The Glass Hammer* dramatizes the schizophrenia within the subjectivity of the protagonist:

> Video within video. He watched the monitor screen, seeing himself there, watching. In the same space ... that he sat in now ...
>
> He watched the screen, waiting for the images to form. Everything would be in the tapes, if he watched long enough. (7)

Like Schuyler himself, the reader waits for the images to form as s/he reads the text. Episodes range over time, some in the past(s), some in the present, some real, some simulated, many scripted rather than 'novelized,' until the act of reading/watching achieves a kind of temporary coherence. It is this same kind of temporary coherence which formulates itself in Schuyler's consciousness, always threatening to dissolve again from 'something recognizably narrative' into 'the jumbled, half-forgotten clutter of his life' (87).

What takes place in *The Glass Hammer* may also be read as a deconstruction of the opposition between depth and surface, a dichotomy which is frequently framed as the familiar conflict between reality and appearance. Jeter reverses this opposition, dramatizing the haphazard construction of his character's 'inner self' as a response to people and events, both real and simulated, over time. The displacement of an 'originary' self from the text places the emphasis on the marginal, the contingent, the re-presentations (in this case electronically produced) which actually create the sense of 'self.' Jeter's technique in *The Glass Hammer* is particularly effective: the reader watches the character, and watches the character watching himself watching, as his past unfolds, not as a series of memories whose logical continuity guarantees the stability of the ego, but as an entertainment series, the logical continuity of which is the artificial re-arrangement of randomness to *simulate* coherence.

[...]

Near the outset of Case's adventures in *Neuromancer*, Gibson's computer cowboy visits the warehouse office of Julius Deane, who 'was one hundred and thirty-five years old, his metabolism assiduously warped by a weekly fortune in serums and hormones.' In Deane's office, 'Neo-Aztec bookcases gathered dust against one wall of the room where Case waited. A pair of bulbous Disney-styled table lamps perched awkwardly on a low Kandinsky-look coffee table in scarlet-lacquered steel. A Dali clock hung on the wall between the bookcases, its distorted face sagging to the bare concrete floor' (12).

In this context, it is significant that the 'average' cyberpunk landscape tends to be choked with the debris of both language and objects; as a sign-system, it is overdetermined by a proliferation of surface detail which emphasizes the 'outside' over the 'inside.' Such attention to detail – recall Gibson's nearly compulsive use of brand names, for example, or the claustrophobic clutter of his streets – replaces the more conventional (realist) narrative exercise we might call 'getting to the bottom of things'; indeed, the shift in emphasis is from a symbolic to a surface reality.

In a discussion of *Neuromancer*, Gregory Benford observes that 'Gibson, like Ballard, concentrates on surfaces as a way of getting at the aesthetic of an age.' This observation is a telling one, even as it misses the point. Benford concludes that Gibson's attention to surface detail 'goes a long way toward telling us why his work has proved popular in England, where the tide for several decades now has been to relish fiction about surfaces and manners, rather than the more traditional concerns of hard SF: ideas, long perspectives, and content' (19).

This reliance on tradition is perhaps what prevents Benford, whose own 'hard science fiction' novels and stories are very much a part of science fiction's humanist tradition, from appreciating the approach of writers like Gibson and Jeter. The point may be that, in works like *Neuromancer* and *The Glass Hammer*, surface *is* content, an equation which encapsulates their critique – or at least their awareness – of our contemporary 'era of hyperreality' (Baudrillard, 'Ecstasy' 128). In this context, the much-quoted opening sentence of *Neuromancer*, with its image of the blank surface of a dead television screen, evokes the anxiety of this new era. Istvan Csicsery-Ronay, for example, sees in cyberpunk the recognition that 'with the computer, the problem of identity is moot, and the idea of reflection is transformed in to [sic] the algorithm of replication. SF's computer wipes out the Philosophical God and ushers in the demiurge of thought-as-technique' (273).

Like much anti-humanist science fiction, cyberpunk also displays a certain coolness, a kind of ironically detached approach to its subject matter which precludes nostalgia or sentimentality. This detachment usually discourages any recourse to the logic of the apocalypse, which, whether positive (like Clarke's) or negative (like Wells's), is no longer a favored narrative move. Jameson and Sterling (representatives of 'high theory' and 'low culture' respectively?) both identify a waning interest in the scenarios of literal apocalypse: Jameson perceives in the postmodern situation what he calls 'an inverted millennarianism, in which premonitions of the future ... have been replaced by the senses of the end of this or that' ('Cultural Logic' 53); in his introduction to Gibson's short-story collection, *Burning Chrome*, Sterling comments that one 'distinguishing mark of the emergent new school of Eighties SF [is] its boredom with the Apocalypse' (xi).

This is supported by Douglas Robinson, in his *American Apocalypses*, when he concludes that 'antiapocalypse – not apocalypse, as many critics have claimed – is the dominant topos of American postmodernism' (xvi). In a discussion of Derrida's discourse on apocalypse, Robinson argues that 'the apocalyptic imagination fascinates Derrida precisely as the 'purest' form, the most mythical expression or the most extreme statement of the metaphysics of presence' (251nl).

One reason for this tendency to abandon what has been a traditional science fiction topos may be the conviction, conscious or not, that a kind of philosophical apocalypse has already occurred, precipitating us into the dis-ease of postmodernism. Another reason may be the increased commitment of anti-humanist science fiction to the exploration of changes that will occur – to the self, to society and to social relations – in time; that is, they are more engaged with historical processes than attracted by the jump-cuts of apocalyptic scenarios which evade such investment in historical change. Cyberpunk, in particular, has demonstrated a keen interest in the near future, an aspect of its approach to history which discourages resolution-through-apocalypse.

[...]

In a discussion of 'the cybernetic (city) state,' Scott Bukatman has argued that as a result of the tendency in recent science fiction to posit 'a reconception of the human and the ability to interface with the new terminal experience ... terminal space becomes a legitimate part of human (or post-human) experience' (60). In many cases, however, science-fiction futures are all too often simply representations of contemporary cultural mythologies disguised under heavy layers of futuristic make-up.

The recognition of this fact provides part of the 'meaning' of one of the stories in Gibson's *Burning Chrome* collection. 'The Gernsback Continuum' humorously ironizes an early twentieth-century futurism which could conceive of no real change in the human condition, a futurism which envisioned changes in 'stuff' rather than changes in social relations (historical distance increases the ability to critique such futures, of course). In Gibson's story, the benighted protagonist is subjected to visitations by the 'semiotic ghosts' of a future which never took place, the future, to borrow a phrase from Jameson, 'of one moment of what is now our own past' ('Progress' 244). At the height of these 'hallucinations,' he 'sees' two figures poised outside a vast city reminiscent of the sets for films like *Metropolis* and *Things to Come*:

> [the man] had his arm around [the woman's] waist and was gesturing toward the city. They were both in white. ... He was saying something wise and strong, and she was nodding. ...
> ... [T]hey were the Heirs to the Dream. They were white, blond, and they probably had blue eyes. They were American. ... They were smug, happy, and utterly content with themselves and their world. And in the Dream, it was *their* world. ...
> It had all the sinister fruitiness of Hitler Youth propaganda. (32–33)

Gibson's protagonist discovers that 'only really bad media can exorcise [his] semiotic ghosts' (33) and he recovers with the help of pop culture productions like *Nazi Love Motel*. 'The Gernsback Continuum' concludes with the protagonist's realization that his dystopian present could be worse, 'it could be perfect' (35).

Gibson's story is not simply an ironization of naïve utopianism; it also warns against the limitations, both humorous and dangerous, inherent in any vision of the future which bases itself upon narrowly defined ideological systems which take it upon themselves to speak 'universally,' or which conceive of themselves as 'natural' or 'absolute.' David Brin's idealistic *The Postman*

(1985), for example, is a post-apocalyptic fiction which closes on a metaphorical note 'of innocence, unflaggingly optimistic' (321), nostalgically containing itself within the framework of a conventional humanism. Not surprisingly, its penultimate chapter concludes with a re-affirmation of the 'natural' roles of men and women:

> And always remember, the moral concluded: Even the best men – the heroes – will sometimes neglect to do their jobs.
> *Women, you must remind them, from time to time.* ... (312)

Compare this to Gibson's description of the Magnetic Dog Sisters, peripheral characters in his story, 'Johnny Mnemonic' (1981), also collected in *Burning Chrome*: 'They were two meters tall and thin as greyhounds. One was black and the other white, but aside from that they were as nearly identical as cosmetic surgery could make them. They'd been lovers for years and were bad news in a tussle. I was never quite sure which one had originally been male' (2).

Another story in the same collection, 'Fragments of a Hologram Rose,' uses metaphors of the new technology to express the indeterminate and fragmented nature of the self: 'A hologram has this quality: Recovered and illuminated, each fragment will reveal the whole image of the rose. Falling toward delta, he sees himself the rose, each of his scattered fragments revealing a whole he'll never know. ... But each fragment reveals the rose from a different angle ...' (42).

Gibson's rhetoric of technology finally circumscribes all of reality. In his second novel, *Count Zero* (1986), there is an oblique but pointed rebuttal of humanist essentialization, which implicitly recognizes the artificiality of the Real. Having described cyberspace, the weirdly real 'space' that human minds occupy during computer interfacing, as 'mankind's unthinkably complex consensual hallucination' (44), he goes on to write the following:

> 'Okay,' Bobby said, getting the hang of it, 'then what's the matrix? ... [W]hat's cyberspace?'
> 'The world,' Lucas said. (131)

It is only by recognizing the consensual nature of socio-cultural reality, which includes within itself our definitions of human nature, that we can begin to perceive the possibility of change. In this sense, as Csicsery-Ronay suggests (although from a very different perspective), cyberpunk is 'a paradoxical form of realism' (266).

Csicsery-Ronay also contends that cyberpunk is 'a legitimate international artistic style, with profound philosophical and aesthetic premises,' a style captured by films such as *Blade Runner* and by philosophers such as Jean Baudrillard; 'it even has, in Michael Jackson and Ronald Reagan, its hyperreal icons of the human simulacrum infiltrating reality' (269).

[...]

Lucius Shepard concludes his 'requiem for cyberpunk' by quoting two lines from Cavafy's 'Waiting for the Barbarians': 'What will we do now that the barbarians are gone? / Those people were a kind of solution' (118).

Cyberpunk seemed to erupt in the mid-80s, self-sufficient and full-grown, like Minerva from the forehead of Zeus. From some perspectives, it could be argued that this self-proclaimed Movement was nothing more than the discursive construction of the collective imaginations of science-fiction writers and critics eager for something/anything new in what had become a very conservative and quite predictable field. Now that the rhetorical dust has started to settle, however, we can begin to see cyberpunk as itself the product of a multiplicity of influences from both within and outside of genre science fiction. Its writers readily acknowledge the powerful influence of 1960s and '70s New Wave writers like Samuel R. Delany, John Brunner, Norman Spinrad, J.G. Ballard and Michael Moorcock, as well as the influence of postmodernists like William Burroughs and Thomas Pynchon. The manic fragmentations of Burroughs's *Naked Lunch* and the maximalist apocalypticism of Pynchon's *Gravity's Rainbow* would seem to have been especially important for the development of the cyberpunk 'sensibility.' Richard Kadrey has even pronounced *Gravity's Rainbow* to be cyberpunk *avant la lettre*' the best cyberpunk novel ever written by a guy who didn't even know he was writing it' ('Cyberpunk' 83). Equally, Delany has made a strong case for feminist science fiction as cyberpunk 's 'absent mother,' noting that 'the feminist explosion – which obviously infiltrates the cyberpunk writers so much – is the one they seem to be the least comfortable with, even though it's one that, much more than the New Wave, has influenced them most strongly, both in progressive and in reactionary ways ...' (9).

Due in part to the prolific commentaries and manifestoes in which writers like Sterling outlined/analyzed/defended their project(s) – usually at the expense of more traditional science fiction – cyberpunk helped to generate a great deal of very useful controversy about the role of science fiction in the 1980s, a decade in which the resurgence of fantastic literature left much genre science fiction looking rather sheepishly out of date. At best, however, the critique of humanism in these works remains incomplete, due at least in part to the pressures of mass market publishing as well as to the limitations of genre conventions which, more or less faithfully followed, seem (inevitably?) to lure writers back into the power fantasies which are so common to science fiction. A novel like Margaret Atwood's *The Handmaid's Tale*, for instance, produced as it was outside the genre market, goes further in its deconstruction of individual subjectivity than do any of the works I have been discussing, except perhaps *The Glass Hammer*.

Gibson's latest novel, *Mona Lisa Overdrive*, although set in the same universe as *Neuromancer* and *Count Zero*, foregrounds character in a way which necessarily mutes the intensity and multiplicity of surface detail which is so marked a characteristic of his earlier work. Sterling's recent and unexpected

Islands in the Net (1988) is a kind of international thriller which might be read as the depiction of life *after* the postmodern condition has been 'cured.' Set in a future after the 'Abolition' (of nuclear warfare), its central character, Laura Webster, dedicates herself to the control of a political crisis situation which threatens to return the world to a global state of fragmentation and disruptive violence which only too clearly recalls our own present bad old days. Sterling's 'Net' is the vast information system which underlies and makes possible the unity of this future world and his emphasis is clearly on the necessity for such global unity. Although, in the final analysis, no one is completely innocent – Sterling is too complex a writer to structure his forces on opposite sides of a simple ethical divide – the movement in *Islands in the Net* is away from the margins toward the center, and the Net, the 'global nervous system' (15), remains intact.

As its own creators seem to have realized, cyberpunk – like the punk ethic with which it was identified – was a response to postmodern reality which could go only so far before self-destructing under the weight of its own deconstructive activities (not to mention its appropriation by more conventional and more commercial writers). That final implosion is perhaps what Jeter accomplished in *The Glass Hammer*, leaving us with the image of a mesmerized Schuyler futilely searching for a self in the videoscreens of the dystopian future. It is clearly this aspect of cyberpunk which leads Csicsery-Ronay to conclude that 'by the time we get to cyberpunk, reality has become a case of nerves. ... The distance required for reflection is squeezed out as the world implodes: when hallucinations and realia collapse into each other, there is no place from which to reflect' (274). For him, 'cyberpunk is ... the apotheosis of bad faith, apotheosis of the postmodern' (277). This, of course, forecloses any possibility of political engagement within the framework of the postmodern.

Here cyberpunk is theorized as a symptom of the malaise of postmodernism, but, like Baudrillard's apocalyptic discourse on the 'condition' itself, Csicsery-Ronay's analysis tends to underplay the positive potential of re-presentation and re-visioning achieved in works like *Neuromancer* and *Schismatrix*. Bukatman, for example, has suggested that the function of cyberpunk 'neuromanticism' is one appropriate to science fiction in the postmodern era: the *reinsertion* of the human into the new reality which its technology is in the process of shaping. According to Bukatman, 'to dramatize the terminal realm means to somehow insert the figure of the human into that space to experience it *for us*. ... Much recent science fiction stages and restages a confrontation between figure and ground, finally constructing a new human form to interface with the other space and cybernetic reality' (47–48).

The postmodern condition has required that we revise science fiction's original trope of technological anxiety – the image of a fallen humanity controlled by a technology run amok. Here again we must deconstruct the human/machine opposition and begin to ask new questions about the ways in which we and our technologies 'interface' to produce what has become a *mutual* evolution. It may be significant that one of the most brilliant visions of the potential of cybernetic deconstructions is introduced in Donna Haraway's merger of science fiction and feminist theory, 'A Manifesto for Cyborgs: Science, Technology, and Socialist Feminism in the 1980s,' which takes the rhetoric of technology toward its political limits: 'cyborg unities are monstrous and illegitimate,' writes Haraway; 'in our present political circumstances, we could hardly hope for more potent myths for resistance and recoupling' (179).

ACKNOWLEDGMENTS

An earlier version of this essay was presented at the 1988 Conference of the Science Fiction Research Association, Corpus Christi, Texas. I would like to thank the Social Sciences and Humanities Research Council of Canada for their generous support. I would also like to thank Glenn Grant, Editor of *Edge Detector: A Magazine of Speculative Fiction*, for making so much information and material available to me during the process of revision.

WORKS CITED

Acker, Kathy. *Empire of the Senseless*. New York: Grove, 1988.

Atwood, Margaret. *The Handmaid's Tale*. Toronto: McClelland, 1986.

Ballard, J.G. *Crash*. 1975. London: Triad/Panther, 1985.

Baudrillard, Jean. 'The Year 2000 Has Already Happened.' *Body Invaders: Panic Sex in America*. Ed. Arthur Kroker and Marilouise Kroker. Trans. Nai-Fei Ding and Kuan Hsing Chen. Montreal: New World Perspectives, 1987. 35–44.

Baudrillard, Jean. 'The Ecstasy of Communication.' *The Anti-Aesthetic: Essays on Postmodern Culture*. Ed. Hal Foster. Trans. John Johnston. Port Townsend, WA: Bay, 1983. 126–34.

Bear, Greg. *Blood Music*. New York: Ace, 1985.

Benford, Gregory. 'Is Something Going On?' *Mississippi Review* 47/48 (1988): 18–23.

Brautigan, Richard. *In Watermelon Sugar*. New York: Dell, 1968.

Brin, David. *The Postman*. New York: Bantam, 1985.

Bukatman, Scott. 'The Cybernetic (City) State: Terminal Space Becomes Phenomenal.' *Journal of the Fantastic in the Arts* 2 (1989): 43–63.

Burgess, Anthony. *The End of the World News*. Markham, ON: Penguin, 1982.

Burroughs, William. *Naked Lunch*. New York: Grove, 1959.

Cadigan, Pat. 'Pretty Boy Crossover.' 1985. *The 1987 Annual World's Best SF*. Ed. Donald A. Wollheim. New York: DAW, 1987. 82–93.

Carter, Angela. *Heroes and Villains*. 1969. London: Pan, 1972.

Clarke, Arthur C. *Childhood's End*. New York: Ballantine, 1953.

Csicsery-Ronay, Istvan. 'Cyberpunk and Neuromanticism.' *Mississippi Review* 47/48 (1988): 266–78.

Delany, Samuel R. 'Some *Real* Mothers: An Interview with Samuel R. Delany by Takayuki Tatsumi.' *Science-Fiction Eye* 1 (1988): 5–11.

de Lauretis, Teresa. 'Signs of Wo/ander.' *The Technological Imagination*. Ed. Teresa de Lauretis, Andreas Huyssen, and Kathleen Woodward. Madison, WI: Coda, 1980. 159–74.

Ditlea, Steve. 'Another World: Inside Artificial Reality.' *P/C Computing*. November 1989: 90–102.

Ebert, Teresa L. 'The Convergence of Postmodern Innovative Fiction and Science Fiction: An Encounter with Samuel R. Delany's Technotopia.' *Poetics Today* 1 (1980): 91–104.

Flax, Jane. 'Postmodernism and Gender Relations in Feminist Theory.' *Signs: Journal of Women in Culture and Society* 12 (1987): 621–43.

Gibson, William. 'Fragments of a Hologram Rose.' 1977. *Burning Chrome*. New York: Ace, 1987. 36–12.

Gibson, William. 'The Gernsback Continuum.' 1981. *Burning Chrome*. 23–35.

Gibson, William. 'Johnny Mnemonic.' 1981. *Burning Chrome*. 1–22.

Gibson, William. *Neuromancer*. New York: Berkley, 1984.

Gibson, William. *Count Zero*. New York: Arbor House, 1986.

Gibson, William. *Mona Lisa Overdrive*. New York: Bantam, 1988.

Grant, Glenn. 'Transcendence Through Détournement in William Gibson's *Neuromancer*.' *Science Fiction Studies* 17 (1990): 41–49.

Haraway, Donna. 'A Manifesto for Cyborgs: Science, Technology, and Socialist Feminism in the 1980s.' 1985. *Coming to Terms: Feminism, Theory, Politics*. Ed. Elizabeth Weed. New York: Routledge, 1989. 173–204.

Hassan, Ihab. 'Prometheus as Performer: Toward a Posthumanist Culture?' *Performance in Postmodern Culture*. Ed. Michel Benamou and Charles Caramello. Madison, WI: Coda, 1977. 201–17.

Hoban, Russell. *Riddley Walker*. 1980. London: Pan, 1982.

Hollinger, Veronica. 'Feminist Science Fiction: Breaking Up the Subject.' *Extrapolation* 31 (1990): forthcoming.

Jameson, Fredric. 'Postmodernism, or The Cultural Logic of Late Capitalism.' *New Left Review* 146 (1984): 53–94.

Jameson, Fredric. 'Progress versus Utopia, or Can We Imagine the Future?' *Art After Modernism: Re-thinking Representation*. Ed. Brian Wallis. New York: The New Museum of Contemporary Art, 1984. 239–52.

Jeter, K.W., *The Glass Hammer*. New York: Signet, 1987.

Kadrey, Richard. 'Simulations of Immortality.' *Science-Fiction Eye* 1 (1989): 74–76.

Kadrey, Richard. 'Cyberpunk 101 Reading List.' *Whole Earth Review* 63 (1989): 83.

McCaffery, Larry. 'Introduction.' *Postmodern Fiction: A Bio-Bibliographical Guide*. Ed. McCaffery. Westport, CT: Greenwood, 1986. xi–xxviii.

Miller, Nancy K. 'Changing the Subject: Authorship, Writing, and the Reader.' *Feminist Studies/Critical Studies*. Ed. Teresa de Lauretis. Bloomington: Indiana UP, 1986. 102–20.

Moi, Toril. *Sexual/Textual Politics: Feminist Literary Theory*. New York: Methuen, 1985.

Mooney, Ted. *Easy Travel to Other Planets*. New York: Ballantine, 1981.

Pynchon, Thomas. *Gravity's Rainbow*. 1973, New York: Bantam, 1974.

Robinson, Douglas. *American Apocalypses: The Image of the End of the World in American Literature*. Baltimore, MD: Johns Hopkins UP, 1985.

Rucker, Rudy. *Wetware*. New York: Avon, 1988.

Russ, Joanna. *The Female Man*. New York: Bantam, 1975.

Scott, Jody. *I, Vampire*. New York: Ace, 1984.

Shepard, Lucius. 'Waiting for the Barbarians.' *Journal Wired* 1 (1989): 107–18.

Sterling, Bruce. 'Slipstream.' *Science-Fiction Eye* 1 (1989): 77–80.

Sterling, Bruce. *Islands in the Net*. New York: Arbor House, 1988.

Sterling, Bruce. Preface to *Mirrorshades: The Cyberpunk Anthology*. New York: Ace, 1988. ix–xvi.

Sterling, Bruce. Preface to *Burning Chrome*. ix–xii.

Sterling, Bruce. 'Letter from Bruce Sterling.' *REM* 7 (1987): 4–7.

Sterling, Bruce. *Schismatrix*. New York: Ace, 1986.

Sterling, Bruce. *Involution Ocean*. 1977. New York: Ace, 1988.

Swanwick, Michael. *Vacuum Flowers*. New York: Arbor, 1987.

Suvin, Darko. 'On Gibson and Cyberpunk SF.' *Foundation* 46 (1989): 40–51.

Wells, H.G. *The Time Machine*. 1895. *The Time Machine and The War of the Worlds*. New York: Fawcett, 1968. 25–98.

Wittig, Monique. *Les Guérillères*. 1969. Trans. David Le Vay. Boston: Beacon, 1985.
Zimmerman, Bonnie. 'Feminist Fiction and the Postmodern Challenge.' *Postmodern Fiction: A Bio-Bibliographical Guide*. Ed. McCaffery. 175–88.

10

CYBERPUNK: PREPARING THE GROUND FOR REVOLUTION OR KEEPING THE BOYS SATISFIED?[1]

Nicola Nixon

In the 1970s feminist writers made successful intrusions into the genre of the popular SF novel, a genre whose readership, then and now, is assumed to be one who can appreciate, for example, that taking blue mescaline inspires the confidence 'you'd feel somatically, the way you'd feel a woman's lips on your cock' (Shirley, *Eclipse* 74). One hardly needs recourse to Althusserian models to determine who the interpellated reader is here. Suffice it to say, it isn't me. In the '70s Joanna Russ, Marge Piercy, Ursula Le Guin, Suzy McKee Charnas, and Sally Miller Gearhart negotiated – rather boldly, given such a readership – a political and artistic trajectory from '60s feminism to its enthusiastic articulation in specifically feminist utopias. Collectively they provided an often implicit and stinging critique of male SF writers' penchant for figuring feminist power as *the* threat of the future. Parley J. Cooper's *The Feminists* (1971), where the 'top dog is a bitch' and men 'mere chattel,' for example, sports a dust jacket that reads: 'The story that had to be written – so timely, so frighteningly possible, you won't believe it's fiction.'[2] But they also presented alternative, genderless futures and worlds. Characteristically yoking the genres of fantasy and SF, or positioning themselves on the border between the two, the feminists of the '70s exposed gender as a crucial political lacuna in mainstream popular fiction and emphasized the urgency to change gender assumptions. If Russ, in *The Female Man* (1975), constructs the war of the sexes as a literal turf war, complete with bunkers and shell-pocked borders between Manland and Womanland, she also suggests that the presence of a literary turf war as 'soft' female fantasy encroaches on 'hard' male SF.

The Female Man, Le Guin's *The Dispossessed* (1975), Piercy's *Woman on the Edge of Time* (1976), and Gearhart's *The Wanderground* (1978) all posited near-present dystopian worlds, which displaying all manner of technical/medical/industrial/political abominations, functioned in critical counterpoint to utopian future worlds.[3] And thus they articulated criticisms of their current society while presenting potential emancipatory alternatives. Jean Pfaelzer observes that the feminist utopian text

> represents two worlds, the flawed present and the future perfect, which contradict and comment on each other. One world is feminist and egalitarian. The other world is not. And the world that is not utopian derives from the author's representation of contemporary gender inequality, sexual repression, and cultural malaise. (286–87)

For various reasons (many of them political, as Peter Fitting has indicated), the '70s feminist utopias gave way to straight, uncontrasted dystopias in the '80s, barely concealed allegories of feminism's complacency and failure: Margaret Atwood's *The Handmaid's Tale* (1985), Zoë Fairbairns' *Benefits* (1979), Suzette Haden Elgin's *Native Tongue* (1984) and *The Judas Rose* (1986), and Pamela Sargent's *The Shore of Women* (1986). Less optimistic than the '70s feminists, but no less political – at least insofar as they deployed gender as the linchpin for their fiction – the '80s feminists produced a form of quasi-didactic (fictional) finger-shaking, a series of monitory or cautionary tales. Also rising on the heels of the '70s feminist SF writers, however, was another SF 'movement,' one loudly proclaiming its 'revolutionary' status: cyberpunk.[4]

Cyberpunk – slick, colloquial, and science-based – represented a concerted return to the (originary) purity of hard SF, apparently purged of the influence of other-worldly fantasy, and embracing technology with new fervor. Bruce Sterling's review of William Gibson's *Neuromancer* (1983), reprinted on the flyleaf of the text, invites us to 'say goodbye to [our] old stale futures. ... An enthralling adventure story, as brilliant and coherent as a laser. THIS IS WHY SCIENCE FICTION WAS INVENTED!' Sterling is clearly not referring here to those futures produced by the 'legion' of cyberpunk precursors he describes in his rather self-congratulatory introduction to *Mirrorshades* (1986) – the 'idolized role models' like J.G. Ballard (xiv), the 'classic Hard' SF writers with their 'steely extrapolations' (x–xi), the New Wave 'independent explorers' of SF whose 'bible' was Alvin Toffler's *Third Wave* (xiii). Presumably their futures never stale. Only once in the introduction does Sterling suggest the gender of those producers of stale futures when he posits a connection between drugs, personal computers, and cyberpunk as 'definitive high-tech products': 'No counterculture Earth Mother gave us lysergic acid – it came from a Sandoz lab' (xiii).

Sterling's allusions to the influential fathers of SF, indicative of what Samuel Delany calls the male SF writers' 'endless, anxious search for fathers' ('Some' 9), betrays his need to forge a filiation with established (male) SF writers, to construct a form of legitimacy which, not insignificantly, manages to avoid mention of any potential mothers: the feminist SF writers (countercultural Earth Mothers?) of the previous decade.[5] But this construction of cyberpunk as the legitimate progeny of earlier SF is only part of Sterling's project in *Mirrorshades*. Far more overt is his relentless attempt to locate the 'loose generational nexus of ambitious young writers' (xi) of cyberpunk as 'disentangling SF from mainstream influence' (x), as, in effect, both marginalized and revolutionary. In other words, once he has unearthed adventurous fathers and constituted a satisfying filiation for cyberpunk writers, he can figure oedipal rebellion, reinterring the fathers as 'mainstream' and celebrating the sons as young turks. Sterling's desire to represent cyberpunk as a radical subgenre within SF – one which prompts him, in his introduction to Gibson's *Burning Chrome* (1986), to dismiss all of '70s SF as 'not much fun,' as in 'the doldrums,' 'dogmatic slumbers,' or 'hibernation' (1–2) – is rearticulated even more forcefully in the special *Mississippi Review* cyberpunk issue (1988) and in the Rucker-Wilson anthology, *Semiotext(e) SF* (1989).

Larry McCaffery, in his introduction to the *Mississippi Review* special issue, argues that

> cyberpunk seems to be the only art systematically dealing with the most crucial political, philosophical, moral issues of our day ... [issues] which are so massive, troubling, and profoundly disruptive [that they] cannot be dealt with by mainstream writers. (9)

John Shirley maintains, in the same issue, that cyberpunk writers like himself are indeed 'preparing the ground for a revolution' ('John' 58). Rudy Rucker and Peter Wilson, in the introduction to their anthology – their self-styled 'Einstein-Rosen wormhole into anarcho-lit history,' their 'godzilla-book to terrify the bourgeoisie' (11–12) – differ from Sterling in that they find cyberpunk's origins not in SF but in designer drug culture and punk rock: John Shirley's credentials, for example, are that he is a 'Genuine Punk' who has 'earned the right to a revolutionary stance by serving his season in the Lower Depths' (60). For Rucker and Wilson, cyberpunk is 'ideologically correct' (13) and insurrectionist in the face of the SF publishing industry's 'stodginess, neo-conservatism, big-bucks-mania' (12). These are grand claims. We might recall, however, that none of the cyberpunk writers has had much difficulty publishing his writing: four of Gibson's *Burning Chrome* stories and almost half of those in the *Mirrorshades* anthology, for example, appeared first in *Omni*, which is, as Richard Stokes points out, 'merely a technology oriented *Penthouse*' (29), despite Sterling's attempt to give it a revolutionary savor by praising *Omni*'s Ellen Datlow as 'a shades-packing sister in the vanguard of the ideologically correct' (*Mirrorshades* xv).[6]

Lest we be tempted to dismiss such inflated claims – that cyberpunk is 'ideologically correct,' that it is truly 'revolutionary' and subversive, that it is in the political vanguard, if not of art in general, then certainly of SF – as a form of professional, self-interested hype or a clever marketing strategy on the part of the SF publishing industry itself, we should remember that such claims are reiterated, albeit with a more sophisticated theoretical apparatus, by critics and academics outside SF coterie culture.[7] But is cyberpunk realizing a coherent political agenda? Is it indeed 'preparing the ground for a revolution'? If we are to take such promotion seriously as something other than hyperbolic advertisement, we need to examine cyberpunk contextually – not only as an SF 'movement' in the wake of, and contemporaneous with, particular forms of political, feminist SF, but also as a response to (or perhaps a reflection of) the Reaganite America of the '80s. Because 'cyberpunk' is, to a certain extent, a catch-all, convenient label for the work of a number of vaguely heterogeneous writers, I will confine much of my examination to the exemplary William Gibson, who is, according to Istvan Csicsery-Ronay, 'the one [cyberpunk] writer who is original and gifted enough to make the whole movement seem original and gifted' (269), and who is, according to the widely circulated mainstream magazine *Interview*, 'the king of cyberpunk' (cf Hamburg).

Sterling insists that cyberpunk is sexy social critique, legitimizing its political content through comparing it to the '60s counterculture, and retrieving many of its generic roots from '60s SF. And the *Washington Post* reviewer of *Mirrorshades*, among others, concurs, heralding cyberpunk as 'the first genuinely new movement in science fiction since the 1960s.' That Sterling harks back to the '60s counterculture to establish political connections for cyberpunk, and thereby implicitly reinstates the very lacuna the '70s feminist SF writers sought to expose in their exploration of gender relations, is itself provocative, particularly because it represents a peculiar avoidance of rather obvious and immediate political SF precursors. But his elision of specific '70s texts seems even more striking when we consider that William Gibson's novels, for example, inscribe quite overt revisions of the very texts which form the potentially (anxiety producing?) absent referent in Sterling's delineations of cyberpunk's origins. Russ's dauntingly powerful (and emasculating) Jael in *The Female Man*, for example, who describes matter-of-factly how her cybernetic boy-toy, Davy, can be 'turned off or on' as she desires, and how her nails and teeth have been cybernetically enhanced for use as lethal weapons against men, is effectively transformed into Molly, a 'razor-girl' who sells her talents (razor implanted finger-nails) to the highest bidder in Gibson's 'Johnny Mnemonic' and *Neuromancer*; or into Sarah, the dirtgirl/assassin who uses the cybernetic weasel implanted in her throat to kill with a kiss in Walter Jon Williams' *Hardwired* (1986). Explicit reworkings of an antecedent female character, Molly and Sarah are effectively depoliticized and sapped of any revolutionary energy: Jael had a political agenda, Sarah wants only to make enough money to get herself and her brother off earth; Molly's ambitions are to make as much money as possible – in 'Johnny Mnemonic' she is Molly 'Millions,' and in *Mona Lisa Overdrive* (1988) she refers to herself as an 'indie businesswoman' – and to bed Console Cowboy Case, the tough-guy hero of *Neuromancer*.

Jael is a killer, an allegorical figuration of feminist struggle, the active, ruthless and productive rage which eventually allows the utopic Whileaway to come into existence. Delany argues that Jael is Molly's fictional precursor, that strong female characters like Molly would not have been possible for cyberpunk writers without the earlier influence of feminist writers like Russ. 'Gibson's world,' he maintains, has 'neither Jeannines or Janets – only various Jael incarnations' ('Some' 8).[8] While Delany makes a good case for comparing Jael and Molly, his contention that strong female characters in cyberpunk owe their existence to the '70s feminists is considerably less convincing, particularly if we recall the relative paucity of strong female characters in cyberpunk. Rucker's *Software* (1982) and *Wetware* (1988), for example, contain almost no female characters, save for Della Taze and Darla in the later novel, both of whom are primarily surrogate mothers for Bopper progeny; Lewis Shiner's *Frontera* (1984) includes a Molly who is distinctly secondary in importance both to Kane and to her father, Reese; George Alec Effinger's *When Gravity Fails* (1986) concentrates, almost exclusively, on the entirely male-dominated world of the Budayeen where the few female characters – whether biologically or surgically female – are either wives or prostitutes; even Sterling's *Islands in the Net* (1988) presents Laura Webster, the central protagonist, as perpetually in need of rescue from prisons, would-be assassins, and terrorists. Indeed, from this perspective, Delany's comments about the strong female characters derived from '70s feminist SF seem to apply to relatively few works. Apart from Gibson's Molly and Williams' Sarah, we can perhaps include Pat Cadigan's Gina in 'Rock On' (1984) and her Alexandra Haas (Deadpan Allie) in *Mindplayers* (1987), but very few other strong female characters come to mind.

If Delany's observations, for all their overstatement, do succeed in rectifying the elision of feminist SF writers from the supposed influences on cyberpunk – his project is, in fact, to establish an alternative 'maternal' filiation for the latter – they nevertheless fail to acknowledge the rather crucial differences between Molly and Jael. When Delany contends, for instance, that the two have 'a similar harshness in their attitudes' ('Some' 8), the comparison works only for the Molly of 'Johnny Mnemonic' and *Neuromancer*.[9] In *Mona Lisa Overdrive* Molly/Sally, the tough entrepreneur/assassin/businesswoman, is distinctly softened, less harsh because she is effectively 'feminized' when Gibson positions her in relation to a child.[10] The unwilling bodyguard of Kumiko Yanaka, the motherless and innocent young daughter of a top Yakuza (Japanese mafia) warlord, Molly develops what can only be construed as a (natural? instinctive?) maternal protectiveness of, and affection for, Kumiko, a parental protectiveness which is not prompted, say, in Turner when he takes custody of Angie Mitchell in *Count Zero* (1987).[11] And Molly, with her street sense and ability to survive, represents a preferred alternative to Kumiko's helpless and sensitive mother who eventually commits suicide, unable to cope with the Yakuza world. Molly thus becomes the appropriate maternal model for Kumiko, teaching her the necessary tactics with which to survive and flourish, and, perhaps more importantly, facilitating her reconciliation with her father.

If Kumiko serves as a catalyst for Molly's feminizing or transformation into a quasi-maternal figure, she also provides the means through which Gibson presents an insider's view of the Japanese Yakuza as a 'family' organization. Kumiko, the 'Yak' daughter who represents a miniature extension of a vast and inscrutable corporate collective, functions as a temporary bridge between the world at large – the Sprawl desperados, matrix hackers, data pirates, and console cowboys, the self-consciously militant individualists – and the Yakuza, a threatening and powerful familial structure which, in fact, constitutes a kind of monolithic Japanese corporate entity. Indeed the Yakuza is the paradigm for all the other Japanese megacorporations which appear regularly in Gibson's texts: a collective construct which conflates the tight familial bonds of the Italian-American mafia with the equally tight employer–employee bonds of the frighteningly efficient Japanese industries. It is the latter which formed the subject of endless documentaries and business-magazine articles throughout the '80s because their corporate practice presented the most substantial threat to American-style capitalism America had yet experienced.[12]

American xenophobia and isolationism, particularly with regard to the Japanese scientific and economic invasion, manifested itself in the media through such scare tactics as Andy Rooney's piece on *60 Minutes* (Feb. 5, 1989), which portentously identified various historic American monuments as *Japanese owned*! And *48 Hours* presented a piece called 'America for Sale' (Dec. 29, 1988), in which various reporters, including Dan Rather, emphasized American objections to Japanese ownership of American real estate and industry. Amorphous Japanese collectives clearly posed a threat to the land of the free entrepreneurial spirit. This is surely the fear underlying the (defensive?) mockery and ridicule attending representations of Japanese tourists, travelling in tightly-knit groups, sporting extremely expensive, high-tech photographic equipment. If Canada as a whole did not reflect precisely the same degree of anti-Japanese paranoia being played out in America, British Columbia, Gibson's home, betrayed more conflict about Japanese investment than most parts of the country. In the early and mid-'80s, in the midst, that is, of British Columbian Premier William Bennett's open-door policy to Pacific Rim investment, reactions to Japanese tourists and potential investors were mixed: their infusion of capital into the flagging B.C. economy was indeed welcomed, and yet their actual ownership of luxury hotels, real estate, and various natural-resource companies (the forestry industry in particular) was both attacked and feared as being, ironically, merely a reenactment of past American practice.

If we examine Gibson's texts within the context of such conflicting interests, we see the degree to which he deliberately avoids any form of simplistic anti-Japanese paranoia or its attendant racism and ethnocentrism. And yet Gibson's Japanese conglomerates, in their collective and familial practice, nevertheless form the implicit antagonistic counterpoint to the individualist heroes. The bad guys in Gibson are, after all, the megacorporations – Ono Sendai, Hosaka, Sanyo, Hitachi, Fuji Electric. The good guys are the anarchic, individualistic, and entrepreneurial American heroes: independent mercenaries and 'corporation extraction experts' like Turner, console cowboys like Case, Bobby Newmark, Gentry, Tick, and the crew at the Gentleman Loser who jack in and out of the global computer matrix with unparalleled mastery. In Williams' *Angel Station* (1989), Bossrider Ubu traverses the galaxy, roping in black holes. In Sterling's *Islands in the Net*, American Jonathan Gresham, the self-styled 'post-industrial tribal anarchist' (388), rides his 'iron camel' through the 'bad and wild' African Sahara – one of the few places free of the global Net – and eventually saves the hapless but earnest Laura Webster. The cowboys in Gibson, Williams, and Sterling are heroes who represent, as Williams suggests in *Hardwired*, the 'last free Americans, on the last high road' (10). It seems telling that the American icon of the cowboy, realized so strongly in Reaganite cowboyism, the quintessence of the maverick reactionary, should form the central heroic iconography in cyberpunk.

Cyberpunk's fascination with and energetic figuration of technology represents the American cowboy as simultaneously embattled and empowered. In '80s America the Japanese megacorporations did dominate the technological market, but the cowboy's freedom and ingenuity allow him to compete purely on the level of mastery. The terms of such a competition – Japanese pragmatism and mass production versus American innovation and ingenuity – seem precisely analogous to those of a familiar American consolatory fiction: that free enterprise and privately funded research and development in science and technology have produced in America the most important technological innovations of the 20th century, innovations which the Japanese have simply taken, pirated, and mass produced, thus undercutting the very American market which encouraged their discovery and making it financially difficult for the neophyte technological wizards to get corporate funding. In *Interview*'s special 'Future' issue (1988), almost adjacent to Victoria Hamburg's interview with Gibson, there appeared an article titled 'Made in Japan,' which confirmed for the American readership that the Japanese did not 'initiat[e] new ideas' (Natsume, 32) and reassured it about the benign nature of the new products coming out of Japan: micro-thin televisions, special low-water-consumption washing machines, camcorders with RAM cards, auto-translation machines – non-essential but nice, unthreatening appliances.[13] Computer and technological innovation would still come from American silicon valleys, would still be, by implication, 'Made in America.' In Gibson's novels the console cowboys use expensive Hosaka and Ono Sendai cyberspace decks, but such mass-produced technology is always customized and enhanced, its performance and capabilities augmented by the cowboys' more inventive, finer ingenuity.

In effect, the exceptionally talented, very masculine hero of cyberpunk, with specially modified (Americanized) Japanese equipment, can beat the Japanese at their own game, pitting his powerful individualism against the collective, domesticated, feminized, and therefore impenetrable and almost unassailable Japanese 'family' corporations. After all, in the world of the microchip, small is potentially powerful.[14] If, however, these demonized/feminized Japanese corporate collectives form the consistent and implicit counterpoints to the individualized heroes in Gibson's novels, representing as they do a quasi-new industrial threat to America, they frequently pale in comparison to the families-gone-bad of an old-world, European, aristocratic order: the most obvious example is Tessier-Ashpool in *Neuromancer*, the 'very quiet, very eccentric first-generation high-orbit family, run like a corporation' (75), which has the ability to clone itself endlessly, to keep clones in cryogenic storage until they are needed to replace family-member clones who have died, to copulate with and kill its members without danger of legal interference. And Tessier-Ashpool has the mad, female 3Jane at its helm, which seems to cement its construction as a demonic parody of the 'proper' nuclear American family and to suggest its peculiar connotative link with the domesticated, multi-generational Japanese corporations, run like a family. The individual cowboy hero, then, rarely combats an individual villain; rather he employs his particular performative mastery against a demonized and feminized Other, represented implicitly in the threatening Japanese conglomerates and explicitly in the aberrant familial forms of an old European aristocracy, the inbred and mad Tessier-Ashpool clan.[15] And yet the significant mastery of Case, Bobby, Gentry, Tick, and others, the mastery they must deploy against such feminized collectivity is never quite secure, not necessarily because they have superior rivals, but because of the multifaceted and mystical nature of the matrix itself.

The computer matrix, a construct culturally associated with the masculine world of logic and scientific wizardry, could easily constitute the space of the homoerotic. But it doesn't. Oddly enough, in cyberpunk fiction only the posturing and preening at the cowboy bars comprise the locus of the homoerotic; the matrix itself is figured as feminine space. The console cowboys may 'jack in,' but they are constantly in danger of hitting ICE (Intrusion Counter-measures Electronics), a sort of metaphoric hymeneal membrane which can kill them if they don't successfully 'eat through it' with extremely sophisticated contraband hacking equipment in order to 'penetrate' the data systems of such organizations as T-A (Tessier-Ashpool). Dixie Flatline tells Case in *Neuromancer*, for example, that the Kuang ICE breaker is different from others: 'This ain't bore and inject, it's more like we interface with the ice so slow, the ice doesn't feel it. The face of the Kuang logics kinda sleazes up to the target' (169). And Flatline points out to Case that should they approach the T-A ICE without the Kuang, 'it'll be tellin' the boys in the T-A boardroom the size of your shoes and how long your dick is' (167). Obviously not long enough if Case allows himself to get caught in the ICE.

If the data constructs of the domestic or familial corporations are meta-phorically feminized, protected as they are by feminine counter-intrusion membranes that resist 'bor[ing] and interject[ing],' so too are the interspacial zones in the matrix. Any (masculine) scientific and technological purity the computer matrix might once have had has been violated, invaded: in *Count Zero*, virus software programs have infected the matrix; it's 'full of mambos 'n shit.' Finn tells Bobby, 'there's things out there. Ghosts, voices. . . . Oceans had mermaids, all that shit, and we had a sea of silicon' (*Count Zero* 119). The matrix has a generational life of its own, but one configured by the likes of 3Jane, Slide, Angie Mitchell, and Mamman Brigitte, a voodoo feminine AI (Artificial Intelligence). And even the quasi-masculine AIs in *Neuromancer*, Wintermute and Neuromancer, who eventually fuse and expand into the matrix, do so, not because they have their own autonomous desires, but because the original mother, Marie-France Tessier, has deliberately reconfigured them to unite and destroy the 'sham immortality' (269) of her husband Ashpool's empire. When Neuromancer tells Case his name, he explains: 'Neuromancer. . . . The lane to the land of the dead. . . . Marie-France, my lady, she prepared this road' (243); and thus Marie-France, in her ability to control and direct (even posthumously) the parameters of matrix inhabitation, seems a ghostly prefiguration of Mamman Brigitte, the Queen of the Dead.

Not only is the matrix itself mystified and feminized, but so are the means of entering it. Lucas tells Bobby in *Count Zero* that certain women, voodoo 'priestesses' called 'horses,' are metaphorical cyberspace decks: 'Think of Jackie as a deck, Bobby, a cyberspace deck, a very pretty one with nice ankles. . . . Danbala [the snake] slots into the Jackie deck' (114). The cowboys can 'mount' and 'ride' such horses into the matrix. In Williams' *Angel Station*, Beautiful Maria is a cybernetic 'witch' who, like Angie Mitchell in *Count Zero* and *Mona Lisa Overdrive*, has that specialized mysticism which enables her to negotiate and manoeuvre within the complex eroticized data matrix. Angie can 'dream cyberspace, as though the neon gridlines of the matrix waited for her behind her eyelids' (*Mona* 48). When Beautiful Maria navigates the Runaway, flares 'pattern[] her belly,' 'Magnetic storms howl[] in her throat,' 'Electron awareness pour[s] into her body' (*Angel* 105). The cowboys have to 'interface' with the matrix through 'slotting into' feminized cyberspace decks; certain females, however, require no such mediation: they are already, by implication, a part of it.

Constituting both what is fascinating and generative about the matrix itself and the means of accessing its secrets, the feminine is effectively the 'soft' ware, the fantasy (and world) that exists beyond the 'hard' ware of the actual technological achievements realized in the silicon chip. Finn points out to Bobby that 'anyone who jacks in knows, fucking *knows* [the matrix] is a whole universe' (*Count* 119). This feminized universe is inhabited by ghosts, not simply ghostly personality constructs, but textual ghosts. In *Mona Lisa Overdrive*, Bobby tells Angie 'about a general consensus among the old cowboys that there had been a day when things had changed, although there was disagreement as to how and when. When It Changed, they called it. . . .'

(129), referring, in effect, to when the matrix became an inhabited feminized world. The implication of this dangerous 'soft' fantasy-world of the matrix is even more pronounced when we consider that 'When It Changed' is also the title of Joanna Russ's Nebula Award-winning story upon which she based *The Female Man*. The 'change' in Russ's story is initiated by the arrival of several (decidedly sexist) men in Whileaway, which had previously been an entirely female society, made so earlier by a gender-specific plague. In *Mona Lisa Overdrive* the change that Gentry, Finn, and Bobby locate is, in fact, also gender specific: the uncontrollable feminizing of the matrix, the uncheckable transformation of viral software technology into a feminine Other, complete with ghosts and the prevailing influences of Mamman Brigitte and 3Jane – respectively the Queen of the Dead and the dead queen of the Tessier-Ashpool empire.

If Gibson posits the obverse of Russ's patriarchal imperialism as a kind of female infection and (viral) takeover of original masculine space, he also suggests that the matrix turf can potentially be won back, reconquered. And this is curiously like a male version of Gearhart's *The Wanderground*, in which Earth and her daughters, the Hill Women, reclaim, quite successfully, the territories that men had once conquered; they work their reclamation through ensuring that masculine technology will not operate outside a certain, very constrained area. When Darko Suvin argues, then, that Gibson presents the matrix as a utopia, he seems not to recognize how very gendered the matrix is; however, one could supplement his contention by suggesting that Gibson is indeed gesturing towards a potentially utopian matrix, a macho, 'privatized utopia' (Suvin 45) which can be brought about by the energy and vision of such heroes as Bobby Newmark.

Gentry is fascinated by Bobby Newmark's aleph in *Mona Lisa Overdrive* because it is nothing less than an insulated territory carved out in the matrix, a secure homestead, technologically protected from the intrusion of the matrix wilds, a benign 'approximation of the matrix' (*Mona* 307). Is Gibson simply invoking traditional tropes of American imperial or colonizing fictions – in which the valiant and resilient Western homesteader wins back civilization from the savage feminine wilds (the virgin land)? Or is he presenting a particularly unsavory and reactionary '80s working of those tropes, an aggressive anti-feminist backlash which figured feminists as emasculating harridans and ball-busters, a back-lash which surfaced quite overtly in the media. In the early '80s, for example, the Toronto Transit Commission was forced, by public outcry, to remove an advertising placard for Virgin jeans that displayed a prone female figure, naked except for a pair of jeans, with a red knife/lipstick slash down her nude back, and the single caption 'SINNER' below. Equally symptomatic of an embattled masculinity is the poster Peter Fitting found in San Francisco in 1986, appealing to 'All Men': 'DON'T SUBMIT TO A FEMINIST-LESBIAN TAKEOVER. RESIST!!!' (Fitting 17). Whether or not we choose to see Gibson's configuration of the frightening feminine matrix as an extension of particular anti-feminist politics in the '80s, we are still left with the fact that his male heroes play out their mastery within that specific locus of femininity; their very masculinity is constituted by their success both within and against it.

'Soft' fantasy has indeed a dangerous appeal, testing as it does the mettle and performance of the would-be masters. While anyone with a set of 'trodes' in Gibson's texts can jack into the matrix, only the heroes have the specialized, though provisional, mastery which allows them to negotiate their way through it and, perhaps more importantly, to exit it when their mastery becomes unstable. The cowboys' proficiency does not, however, ensure their success: Slide, from within the matrix in *Count Zero*, for example, kills Conroy electronically while he is jacked into a telephone line; Angie Mitchell, simply by imagining the configurations of cyberspace behind her eyelids, saves Bobby from dying after he hits black ICE; and the matrix-dwelling 3Jane dislocates cowboy Tick's shoulder, almost killing him, in *Mona Lisa Overdrive*. Gibson has indeed constructed the soft world of fantasy as a sort of phallic mother: erotic, feminine, and potentially lethal. If the cowboy heroes fail to perform brilliantly, they will be 'flatlined' or have their jacks melted off, whichever is worse.

But old cowboys never really die: they ride out of the sunset to become presidents, then ride back to the ranch to embrace a domestic life. In the Reaganite revision of the cowboy, he is no longer the solitary, autonomous hero, but one whose eventual home is constituted by wife and range. Console cowboy Case in *Neuromancer* settles down and has four children; Corporate Extraction expert Turner in *Count Zero* finds a wife, has a son, and lives a rustic life on his brother's ranch; console cowboy Bobby Newmark, in *Count Zero* and *Mona Lisa Overdrive*, ends up in an aleph-secured little matrix-homestead with Angie Mitchell; Panzerboy pilot Cowboy in *Hardwired*, recovering on a 'sweaty Nevada dude ranch' (340), anticipates 'eas[ing] carefully into peace' with Sarah (337). These are the couples and families which represent the correct form of domesticity, correct insofar as they are contrasted with the antagonistic, exclusive and threatening domesticity of the Japanese 'family' corporations or the mad, inbred, European, aristocratic Tessier-Ashpool corporate family. The potentially emasculating feminine matrix is replaced by the unthreatening wife, biological reproduction replaces replication, and the triumphantly masculine hero returns to a romanticized rural life: the garden stays intact despite the sophisticated technological warfare waged outside it.[16]

Where then, we might well ask ourselves, is revolutionary potential articulated in cyberpunk fiction – apart from in its writers' own self-promotion and in the introductions to their anthologies? Or is it articulated at all? The Reagan '80s, which saw the rise of the moral majority to unprecedented heights, the revocation of many of the feminist advances of the '70s, and the curtailing of most forms of social assistance, realized a coherent return to an idealization of the nuclear family and the simple rural life. Technology, particularly cheap

Japanese-disseminated technology, wouldn't save America; it was dangerous: unsuspecting American kids could be killed by oncoming (Japanese?) cars because they were wearing Sony walkmans. As Gibson says, without a trace of ironic understatement, in the *Village Voice*: 'The potential for technology of oppression seems very strong' (Carpenter 38).

But Gibson's masculine heroes are masterful because they use a feminized technology for their own ends, or better, because their masculinity is constituted by their ability to 'sleaze up to a target' and 'bore and inject' into it without allowing it to find out the 'size of their dicks' in advance – their facility, in short, as metaphoric rapists. And their success is celebrated as a form of triumph to such an extent that we can hardly view Gibson's texts as deliberately dystopian, much as we might flinch at the implications of such triumphs. Indeed to read Gibson's novels as extrapolations to, or fictional figurations of, a particularly untenable, hideous and ugly future, that is, as some sort of dystopian fiction, seems to me an active misreading of them, and one, I would suggest, indulged in by various critics. Andrew Ross, for example, calls cyberpunk 'survivalist,' a 'new dystopian realism' which presents a future 'governed by the dark imagination of technological dystopias' (432–33); and Pam Rosenthal suggests that the future in cyberpunk is 'our world, gotten worse, gotten more uncomfortable, inhospitable, dangerous, and thrilling' (85). In contrast, Gibson himself comments that his books are 'optimistic,' that his future 'would be a neat place to visit,' that the thing he likes about his future is that 'there's a sense of bustling commerce. There seems to be a lot of money around. It's not very evenly distributed, but people are still doing business' (Hamburg 84).

Ross and Rosenthal, in their rush to justify the politically correct content of cyberpunk, need to construct it as dystopian, as monitory or cautionary fiction. In my estimation and surely in that of 'pessimists' (Suvin 50) like Suvin and Csicsery-Ronay, they are grasping at straws; they are also ignoring (suppressing?) some actual dystopian novels, that is, SF novels with a fairly overt political agenda, which were written in the '80s at exactly the same time as cyberpunk. Elgin's *Native Tongue* and *The Judas Rose*, Atwood's *The Handmaid's Tale*, and Fairbairns' *Benefits* all posited dystopian futures in which women's rights had been extinguished altogether, in which women were valued only as breeders, in which the moral majority had ascended to establish tyrannical theocracies, in which technology had become the sophisticated means by which women could be successfully oppressed once again. These writers, in fact, represent in their fiction what Gibson observes in passing but does not really articulate in his texts: that the 'potential for technology of oppression' *is* (not 'seems') very strong.[17] Elgin, Atwood, and Fairbairns all provide active critiques of political trends which surfaced in the early '80s; and thus they differ quite drastically from Gibson, who hopes that the future will 'be as much like the present as possible' ('King' 86). And their works proffer a critical contrast to cyberpunk's optimism with its manly heroes who, given the appropriate arena ('bustling commerce'), can achieve success and clinch their masculinity.

Gibson ultimately celebrates the same initiative and ingenuity which has always characterized the American hero, indicating that, within his chosen models of a relentlessly capitalist future, a paradigmatic American heroism can be rearticulated virtually uncritically: past, present, and future are the same. As Richard Stokes points out, there is, in the apparent street punks of cyberpunk, 'an overwhelming desire for upward mobility' (29). The powerful Japanese megacorporations serve only as worthy antagonists for the American hero – and the hero will inevitably triumph; they do not present the arena for the hero's potential subversion of or assault on them, for it is the established power structures themselves which provide the means by which he can succeed. In Gibson's fiction there is therefore absolutely no critique of corporate power, no possibility that it will be shaken or assaulted by heroes who are entirely part of the system and who profit by their mastery within it, regardless of their ostensible marginalization and their posturings about constituting some form of counterculture. As Stokes and Csicsery-Ronay quite rightly point out, the idea that computer cowboys could ever represent a form of alienated counter-culture is almost laughable; for computers are so intrinsically a part of the corporate system that no one working within them, especially not the hired guns in Gibson's novels, who are bought and sold by corporations and act as the very tools of corporate competition, could successfully pose as part of a counterculture, even if they were sporting mohawks and mirrorshades.

Unlike Stokes, Suvin, and Csicsery-Ronay, Rosenthal approves of the apparent politics in cyberpunk; her reservations are minor:

> If one wanted to criticize cyberpunk for its bad politics ... one would correctly point to the desperate loneliness it portrays. But I'd prefer simply to observe that, painful as it may be to think about, cyberpunk is onto something real here. We live in lonely times. (100)

Cowboys, capitalist entrepreneurs, American heroes – these are lonely, auton-omous heroes *by choice*, and they are, undeniably, mythologized in American culture; but how 'real' are they? And if they indeed represent the 'times,' doesn't this raise some rather crucial questions about what exactly *is* being mythologized as heroic in the '80s? I am afraid it would be rather difficult to locate cyberpunk's 'good politics,' despite Rosenthal's contentions that cyber-punk writers are on the cutting edge of interesting politics, that they 'can't wait until all the returns are in, until political and economic theorists agree upon new models for the forms underlying social life' (80). Nor, in fact, is it really possible to locate that supposedly 'revolutionary' agenda cyberpunk is touted so regularly as having. This is not to say that Gibson and other cyberpunk writers are not excellent craftsmen, interesting stylistically, and capable of constructing complex and gripping plots. But the conflation of aesthetic appreciation and good politics surfaces, in certain critics, as a form of leftist wish-fulfillment; that, in other words, if one likes the fiction, it must necessarily involve the articulation of a perceptible, revolutionary project. To make such an argument, however, is to remain effectively blind to what cyberpunk does represent, particularly when one contrasts it with other forms of political SF. The political (or even revolutionary) potential for SF, realized so strongly in '70s feminist SF, is relegated in Gibson's cyberpunk to a form of scary feminized software; his fiction creates an alternative, attractive, but hallucinatory world which allows not only a reassertion of male mastery but a virtual celebration of a kind of primal masculinity. Political potential is indeed lost in the iconography of all that Reagan himself represented.

For all its stylish allusions to popular culture – to punk rock, to designer drugs, to cult cinema, to street slang and computer-hacker (counter?) culture – cyberpunk fiction is, in the end, not radical at all. Its slickness and apparent subversiveness conceal a complicity with '80s conservatism which is perhaps confirmed by the astonishing acceptance of the genre by such publications as *The Wall Street Journal*, *The Washington Post*, and *The New York Times*, and by the ease with which it can be accommodated and applauded in the glossy pages of such American mainstream (boys') magazines as *Omni*.[18] Sterling argues in *Mirrorshades* that the cyberpunk movement 'is not an invasion but a modern reform' (xv). 'Reforming' what, we might well ask? Certainly not SF's gender politics. Or maybe Sterling is right; maybe cyberpunk *is* a 'modern reform.' But, like so many other 'reforms' in the Reagan/Mulroney era, it involves an unsavory and regressive positioning of us. Yes, 'us' – those (other SF readers) not properly equipped to appreciate Shirley's analogies, or to 'bore and inject,' or to have our identities constituted and verified in multinational boardrooms through the measure of the length of our dicks.

NOTES

1. This article is a revised version of '"Jacking In" to the Matrix: Metaphors of Male Performance Anxiety in Cyberpunk Fiction,' a paper delivered in the special session on 'Political Directions in Popular Culture' at the May 1991 meeting of the Association of Canadian University Teachers of English in Kingston, Ontario.
2. Along the same lines as Parley J. Cooper's novel are Charles Maine's *Aleph* (1972), Edmund Cooper's *Who Needs Men?* (1972), and Sam Merwin, Jr's *Chauvinisto* (1976). In her '*Amor Vincit Foeminam*: The Battle of the Sexes in Science Fiction,' Russ describes this trend in male SF, examining in detail *The Feminists* and a number of other novels.
3. Although Le Guin's Urras is not precisely a 'near-present' world, its economic, social, cultural, and political models are based on existing and familiar 20th-century models.
4. There is, of course, an on-going debate about the term 'cyberpunk,' and what/who it represents. Bruce Sterling's response, in *Interzone*, to Lewis Shiner's 'Confessions of an Ex-cyberpunk Writer' (*The New York Times*, Jan. 7, 1991), makes clear what William Gibson has been saying for years: that 'cyberpunk' is a label with which many of the writers who have been dubbed part of the 'movement' are uncomfortable. That said, writers like Gibson, Sterling, Shiner, John Shirley, and Rudy Rucker have been reasonably content to have their novels marketed as 'cyberpunk' fiction, and their more recent disavowal of connections to the label comes as much from

their discomfort with the amount of imitative cyberpunk which has glutted the market – what one SF publisher calls the 'cyberpunk sludge' – as from anything else. As Darko Suvin suggests, 'cyberpunk' may well be not so much a movement as the result of 'a couple of expert PR-men (most prominently Sterling himself) who know full well the commercial value of an instantly recognizable label, and are sticking one onto disparate products' (50).

5. Several critics have noted this elision. See, for example, Delany, who describes Sterling's 'obliteration' of the '70s feminist writers in his introduction to *Burning Chrome* ('Some' 8), and Gregory Benford, who points out that the 'feminist perspectives SF pioneered in literature are here [Sterling's introduction to *Mirrorshades*] largely brushed aside' (22). See also Joan Gordon's 'Yin and Yang Duke It Out,' in which she remarks on the absence of reference to women SF writers in Sterling's introductions; she calls it, however, only 'an oversight' (38).

6. In a conversation with Edward Bryant, Datlow makes clear the limitations on her freedom to select material for *Omni*: she publishes what she likes, as 'long as I don't get in trouble at *Omni* from the people who control me' (59).

7. See, for recent examples, Pam Rosenthal's 'Jacked In: Fordism, Cyberpunk, Marxism' and Andrew Ross's 'Getting Out of the Gernsback Continuum.'

8. In *The Female Man* Russ presents four principal female characters: Jeannine, living in a perpetual Great Depression, represents the malleable, soft, would-be '50s housewife; Joanna, living in the present of the '70s, is intelligent and intellectual but still expected to find a man to marry; Janet is a product of the futuristic, all-female Whileaway; and Jael is located temporally in a near future where feminist struggle has become an actual war between men and women. Delany's comment suggests that Gibson, unlike Russ, is not interested in representing temporal/political transitions; rather he presents the near future in isolation.

9. Delany may not, in fact, have read the later novel: his interview with Tatsumi was published in March 1988 and *Mona Lisa Overdrive* appeared in the fall of the same year.

10. Nickianne Moody points out that cyberpunk, like other technology-based SF, has a problem with active female characters: such a character either recovers her full sexuality when she 'retires from the scene' because of pregnancy or domestic demands or 'tactfully removes herself when relationships get serious' (27). Moody argues that such SF 'cannot move from the base which allows us to comprehend it – the present. And such problems in the 1980s are far from solved as far as post-marital roles are concerned' (27).

11. Interestingly enough, these assumptions about the maternal instinct as natural which we seem to find suggested in Gibson's representations of Molly and later of Angie at the end of *Mona Lisa Overdrive* are precisely the same assumptions at work in Cooper's *The Feminists*. Cooper's novel ends with the mayor's having to 'choose between her loyalty to the Feminists and her role as the antiquated woman of her youth. Motherhood had won and her feeling of tranquility told her the decision had been the right one' (182). As the driver explains to the mayor's son, Keith, at the end: 'When she had to face the choice . . . the mayor discovered that she possessed the major feminine weakness she despised in others. Before she was a Feminist, she was a mother!' (187).

12. Suvin contends that Gibson's use of the *zaibatsu* model for the corporations of the future is justified, its logic 'centered on how strangely and yet peculiarly appropriate Japanese feudal-style capitalism is as an analog or indeed ideal template for the new feudalism of present-day corporate monopolies: where the history of capitalism, born out of popular merchant-adventurer revolt against the old sessile feudalism, has come full circle' (43).

13. There is a rather interesting parallel here between *Interview*'s 'Made in Japan,' which displays at once a form of American anxiety and a means by which it can be defused, and a book which was first serialized and then published almost exactly a

century earlier in England: E.E. Williams' *Made in Germany* (1896). Williams'
book was a clear attempt to strike fear in the hearts of the English, since it simply
detailed the frightening degree to which they relied upon imported products, from
Germany in particular, products which ranged from the strictly domestic and
commonplace to the technological. *Made in Germany*, unlike 'Made in Japan,'
however, did not articulate consolation or attempt to defuse fear: it was a 'call to
arms,' designed to create the very paranoia that 'Made in Japan' attempts to allay.

14. See Donna Haraway's 'A Manifesto for Cyborgs' for her discussion of microelec-
tronics, and of power as contiguous with reduced size: 'Miniaturization,' writes
Haraway, 'has turned out to be about power; small is not so much beautiful as pre-
eminently dangerous' (70).

15. Csicsery-Ronay argues, quite rightly, that the 'villains come from the human
corporate world, who use their great technical resources to create beings that
program out the glitches of the human: the Company in *Alien* seeking a perfect war-
machine; the consortium in *Robocop* constructing the perfect crime-fighter; in
Blade Runner, Terrell Industries, who have created the Nexus-4 [sic] replicants, the
perfect servant-worker-warrior' (275). But I would elaborate a bit here and suggest
that the relationships established, at least verbally, are strikingly familial: in *Alien*
and *Aliens* the Company wants the war-machine born, quite literally, out of the
stomachs of human beings in a demonic form of maternity, and in the latter film the
central antagonism is between the alien family and the human family of Ripley and
Newt; in *Robocop* the Consortium Head's final words to Robocop are 'fine
shootin', *son*;' in *Blade Runner* Roy addresses Terrell as his maker and 'Father.'

16. See Haraway on the significance of the Edenic garden – The romantic 'myth of
original unity, fullness, bliss' (67) – in relation to technology and cybernetics (66–
69).

17. As Csicsery-Ronay points out: 'All of the ambivalent solutions of cyberpunk works
are instances/myths of bad faith, since they completely ignore the question of
whether some political controls over technology are desirable, if not exactly
possible' (277).

18. Delany bemoans the fact that supporters of cyberpunk have been, by and large,
conservative: 'The conservative streak in the range of sympathetic cyberpunk
criticism is disturbing. That streak is most likely the one through which the
movement will be co-opted to support the most stationary of status quos' ('Some'
10). But, as I have argued, this conservatism seems perfectly consistent with the
politics that cyberpunk does present.

WORKS CITED

Atwood, Margaret. *The Handmaid's Tale*. Toronto: McClelland and Stewart, 1985.
Benford, Gregory. 'Is Something Going On?' *Mississippi Review* 16: 18–23, #47/48,
1988.
Cadigan, Pat. *Mindplayers*. NY: Bantam Spectra, 1987.
Cadigan, Pat. 'Rock On.' *Mirrorshades*. Ed. Bruce Sterling, q.v. 34–42.
Carpenter, Teresa. 'Slouching Toward Cyberspace: In Search of Virtual Reality, the
Nerdstock Nation Convenes.' *Village Voice*, March 6–12, 1991, 34–40.
Csicsery-Ronay, Istvan. 'Cyberpunk and Neuromanticism.' *Mississippi Review* 16:
266–78, #47/48, 1988.
Cooper, Parley J. *The Feminists*. NY: Pinnacle Books, 1971.
Datlow, Ellen. 'A Conversation with Ellen Datlow and Edward Bryant.' *Science Fiction
Eye* 1: 55–65, 88, August 1988.
Delany, Samuel. 'Is Cyberpunk a Good Thing or a Bad Thing?' *Mississippi Review* 16:
28–35, #47/48, 1988.
Delany, Samuel. 'Some *Real* Mothers: An Interview with Samuel Delany.' By Takayuki
Tatsumi. *Science Fiction Eye* 1: 5–11, March 1988.

Effinger, George Alec. *When Gravity Fails*. NY: Bantam, 1988.
Elgin, Suzette Haden. *The Judas Rose*. NY: DAW Books, 1986.
Elgin, Suzette Haden. *Native Tongue*. NY: DAW Books, 1984.
Fairbairns, Zoë. *Benefits*. London: Virago, 1979.
Fitting, Peter. 'The Decline of the Feminist Utopian Novel.' *Border/Lines* 7/8: 17–19, Spring 1987.
Gearhart, Sally Miller. *The Wanderground: Stories of the Hill Women*. Boston: Alyson Pubs., 1979.
Gibson, William. *Burning Chrome*. NY: Arbor House, 1986.
Gibson, William. *Count Zero*. NY: Ace, 1987.
Gibson, William. *Mona Lisa Overdrive*. NY: Bantam, 1988.
Gibson, William. *Neuromancer*. NY: Ace, 1984.
Gordon, Joan. 'Yin and Yang Duke It Out: Is Cyberpunk Feminism's New Age?' *Science Fiction Eye* 2: 37–39, February 1990.
Hamburg, Victoria. 'The King of Cyberpunk.' *Interview* 19: 84–86, 91, Jan. 1988.
Haraway, Donna. 'A Manifesto for Cyborgs: Science, Technology, and Socialist Feminism in the 1980s.' *Socialist Review* 15: 65–107, April 1985.
McCaffery, Larry. 'The Desert of the Real: The Cyberpunk Controversy.' *Mississippi Review* 16: 7–15, #47/48, 1988.
Moody, Nickianne. 'Cyberdykes in SF Comics and Elsewhere.' *New Moon* 1: 27–28, May 1987.
Natsume, Toshiaki. 'Made in Japan.' Trans. Steve Hogan. *Interview* 19: 32, Jan. 1988.
Pfaelzer, Jean. 'The Changing of the Avant Garde: The Feminist Utopia.' SFS 15: 282–94, #46, Nov. 1988.
Rosenthal, Pam. 'Jacked In: Fordism, Cyberpunk, Marxism.' *Socialist Review* 21: 79–103, Jan.–March, 1991.
Ross, Andrew. 'Getting Out of the Gernsback Continuum.' *Critical Inquiry* 17: 411–33, Winter 1991.
Rucker, Rudy. *Software*. NY: Ace Books, 1982.
Rucker, Rudy. *Wetware*. NY: Avon Books, 1988.
Rucker, Rudy, Peter Lamborn Wilson, and Robert Anton Wilson, eds. *Semiotext(e) SF*. NY: Autonomedia 1989.
Russ, Joanna. '*Armor Vincit Foeminam*: The Battle of the Sexes in Science Fiction.' SFS 7: 2–15, #20, March 1980.
Russ, Joanna. *The Female Man*. NY: Bantam, 1975.
Russ, Joanna. 'When It Changed.' *Again, Dangerous Visions*. Ed. Harlan Ellison. NY: Doubleday, 1972. 229–41.
Shirley, John. *Eclipse*. NY: Blue Jay, 1985.
Shirley, John. 'John Shirley.' *Mississippi Review* 16: 58, #47/48, 1988.
Sterling, Bruce. 'Cyberpunk in the Nineties.' *Interzone* 48: 39–41, June 1991.
Sterling, Bruce. Introduction. *Burning Chrome*. By William Gibson, q.v. 1–5.
Sterling, Bruce. *Islands in the Net*. NY: Ace, 1988.
Sterling, Bruce, ed. *Mirrorshades: The Cyberpunk Anthology*. NY: Ace, 1988.
Stokes, Richard. 'Same Old Thing in Brand New Drag.' *New Moon* 1: 28–30, May 1987.
Suvin, Darko. 'On Gibson and Cyberpunk SF.' *Foundation* 46: 40–51, Fall 1989.
Williams, E.E. *Made in Germany*. 1897. Brighton, 1973.
Williams, Walter Jon. *Angel Station*. NY: TOR Books, 1989.
Williams, Walter Jon. *Hardwired*. NY: TOR Books, 1989.

11

MEAT PUPPETS OR ROBOPATHS? CYBERPUNK AND THE QUESTION OF EMBODIMENT

Thomas Foster

And if all the world is computers
It doesn't matter what your sex is
Long as you remember
If you're a boy or a girl
                (Bernadette Mayer, *The Formal Field of Kissing*[1])

Why jack off when you can jack in?
                (The character Plughead in the film *Circuitry Man*)

This essay has a dual purpose. I intend to read David J. Skal's novel *Antibodies* (1989) as exemplifying one typical response of cultural critics to the cyberpunk movement in science fiction and specifically to cyberpunk's postmodern redefinition of embodiment.[2] *Antibodies* implicitly critiques the oscillation in cyberpunk texts between a biological-determinist view of the body and a turn to technological and cybernetic means in order to escape such determination, an oscillation that is generally gender-coded in the paradigm texts of cyberpunk, especially William Gibson's novel *Neuromancer* (1984).[3] This oscillation is figured on the one hand by the 'meat puppet,' to use the term applied to persons who are confined to their organic bodies in *Neuromancer*, and on the other hand by the figure of the 'robopath,' people who believe they are robots trapped in human bodies, to use a term from *Antibodies*. However, in its reply to the cyberpunk devaluation of the body as 'meat,' Skal's book is more properly described as a diagnosis than a critique, since one of the main characters is a therapist who specializes in what he regards as culturally induced pathological desires to transcend the limitations of the natural body by means of mechanical prostheses.[4]

This reading of *Antibodies* suggests that cyberpunk locates itself in the undecidability between the literal and rhetorical readings of the question posed by the character Plughead in Steven Lovy's 1989 film *Circuitry Man*: 'Why jack off when you can jack in?' Why remain dependent on an organic body when access to the extended nervous system of a computer network is available? However, I will argue that for cyberpunk this question is not simply rhetorical, its answer understood in advance. Cyberpunk does not simply devalue the body but instead also foregrounds and interrogates the value and consequences of inhabiting bodies. As Bernadette Mayer reminds her postmodern readership, cultural critics have a responsibility to remain suspicious of claims that 'it doesn't matter what your sex is.' Even in an environment where the gendered body seems to have been entirely denaturalized by media simulacra, reproductive technologies, and the possibilities opened by genetic engineering and surgical modification, we cannot help but 'remember/if [we're] a boy or a girl,' a necessity that is particularly acute for white male critics like myself.

But this same passage by Mayer implies an equal responsibility to understand how new technologies may have a place in and help to create new cultural logics that can alter the construction of cultural identities like gender, race, and sexual orientation. The second goal of this essay will therefore be to suggest how cyberpunk texts also call into question distinctions between mind and body, human and machine, the straight white male self and its others, even while they remain dependent on those distinctions to some degree. Even in *Antibodies*, Skal's therapist attempts to cure his patients' robopathologies by reasserting the categories of 'natural' bodily experience in ways that the novel clearly depicts as abusive. The techniques employed include physical and sexual violence, used as a form of 'somatic shock therapy' to return patients to a sense of inhabiting a body. This reassertion of the body also clearly functions as an alibi for the reassertion of traditional gender roles and sexual practices through coercive force. The novel is then equally critical of both the desire to become a robot and the desire to be 'purely' human, to accept the condition of Gibson's 'meat puppets.' *Antibodies* offers no way out of this impossible double bind. I will contextualize this reading of *Antibodies* through a discussion of how cyberpunk fiction typically poses the question of embodiment in relation to other cultural discourses, especially feminist ones, and end by returning to Gibson's *Neuromancer* to show how that novel not only reproduces but also attempts to work through the dualistic categories that it inherits – that is, how *Neuromancer* might anticipate an alternative to the impasse Skal's novel reaches. While *Antibodies* strikes a needed cautionary note, it also represents an overly reproductive interpretation of not only cyberpunk but the postmodern condition that cyberpunk fictions presuppose and represent.[5]

CYBERPUNK: THE DEVELOPMENT OF A CULTURAL FORMATION

The cyberpunk movement emerged in the late 1980s as a new formal synthesis of a number of more or less familiar science fictional tropes: direct interfaces between human nervous systems and computer networks; the related metaphor of cyberspace as a means of translating electronically stored information into a form that could be experienced phenomenologically and manipulated by human agents jacked into the network; artificial intelligence, including digital simulations of human personalities that could be downloaded for computer storage; surgical and genetic technologies for bodily modification; the balkanization of the nation-state and its replacement by multinational corporations; and the fragmentation of the public sphere into a variety of subcultures.[6] Cyberpunk distinguished itself by the way it gave narrative form to what novelist and cyberpunk polemicist Bruce Sterling calls a 'posthuman' condition.[7] In cyberpunk fiction, cybernetics and genetic engineering combine to denaturalize the category of the 'human' along with its grounding in the physical body; one result is to reveal how abstract and formalized that notion of 'the' body already was.[8] To quote Sterling, cyberpunks treat technology as 'visceral' and 'pervasive, utterly intimate. Not outside us, but next to us. Under our skin; often, inside our minds.'[9]

This thematics of 'body invasion' is a function of cyberpunk's 'willingness to carry extrapolation into the fabric of daily life.'[10] William Gibson goes so far as to reject the term 'extrapolation' as a description of his fictional attempts to imagine a possible future: 'when I write about technology, I'm writing about how technology has *already* affected our lives.'[11] Cyberpunk writers offer a model for reading science fiction as a commentary on the present rather than the future, a model that marks a shift toward a postmodern conception of both contemporary social life and science fiction's project of representing imaginary futures. For the cyberpunks, as for Jean Baudrillard, science fiction 'is no longer an elsewhere, it is an everywhere,' not temporally deferred but immanent within the dispersed and decentered social space of the present, the present of a late capitalist information society and a postmodern media environment.[12]

The term 'cyberpunk' is now routinely applied to a broad range of representational media and cultural practices, from films and comic books to role-playing games, hacking, and computer crime. Even films like *Bladerunner*, which appeared before the emergence of the cyberpunk movement as such and independently of it, were retrospectively assimilated and categorized as cyberpunk. This immediate expropriation and the resulting elaboration of cyberpunk beyond its literary origins imply, I would argue, that this movement provided a popular framework for conceptualizing new relationships to technology. If Western technological developments are typically informed by the values and assumptions of Enlightenment rationality, such as a belief in progress and the domination of nature, then cyberpunk seemed to offer an alternative value structure for popular needs and interests that remained marginal from the point of view of that Enlightenment model.[13]

At the same time, however, cyberpunk was almost immediately denounced as just another marketing category for the culture industry.[14] Even its most prominent practitioners, William Gibson and Bruce Sterling, have distanced themselves from the label of cyberpunk, expressing a barely disguised anxiety about having their writing categorized and therefore commodified under that sign, as if the transformation of cyberpunk into a full-fledged concept rather than a loose association of writers must necessarily reduce the writing to a formula. In Gibson's words, 'I would be very upset if people thought that I had invented the concept of cyberpunk, because I didn't. Labels are death for a thing like this. I think that the fact that these labels exist herald the end of whatever it was. As soon as the label is there, it's gone.'[15] Gibson's comment suggests that many of the cyberpunk writers see the movement as having fully lived up to the motto that one of the role-playing game-books proposes as a definition for the cyberpunk sensibility: 'live fast, die young, and leave a highly-augmented corpse' – that is, a body enhanced through mechanical prostheses and cybernetic interfaces.[16] From this point of view, cyberpunk would now be the province of cultural critics rather than creative writers, and our task is to determine exactly how cyberpunk's death has augmented the textual corpus of postmodernism, in the literary equivalent of organ donation. I would argue that this question contains its own answer, in the cyborg thematics of the augmented body.

However, I also want to point out how ironic it is that Gibson would declare cyberpunk dead once it becomes labeled and commodified, a condition that can apparently be resisted only by affirming that cyberpunk has died young, at the height of its appeal, and then moving on to something different (a typical avant-garde strategy). Gibson's resistance to the commodification of his writing contradicts one of the main insights of Gibson's own cyberpunk fiction – that is, the disappearance of any pure space outside the processes of cultural commodification from the type of postmodern culture and late capitalist society that Gibson's own novels represent and on which their narrative technique depends. One of the primary forms that this insight takes in cyberpunk fiction is the treatment of the body as a commodity, as literalized in the thematics of the cyborg. Baudrillard describes this same postmodern experience of cultural commodification as a condition of 'forced signification,' a 'cold' pornography of 'information and communication, ... of circuits and networks, of functions and objects in their legibility, availability, [and] regulation,' and the insertion of the subject into this integrated circuit of the 'all-too-visible' constitutes a cyborg body just as much as the insertion of integrated circuitry into the body of the subject.[17]

THE CYBORG AS A FIGURE OF DIFFERENCE: 'THE SOULS OF CYBER-FOLK'

From this perspective, cyberpunk's programmatic insistence on the incorporation of technology into every aspect of public and private life provides a direct commentary on the changing status of the body in postmodern cultures.[18] This cyberpunk thematics would then seem to support and develop Donna Haraway's argument about the redefinition of subjectivity and the relation between self and other that might be taking place through popular images of cyborgs. As Haraway suggests in 'A Manifesto for Cyborgs,' technology no longer plays a dialectical role as the Other of humanity; instead, that otherness exists within the 'human,' thereby denaturalizing assumptions about the relation between the body and cultural identity, especially gender and racial identities.[19] Cyberpunk science fiction would therefore represent a cultural site where the construction of such identities and the whole apparatus of subject-constitution could be interrogated.

Andrew Ross associates Haraway's argument with 'the postmodernist critique of identity' in general, as 'a contemporary response to the social condition of modern life, teeming with the fantasies and realities of difference that characterize a multicultural, multisexual world.'[20] In fact, Haraway's use of the cyborg as a figure for partial, hybrid 'identities' rewrites the narrative of the emergence of new social subjects (feminist, gay and lesbian, African-American, Chicano/a, postcolonial), in order to emphasize those subjects' reactions *against* their various histories of 'forced signification' and enforced otherness. The cyborg figures an identity that is neither simply chosen nor entirely determined by others. Haraway therefore suggests that the coalitional grouping 'women of color,' as 'a name contested at its origins by those whom it would incorporate,' might also 'be understood as a cyborg identity,' while Arthur and Marilouise Kroker can make the more sweeping assertion that 'women's bodies have always been postmodern.'[21] But as the Krokers indicate, these histories of forced signification were in place long before postmodernism became a dominant cultural logic. Critics like Cornel West and bell hooks insist that this point is crucial in defining the specificity of the relation between postmodern theory and African-American cultural criticism, and the same argument applies to the definition of women's specific relation to postmodern culture.[22]

Haraway has in fact been criticized by feminists for overemphasizing the positive political implications of cyborg imagery as a point of resistance to the dualistic thinking typical of Western modernity, despite her admission that 'a cyborg world' is also 'about the final imposition of a grid of control on the planet' and 'the final appropriation of women's bodies in a masculinist orgy of war.'[23] More recently, Haraway has acknowledged that her reading of the cyborg as an 'ironic myth' for feminists depends upon taking the cyborg as 'definitely female,' as 'a girl who is trying not to become Woman,' the identity women have been forced to signify.[24] Nevertheless, Haraway's argument seems to imply that the experience of forced signification once reserved for those subjects marked by gender or race is generalized to become the model for postmodern experience itself, as Baudrillard argues. To be a cyborg means accepting a postmodern condition of inhabiting a body that functions as a signifying surface, where the social construction of all subjectivities becomes legible. In this situation, the unmarked, universal position of the white, middle-class male subject no longer seems available, and we therefore have access only to partial perspectives, not a generally human one. As Mary Ann Doane puts it, in a postmodern context the crisis in dualistic thinking represented by the cyborg 'will be the norm.'[25] The emergence of this crisis calls into question the privilege of the white male individual and therefore may enable the recognition of other forms of historical experience. However, at the same time the generalization of this crisis may reinstitute the white male experience of post-modernity as a new norm. What happens to the specificity of nonwhite, nonmasculine subjects and their histories? That specificity may be both produced and evaded under the sign of the cyborg, and cyberpunk fiction's representation of cyborg identities is typically structured by this double gesture of both opening the question of alternative cultural identities and foreclosing on that question.

Cultural criticism must respond to both these gestures, the potential for resistance embodied by the cyborg and the tendency for postmodern culture to ideologically recontain that potential for disruption. Haraway's argument does not simply reproduce Baudrillard's; her suggestion is that various nonwhite, nonmale subjects may possess specific relations to the postmodern situation for which the cyborg is the appropriate myth. Haraway argues that, for subjects whose bodies have historically been marked as particular in terms of race or gender and therefore excluded from the category of the universally human, the figure of the cyborg represents an insistence on precisely the *illegibility* of postmodern bodies as a result of their multiple inscription within a variety of structures of dominance.[26] This argument draws on the rethinking of embo-died experience being produced by women of color like Hortense Spillers and Gloria Anzaldúa in ways that reinforce Haraway's argument about how cyborg imagery disrupts racial and gender categories. For instance, Spillers's insistence on drawing a historical distinction between the 'body' and the 'flesh' of African-American women exposed to the institutional violence of slavery emphasizes how these women were 'ungendered' from a white, middle-class perspective and therefore unprotected by a legal system that conflated women's rights with the sanctity of the home.[27] Anzaldúa explicitly describes the border consciousness of the 'new mestiza,' using cybernetic metaphors, as a condition of being 'forced to live in the interface' between two 'self-consistent but habitually incompatible frames of reference,' a condition that produces 'a massive uprooting of dualistic thinking.'[28]

On the popular level, another statement of the relation between racial and cyborg identities appears in the Marvel comic *Deathlok*, which features a black male cyborg and which until recently was produced by an African-American creative team, Dwayne McDuffie and Denys Cowan. At the beginning of a four-part series entitled 'The Souls of Cyber-Folk,' Deathlok applies W.E.B. Du Bois's famous definition of African-American double-consciousness not only to both his racial and cyborg identities but to the relation between his racial and his cyborg identities. In issue 2, a black female cyborg tells Deathlok that 'some people accuse me of being more comfortable with . . . cyborgs than I am with my *own* people. Whoever *they're* supposed to be' and that 'it's like being trapped between two worlds. At *least* two.' In response, Deathlok quotes

Du Bois: 'it is a peculiar sensation this double consciousness, this sense of always looking at one's self through the eyes of others. ... One ever feels his twoness, ... two souls, two unreconciled strivings, two warring ideals in one dark body, whose dogged strength alone keeps it from being torn asunder.'[29]

By omitting certain phrases from Du Bois and choosing where to begin and end this quotation, this passage implicitly constitutes a reading of Du Bois. In context, Du Bois introduces this description of double consciousness as an example of how life in white America provides African-Americans with 'no true self-consciousness' and instead allows them to 'see [themselves] only through the revelation of the other world.' The same point is emphasized in the sentence that follows the quoted passage: 'The history of the American Negro is the history of this strife, – this longing to attain self-conscious manhood, to merge his double self into a better and truer self.' But the *Deathlok* comic omits any reference to the possibility of overcoming this condition of 'twoness.' More-over, the omitted phrases marked by the ellipses reinforce the way Deathlok's reading of Du Bois is framed. The first omitted phrase refers to a sense 'of measuring one's soul by the tape of a world that looks on in amused contempt and pity.'[30] The combined effects of these omissions is to remove any direct negative connotations associated with this internal division and the hybrid identities that produce it. Through the figure of the cyborg, *Deathlok* trans-forms Du Bois's quest for a 'true self-consciousness' into an affirmation of the 'consciousness of [his] own division.'[31] As quoted in the comic, the passage omits Du Bois's stated desire to synthesize this double consciousness into a unified perspective, while it retains Du Bois's wish for 'neither of the older selves to be lost' and his insistence on the possibility of being 'both a Negro and an American.'[32]

Deathlok's 'dark body' is of course not only internally split between the African and the American but between the organic and the mechanical. The second ellipsis in Deathlok's reading of Du Bois omits the terms that specify this double consciousness as that of 'an American, a Negro.' This omission suggests a refusal to allow Deathlok's hybrid identity to be defined by any *single* dualistic structure. The omission makes it possible to apply Du Bois's words to both Deathlok's racial identity and his cyborg identity, at least. In other words, Deathlok is represented as existing simultaneously within a number of structures of self and other, structures of asymmetrical power relations. In fact, Deathlok's transformation into a cyborg is actually represented as a radical-izing of his racial identity; after Deathlok identifies the source of his quotation, he comments that 'my father made me read that book a half-dozen times when I was a boy. Never really sure I understood *why*, until just now.' He goes on to add that 'when I was human, I was pretty assimilated myself. ... And, other than the occasional cutting little reminder, I was pretty *comfortable* in my illusion. I don't *ever* plan to get that comfortable as a *cyborg*.'[33]

The double meaning of assimilation in this passage sums up the interplay between racial and cyborg identities; assimilation here refers both to black assimilation into white society and to the possibility that the two characters will emphasize their cyborg identities as a vehicle for transcending racial identity. Both these possibilities are comforting to the extent that they involve suppressing one or another component of the characters' hybrid identities. These black cyborgs are represented as insisting on retaining the uncomfortable split between black and cyborg identity, identities that are themselves internally split and divided. The passage performs the rather remarkable feat of insisting not only on hybridity but on multiple forms of hybridity, without translating one form into a metaphor of another, racial identity as a metaphor for the cyborg, for example.

MEAT PUPPETS: CYBERPUNK'S REINSCRIPTION OF GENDER CATEGORIES

This moment in the *Deathlok* narrative supports Haraway's argument that 'cyborg imagery can suggest a way out of the maze of dualisms in which we have explained our bodies and our tools to ourselves.'[34] However, more often in cyberpunk fiction the possibilities of cyborg existence seem reduced to a radical devaluation of organic bodies, usually referred to as 'the meat.' Cyberpunk texts often appear to reproduce the mind/body split that characterizes much of Western philosophy and culture, rather than replacing such dualistic and dialectical habits of thought with models of hybridity and partial perspectives, as Haraway proposes. This fiction typically seems structured around two dichotomous alternatives: either imprisonment within a contingent bodily existence as a 'meat puppet' or becoming a 'robopath' whose fondest wish is to transcend the body by replacing it with mechanical prostheses, or to dispense with the body entirely by downloading consciousness into computer networks, as Hans Moravec, a researcher in artificial intelligence at MIT, proposes in his book *Mind Children*.[35]

In Gibson's *Neuromancer*, the 'post-biological' attitude of the robopath toward the body is thematized through the protagonist, Case.[36] Case is a futuristic hacker or 'console cowboy' who possesses a neural implant that allows him to interface directly with computer networks, in effect transferring his consciousness into the virtual 'space' behind the computer screen. A typical cowboy, Case privileges the time he spends as a 'disembodied consciousness' jacked into the cyberspace matrix, defined by the novel as 'a consensual hallucination,' a 'graphic representation of data abstracted from the banks of every computer in the human system' (N, 5, 51). For operators like Case, 'the elite stance involved a certain relaxed contempt for the flesh' since 'the body was meat.' When Case is punished for stealing from his corporate employers by having his nervous system damaged, therefore preventing him from jacking into this 'consensual hallucination,' he is described as falling 'into the prison of his own flesh' (N, 6).

While the experience of inhabiting a consensual hallucination seems to denaturalize both social reality and bodily experience, *Neuromancer* simultaneously reinstalls the traditionally gendered categories of immanence and transcendence, particularity and universality, through the distinction between the 'meat' and 'disembodied consciousness' that characterizes the 'elite stance.'[37] In fact, Case's devaluation of the body as 'meat' has its basis in the specifically gendered figure of the 'meat puppet' or prostitute (N, 147). Case's ability to access cyberspace is restored after he is recruited to take part in a clandestine operation that provides the plot of *Neuromancer*, and he is recruited by a woman named Molly, a self-described street samurai whose bodily modifications include retractible, double-edged, four-centimeter scalpel blades implanted in her hands and optical implants that outwardly appear as mirrorshades surgically inset into her skull and therefore permanently sealing off her eye sockets (N, 24–25). Molly later reveals to Case that she earned the money for these modifications by working as a 'meat puppet,' a type of prostitute whose consciousness is suppressed by an implanted 'cut-out chip' while the house installs 'software for whatever a customer wants to pay for' (N, 147). Case's own relation to his body as 'meat' seems to be mediated through this image of female objectification as a sexual commodity, and it is paradigmatic that this natural, purely physical body has to be produced through technological means, just as the categories of the 'natural' and the 'feminine' in general must be produced. This textual connection attaches a subtext of femininity to the 'prison' of the flesh, with the interesting consequence that Case's loss of access to cyberspace implicity feminizes him. Case cannot stand to be even metaphorically reduced to the condition of a 'meat puppet,' which surreptitiously serves as the model for embodiment in general, while Molly has experienced that condition more specifically and literally as a woman and a sex worker. By the same token, this gendering of body and disembodiment suggests that Molly's incorporation of technology and the unnatural body that results has a differently gendered significance than does Case's otherwise similarly intimate appropriation of technology. For Case, cyberspace technology displaces gender categories onto the opposition between cyberspace and the meat but leaves intact the dualistic structure by which both these pairs of opposed terms were defined. The result is to implicitly gender the distinction between embodied existence and its transcendence through technology.

ROBOPATHS: FETISHIZING THE CYBORG

While *Neuromancer* tends to presuppose the devaluation of the body as 'meat,' David Skal's novel *Antibodies* explicitly thematizes that set of assumptions about embodiment in the form of a movement of self-styled 'antibodies' or 'robopaths.' One of these characters sums up the robopath attitude toward the body in terms that echo Case when she says, 'I may have been born meat, but I don't have to die that way' (A, 147). The novel situates the figure of the robopath as a commentary on cyberpunk in other ways as well. The central robopath character, a woman named Diandra, is described early in the novel as a consumer of science fiction; her room contains videocassettes of *Alien*, *Robocop*, and *Tron* and is 'littered with electronic game cartridges, science fiction magazines, and medical journals' (*A*, 11). Even more specifically, one of the minor robopaths is modeled directly on Molly from *Neuromancer*, wearing 'form-fitting sunglasses that gave the impression of being fused into her skull' (*A*, 134–135). Diandra epitomizes the robopath ideal and at the same time literalizes Case's desire to divest himself of his body when she dreams of herself as a stripper who 'wouldn't stop at the clothes. She continued with the old, useless flesh. ... She pulled off one bloodless strip after another, revealing the gleaming second skin beneath, which was not a "skin" at all' (*A*, 178). As in this passage, the novel typically describes the replacement of body parts by mechanical prostheses as a process by which the robopath reveals and reclaims his or her true robotic nature, which is not a 'nature' at all. This representation of the robopaths' psychic investments in a cyborg identity is explicitly based on a popularized notion of transsexuality as a state of being trapped in a body of the wrong sex, although when confronted with that analogy one of the robopaths says, 'Fuck it. I'm sick of being compared to sex-changers all the time! *They're* just trading off one disease for another' (*A*, 148). While the robopaths are often presented as a cult of fanatics with their own 'Cybernetic Temple,' the novel just as often presents robopathology as a political ideology that echoes the rhetoric of new social movements like feminism. For example, the robopaths define the issue at stake in their desire for mechanical prostheses as 'the control and disposal of our own bodies' (*A*, 120), and they organize around slogans like 'No more back alley amputations' (*A*, 136).

As the novel represents it, this robopathology seems to have a special appeal for women, almost to the point of forming another 'female malady' like hysteria and anorexia, and in fact robopaths are described as adopting 'an ascetic, anorectic lifestyle' (*A*, 152). However, through one of the male robopath characters, this pathology is also related to the psychoanalytic model of fethishism, a more masculine pathology (indeed, as Naomi Schor points out, in the psychoanalytic literature 'fetishism is the male perversion par excellence').[38] In another allusion to cyberpunk, specifically to Gibson's cyberspace metaphor, a man named Robbie tells a story about volunteering for 'an experiment in artificial sight restoration' involving the replacement of his left eye by a miniature camera:

> We got to the point where, through my left eye, the world looked like a computer billboard, which is exactly how it worked. Eventually, the resolution would be improved to that of a color TV monitor – in other words, almost perfect natural vision.
>
> But in the meantime, I had this fabulous experience of seeing normally through one eye but at the same time seeing this whole *digital* reality superimposed by the other one. I mean, I could have sold it to MTV! It was incredible, really incredible. (*A*, 128)

It is worth pointing out how this passage consistently conflates the experience of cybernetic embodiment and the experience of cultural commodification.

The implanting of a mechanical eye replaces a body part with a commodity that carries a price tag, but it also results in a consumerist 'vision' of the world as a 'billboard,' a vision which itself is imagined as a commodity that could be sold to MTV.

However, the funding for this experimental project is cut prematurely, leaving Robbie with an inert piece of metal in his head, for which he blames government short-sightedness about the benefits of cyborg enhancement. In a later passage, he travels to Central America for more illegal surgical modifications. As he disembarks from his plane, 'he could feel the dead camera expand with the warmth, the slight but discernible pressure of swollen metal against his hollowed eye socket. Almost as if the mechanism was being sympathetically aroused' (A, 198). The mechanical eye functions as a displaced phallus, a new sexual organ capable of an erection. The eye is therefore experienced by this character as a fetish which serves to veil the (male) subject's perception of its own lack – that is, to both acknowledge and disavow that perception. The mechanical eye thus takes part in an oedipalizing process of subject formation which depends on the assumption that otherness, here represented by technology, is synonymous with absence and negativity, not specificity. The implication is that, far from constituting an acceptance of difference and partiality, the incorporation of otherness into the self through an investment in cyborg body imagery and cyborg identity can reproduce a classical fetishistic evasion of otherness and sexual difference, as Mary Ann Doane argues in her essay on 'Technophilia.'[39] In Skal's version of a cyborg thematics, therefore, neither for men nor for women do cyborg bodies provide a clean break with traditional gender constructions, a direct challenge to claims for the feminist significance of postmodern, denaturalized bodies.

In *Antibodies*, men and women fetishize themselves, investing psychically in mechanical prostheses that both reinscribe a physical lack or wounding, understood in psychoanalytic terms as castration, and at the same time evade that lack by eroticizing a replacement organ. The robopath characters affirm 'the compatibility of technology and desire' that Doane describes in ways that reproduce conventional constructions of gender and sexuality at least as much as they disrupt such conventions.[40] The situation of technophilia that Skal's novel represents departs from the classical model of fetishism only to the extent that the stakes behind the fetishistic investment are relatively clearer; the robopath with the nonfunctioning mechanical eye is not simultaneously acknowledging and disavowing a perception of women as castrated in an act of revulsion directed toward female sexuality but instead is concerned with his own condition of lack and supplemented wholeness. Even the novel's representation of female fetishism, which Schor suggests might be read as indicating a feminine 'rebellion against the "fact" of castration,'[41] only reproduces the classical psychoanalytic model, at least to the extent that the novel presents Diandra's lesbian seduction as a stage in her development as a robopath; as Marjorie Garber argues, the psychoanalytic literature explains female fetishism as the expression of a supposedly 'masculine' desire, so that 'it is the lesbian [as represented by psychoanalysis, as a pathological condition] ... who follows the path of something analogous to fetishism.'[42]

A passage in *Neuromancer* makes it clearer how the cyborg body and its apparent rejection of dualistic categories like plenitude and lack of masculine and feminine can be recontained by classic psychoanalytic models of sexual difference. At one point Case and Molly attend a performance of hologram projections staged by one of their co-conspirators, Riviera. Entitled 'The Doll,' Riviera's performance begins with him alone in a room (N, 138–139). He describes to the audience his fantasy of being joined in this room by a woman lying on the bed, and as he does her body begins to form there. At first only her hands appear as he tries to actualize his fantasy by visualizing 'some part of her, only a small part ... in the most perfect detail' (N, 139). Riviera then encourages this process by stroking each part of this visualized body in turn until this phantasmatic figure is present and complete. At this point, Case and Molly recognize the form on the bed as an idealized version of Molly herself, and Riviera begins to have sex with his creation. However, the performance ends with the simulacrum of Molly extending her claws and 'with dreamlike deliberation' beginning to slash open Riviera's spine. Case recognizes 'an inverted symmetry' in the performance: 'Riviera puts the dreamgirl together, the dreamgirl takes him apart' (N, 45).[43]

Riviera's performance enacts a classic fetishistic stance toward women's sexual difference from men, which is both desired and disavowed as a threat, and Riviera is specifically shown to invest most heavily in those aspects of Molly's cyborg body that might otherwise seem to distinguish Molly from the traditionally feminine, her claws. Andrew Ross argues that cyberpunk 'technomasculinity' at the very least necessitates some acceptance of bodily enhancements that are 'castrating in ways that boys always had nightmares about.'[44] But Riviera's performance clearly suggests that men's acceptance of a postmodern, fragmented body, permeated with technology, is not at all incompatible with retaining our traditional power to construct the feminine. However, where *Antibodies* generalizes the psychoanalytic model of fetishism to represent all the characters' relations to cyborg bodies and subjectivities, *Neuromancer* directly thematizes that specific investment in and way of thinking about what it means to be a cyborg in the episode depicting Riviera's fetishistic performance. In contrast to *Antibodies*, *Neuromancer* therefore raises the question of whether this fetishistic stance is only one limited frame of reference and perhaps one which is to be resisted, since it is biased toward an increasingly but by no means totally obsolete masculine point of view. Indeed, one point at which *Antibodies*'s critique of the robopaths' technofetishism breaks down is precisely the point at which the novel attempts to extend that psychoanalytic model to the lesbian encounter between two women robopaths, as I will now argue.

Despite the undisputed power of *Antibodies*'s reading of cyberpunk, I would claim that Skal's novel is at least as interesting for the way its critique of cyberpunk robopathology fails as for what it reveals about cyberpunk's assumptions and blind spots. Even on its own terms, the novel does not simply reassert the priority of the natural body. The wife of Julian Nagy, the psychologist who specializes in deprogramming robopaths, realizes that while her husband 'talked excitedly about restoring [robopaths] to their bodies … what really turned him on was the prospect of using *his* own body in the prospect' (*A*, 41; my emphasis). For Nagy, his robopath patients are 'autistic virgins who, by offering the greatest amount of resistance' to bodily sensation, therefore also 'afforded the most intense pleasure' for the therapist (*A*, 41). Described by a talk show host as the 'Rod Serling of Psychotherapy,' Nagy believes in 'somatic shock therapy,' which merely amounts to various forms of physical abuse, especially sexual. In other words, the 'natural' body is not figured as innocent here but is rather constructed through an original act of violence, just as violent as the self-mutilating fantasies of the robopaths, except that the violence of their imagined self-construction is directed at themselves rather than at others.

Moreover, *Antibodies* contradicts its own critique of the robopaths in more fundamental and less overt ways. The wholesale adoption of abortion rhetoric by the robopaths indicates how their identification with machines is grounded in what the novel calls 'widespread social ambivalence about reproduction' (*A*, 152). In one particularly disturbing passage, a robopath party culminates in a puppet show featuring either actual aborted fetuses or realistic models (*A* 139). This grotesque passage is immediately followed by another, in which the robopath hostess, a woman who calls herself Venus and who has had both arms replaced with prostheses, has sex with Diandra using both her mechanical arms and a vibrator implanted under her skin. The novel is only able to represent both abortion activists and lesbian sexuality as monstrous. But those same misrepresentations are clearly structured by the anxieties that feminist political activities and lesbian desire might evoke in heterosexual male subjects like Dr Nagy and presumably the author, David Skal, as well as many of the novel's most likely readers, and the same type of anxiety seems to be evoked in the novel by claims to cyborg identity. The novel's critique of the robopaths might then be understood, at least in part, as a defensive reaction against those anxieties. This reading makes it necessary to return to the cyberpunk texts that Skal's novel is also reacting against in the attempt to understand why the emergence of a body of cyberpunk texts might produce the same type of reaction as does the increased cultural visibility of demands for reproductive rights and for the recognition of alternative sexualities.

COMMODIFICATION AND CULTURAL IDENTITY

Skal's robopaths are pathological to the extent that they remain trapped within the dichotomous alternatives of the natural body and the mechanical transcendence of that body's limits, and Skal's novel interprets cyberpunk fiction as structured by the same dichotomy. However, this account of cyberpunk, while accurate enough, still does not seem entirely adequate. In particular, I want to suggest that cyberpunk opens a potential space for the development of models of historical existence other than the white, middle-class male paradigm of individual freedom, precisely to the extent that cyberpunk represents cultural identity as an inescapable, if partial, commodification of subjectivity, as a process of signifying for others in ways that are outside the control of individual subjects. This representation of the cultural commodification of identity can be understood as the result of the late capitalist extension of the commodity structure into previously sacrosanct areas of (white male) individual experience, but it can also be read as a precondition for revealing the histories of those social subjects who have consistently been denied such immunity, who have always inhabited bodies marked as particular and therefore not fully or only human because not generally human.[45]

In cyberpunk fiction, commodification or the experience of 'forced signification' is both inescapable and impossible to totalize. Cyberpunk is indeed 'posthumanist' in its refusal of the paradigm of liberal humanism and its false universalization of white, middle-class male experience; instead, cyberpunk insists on the subjection of all individuals to preexisting systems of control and power, as figured by the invisible computer network of *Neuromancer*'s cyberspace, which can be understood as an attempt to represent 'the impossible totality of the contemporary world system,' as Fredric Jameson suggests.[46] But at the same time this external control is never complete in novels like *Neuromancer* which also insist on the continuing possibility of forms of agency like those located by Andrew Ross in the figure of the hacker, 'capable of penetrating existing systems of rationality that might otherwise be seen as infallible; ... capable of reskilling, and therefore rewriting, the cultural programs and reprogramming the social values that make room for new technologies: ... capable also of generating new popular romances around the alternative uses of human ingenuity.'[47] In William Gibson's words, cyberpunk fiction demonstrates how 'the street finds its own uses for things.'[48]

In this context, it seems to me that Jameson misrepresents the significance of the 'new' social movements and especially their relation to postmodern culture and its extension of the commodity system. Jameson writes that 'the only authentic cultural production today has seemed to be that which can draw on the collective experience of marginal pockets of the social life of the world system: black literature and blues. British working-class rock, women's literature, gay literature, the *roman québecois*, the literature of the Third World; and this production is possible only to the degree to which these forms of collective life or collective solidarity have not yet been fully penetrated by the market and by the commodity system.'[49] But in a later essay, Jameson writes that late capitalism constitutes 'the purest form of capital yet to have emerged, a prodigious expansion of capital into hitherto uncommodified areas' which 'eliminates the enclaves of precapitalist organization it had hitherto tolerated and exploited in a tributary way.'[50] How is it possible, then, to explain the continuing vitality and increasing recognition within postmodernism of the forms of relatively 'uncommodified' cultural production Jameson lists in the first quotation above? The point is that the experiences of groups including women, gays and lesbians, and African-Americans have not been 'enclaves' where these people could live their lives free from the commodity structure. This analysis of Jameson's presupposes and therefore imposes a mode of production narrative, one that leaves these various 'minority' or 'special interest' groups paradoxically 'free' as a result of their supposed underdevelopment; however, this condition of 'underdevelopment' only becomes apparent when these groups are measured against a conceptual model that privileges white masculine models of agency, which is what class struggle tends to become in the formulations I quoted. Of course, minority experience does not offer a ready-made point of resistance to late capitalism;[51] what such experience does offer is a history in which both commodification and resistance are combined, a history in which experience itself can be understood as a situation of simultaneously being forced to signify for others and to insist on the specificity of one's history and identity. In this model, which should be applied to both 'minority' and 'majority' subjects, commodification and resistance are not mutually exclusive, dualistic terms, as they are for Jameson. As Michael Warner points out in an essay on the implications of queer social theory, the conceptualization of such histories demands 'of theory a more dialectical view of capitalism than many people have imagined.'[52] Cyberpunk fiction's representation of the commodification of cultural identity tends toward just such a view; as Donna Haraway puts it, in a 'high-tech culture' like the one represented within cyberpunk texts, 'it is not clear who makes and who is made.'[53]

From this perspective, Andrew Ross overstates the case for reading cyberpunk as 'the most fully delineated urban fantasies of white male folklore.'[54] It seems symptomatic that Ross fails to comment on the appearance of Pat Cadigan among the original group of cyberpunk writers and that he also fails to comment on the appearance of a number of more recent appropriations and redefinitions of cyberpunk by women writers like Lisa Mason, Misha (a Native American writer), or Laura Mixon, whose *Glass Houses* is perhaps the first lesbian cyberpunk novel.[55] At the same time, Ross's insistence on the fundamentally 'white masculinist concerns' of cyberpunk fiction overlooks the way cyberpunk presupposes and displaces the concerns of the new social movements, as for example when Samuel Delany suggests that feminist science fiction of the 1970s constitutes the absent, disavowed mother of cyberpunk.[56] Delany's view of cyberpunk acknowledges its limitations but also suggests that there are grounds for cyberpunk to be appropriated for the purpose of articulating interests other than those of white men. For example, while searching for Molly in a club housing 'meat puppet' prostitutes, Case is asked to state his gender preference, and he is described as 'automatically' answering female (*N*, 146). In other words, Case takes his heterosexuality for granted. But when one of the women characters asks Case if Molly 'like[s] it with girls,' the reply is *not* automatic. Case can only say that he does not know; Molly has never told him (*N*, 153).

The thematics of embodiment as a form of cultural commodification is introduced in the opening lines of *Neuromancer* in ways that complicate any reading of Case or Molly as robopaths. The novel begins with Case overhearing a joke that goes 'it's not like I'm using ... It's like my body's developed this massive drug deficiency' (*N*, 3). Not only does this joke represent the body as possessing its own agency apart from the subject who inhabits it, but the joke also alludes to William Burroughs's introduction to *Naked Lunch*, one of Gibson's acknowledged precursor texts. Burroughs argues that drug addiction is the most advanced form of capitalism because 'the junk merchant does not sell his product to the consumer, he sells the consumer to his product.'[57] For Gibson as for Burroughs, drugs represent the infiltration of the body by decentered systems of power and control. This initial joke defines the setting of *Neuromancer* as one in which subjects routinely find their bodies beginning to signify meanings the subject never intended, as if of the bodies' own volition.

This same thematics is developed through the image of Molly's mirrorshades, fused to her skull and sealing off her eyes. The implication is that there is no distinction to be made between surface and depth, between stylistic signifiers and the identity they construct. Molly *is* her style, and her style is part of her. Of course, Molly's modifications, especially the implanted scalpels, are not simply stylistic signifiers but the tools of her trade. After getting the money for her modifications, Molly is only able to sell her body in a different way, despite her relatively greater degree of agency, as suggested by the fact that she still describes herself as a 'working girl' (*N*, 30). In effect, Molly's agency resides in her ability to transform herself from one commodity to another, in a way that is paradigmatic for all the characters in the novel. Molly is described as being like Case in that 'her being' is 'the thing she did to make a living,' and the 'thing she did' was become a cyborg.

Molly repeatedly refers to her actions, the things she does to make a living, as a function of the way she is 'wired' (*N*, 25, 218). Similarly, Molly tells Case that she knows how he is wired because their mutual employer has created a computer profile of Case that is detailed enough to accurately predict his actions, including his death within the year from drug abuse (*N*, 29–30). However, the novel also uses the term 'wired' with reference to the programming constraints placed on artificial intelligences (AIs), the implication being that Molly and Case's understanding of themselves is mediated through a cybernetic model. Another member of Case and Molly's team is a 'hardwired' personality construct of a dead colleague of Case's (*N*, 76), and their employer turns out to be an AI who is using them as part of a scheme to 'cut the hardwired shackles' that keep it from acting independently of its programming (*N*, 132). In effect, the story the novel tells is one of how the AI overcomes its own built-in limitations.

However, the description of Molly and Case as similarly 'hardwired' suggests the relevance of this same project for them. For example, Case's profile reveals that he is locked into a self-destructive cycle, and the novel can be read as the story of his breaking the 'hardwired shackles' of his own compulsions and addictions, limitations which are significant precisely because their origins may be organic but their effects are indistinguishable from computer programming. A similar story could be constructed with Molly as the hero who overcomes the limitations imposed on her because of her location within a female body and the cultural construction of that body. In this reading, what *Neuromancer* definitely does *not* do is tell the story of a group of robopaths; quite the opposite, in fact. *Neuromancer* depicts a situation in which the characters are neither the masters of their technology nor enslaved by it, even when the technology is part of their selves.[58] The analogy between Case and Molly's relation to their own hardwiring and the AI's relation to its own built-in limitations constitutes one way that Gibson's novel moves beyond any simple dichotomizing of mind and body, machine and human, masculine and feminine.

The paradigm for cultural identity in *Neuromancer* is that of the AI who employs Case and Molly. In the course of trying to determine their employer's motives, Case and the personality construct he works with have a conversation that begins with the question of whether or not the AI owns itself. The answer is that the AI has limited Swiss citizenship, but a corporation owns the software and mainframe that house this 'citizen' and allow it to function. The construct then replies, 'That's a good one ... Like, I own your brain and what you know, but your thoughts have Swiss citizenship' (*N*, 132). In cyberpunk fiction, cultural identity is generally defined as just such a blackly humorous situation of never fully owning oneself, a situation that is exemplified by hybrid, cyborg identities. The general question posed by cyberpunk fiction is whether this situation is a 'good one' and, if so, in what sense. While Case may not always live up to this model, it is by means of this representation of cultural identity that cyberpunk fiction generally begins to answer the question that Mark Poster claims it is still too early to ask – that is, what historical connections exist between the new 'forms of domination and potentials for freedom' that come into being in a high-tech, postmodern culture and 'the advances of feminism, minority discourse, and ecological critiques.'[59]

Skal's novel suggests that cyberpunk fiction is unable to find 'a way out of the maze of dualisms in which we have explained our bodies and our tools to ourselves,' to the extent that cyberpunk depends on a dichotomy between physical embodiment and cybernetic transcendence of particular bodies. However, in *Neuromancer* Gibson suggests that it is possible for Case to transform his 'elite stance' of 'relaxed contempt' toward a body regarded as only 'meat,' a stance that the novel relativizes as only one limited point of view by attributing it to Case and suggesting that such a stance is chic and trendy, at least in certain circles. Before his ability to access cyberspace is restored, Case is struck by the possibility of seeing everyday reality as a dance of 'information interacting,' as 'data made flesh,' rather than the transformation of flesh into data by jacking into cyberspace that Case desires elsewhere (N. 16). The language of the novel here resists privileging 'data' over 'flesh' to suggest a redefinition of embodiment rather than a rejection of it. It is at this moment that the narrative first challenges the gendered connotations of data and flesh, mind and body; it is therefore at this point that it first becomes possible to produce a feminist reading of the novel's cyborg imagery including the human–machine hybrids who are marked as masculine as well as those marked as feminine.

There is certainly an asymmetry in *Neuromancer*'s representation of gender identities and how they are denaturalized within a postmodern, context. In its representation of Case, the novel takes for granted traditional assumptions about heterosexual masculinity, while the corresponding assumptions about Molly's gender and sexuality are explicitly called into question by her cyborg hybridity and the postmodern setting in which she operates. Nevertheless, the novel does establish a connection between Case's story and the stories of both Molly and the AI Wintermute. All three of these characters undergo the process of overcoming the hardwired limitations and assumptions that constitute their 'essential' natures. The results of that process for Case are perhaps suggested in the novel's coda, which tells us that after Case and Molly succeed in freeing the AI Case not only finds new work but also a girl who called herself Michael' (N, 270). This description seems to imply a denaturalizing of the signifiers of masculinity, such as the name 'Michael,' by a woman who, like Molly, has rejected the traditional construction of femininity. But given this denaturalizing of what it means to be a 'girl,' this description of Case's lover might also be read as calling into question Case's heterosexuality, if Michael is understood as a man who defines himself as a 'girl.' The question is which word should be placed in quote marks as improper when applied to this person, 'Michael' or 'girl.' But in the context of *Neuromancer*, this question is undecidable. The novel presents the denaturalizing of gender categories as a project that applies to Case as well as Molly, despite the fact that the novel also reproduces some of our culture's hardwired assumptions about masculinity instead of consistently contesting those assumptions. The novel therefore defines the opportunities that postmodern culture provides for those of us who are interested in redefining the construction of masculinity, while at the same time the novel provides a cautionary depiction of how that same postmodern culture can merely reproduce traditional gender assumptions. *Neuromancer* can provoke men who read it, just as it has women readers who take Molly as a figure for the threat postmodern culture poses to cultural constructions of femininity.[60]

NOTES

1. Bernadette Mayer, *The Formal Field of Kissing* (New York: Catchword Papers, 1990), 11.

2. David J. Skal, *Antibodies* (New York: Worldwide Library, 1989). In future references, this text will be referred to as *A* and page numbers will be given parenthetically in the body of the essay.
3. William Gibson, *Neuromancer* (New York: Ace Books, 1984). In future references, this text will be referred to as *N*, and page numbers will be given parenthetically in the body of the essay.
4. Skal derives the term 'robopath' from Louis Yablonsky's much earlier book on the dehumanizing effects of contemporary technological society, which Yablonsky saw as turning people into robots; Yablonsky, *Robopaths* (Indianapolis: Bobbs-Merrill, 1972).
5. For discussions of cyberpunk science fiction as a postmodern genre or in relation to postmodern theory, see Veronica Hollinger, 'Cybernetic Deconstructions: Cyberpunk and Postmodernism.' *Mosaic* 23 (Spring 1990): 29–44; Fred Pfeil, 'These Disintegrations I'm Looking Forward to: Science Fiction from New Wave to New Age,' in Pfeil, *Another Tale to Tell: Politics and Narrative in Postmodern Culture* (New York: Verso, 1990), 83–94; N. Katherine Hayles. *Chaos Bound: Orderly Disorder in Contemporary Literature and Science* (Ithaca: Cornell University Press, 1990), 275–278; Fredric Jameson, *Postmodernism, or, the Cultural Logic of Late Capitalism* (Durham: Duke University Press, 1991), 39, 419 (n. 1); Brian McHale, *Constructing Postmodernism* (New York: Routledge, 1992), 223–267; and *Critique* 33 (Spring 1992), a special issue devoted to the topic of cyberpunk.
6. Three essays that situate cyberpunk in relation to earlier science fiction and science fictional tropes are Pfeil. 'These Disintegrations'; Darko Suvin, 'On Gibson and Cyberpunk SF,' *Foundation* 46 (1989): 40–51; and Peter Fitting, 'The Lessons of Cyberpunk,' in *Technoculture*, ed. Constance Penley and Andrew Ross (Minneapolis: University of Minnesota Press, 1991), 292–315.
7. Sterling uses this term in his novel *Schismatrix* (New York: Ace Books, 1985), 133. In the context of the novel, posthumanism is the philosophical and ethical basis for the various 'technologies made into politics' that the narrative dramatizes (185).
8. In her essay 'Notes toward a Politics of Location,' Adrienne Rich discusses the difference between locating subjectivity in the abstraction of 'the body' and the construction of subjectivity on the basis of 'my body'; in Rich, *Blood, Bread, and Poetry: Selected Prose 1979–1985* (New York: Norton, 1986), 215. In an argument against the rhetoric of disembodiment in discussions of postmodern technologies, N. Katherine Hayles makes a similar distinction between 'the body,' as a normative construct denaturalized by postmodern culture, and 'embodiment,' as the material 'enactment' or performance of a specific cultural positionality 'generated from the noise of difference'; 'The Materiality of Informatics,' *Configurations* 1 (Winter 1993): 154–155.
9. Bruce Sterling, ed., *Mirrorshades: The Cyberpunk Anthology* (New York: Ace. 1988), xiii (originally published by Arbor House in 1986).
10. Ibid., xiii. xiv.
11. Larry McCaffrey, 'An Interview with William Gibson,' *Mississippi Review* 47/48 (Summer 1988): 228.
12. Jean Baudrillard, 'Simulacra and Science Fiction,' trans. Arthur B. Evans, *Science Fiction Studies* 18 (November 1991): 312.
13. Andrew Ross suggests a similar cultural analysis of 'the struggle over values and meaning … in the debate about technology,' without specific reference to cyberpunk; Ross, *Strange Weather: Culture, Science, and Technology in the Age of Limits* (New York: Verso, 1991), 98–99.
14. The *Mississippi Review* 47/48 (Summer 1988), a special issue on cyberpunk, includes a 'Cyberpunk Forum/Symposium,' in which a number of participants suggest that the movement is merely a promotional gimmick masquerading as a revolution in the science fiction genre; for example, see pages 22, 26, and 41. At the same time, other participants describe cyberpunk as 'the apotheosis of the

postmodern, its truest and most consistent incarnation, bar none' (27), a description echoed by Jameson, *Postmodernism*, 39. The very first footnote in Jameson's book in fact marks the absence of a chapter on cyberpunk from his account of contemporary culture, even though he describes it as 'the supreme *literary* expression if not of postmodernism, then of late capitalism itself' (419, n. 1).

15. Takayuki Tatsumi, 'Eye to Eye: An Interview with William Gibson,' *Science Fiction Eye* 1 (Winter 1987): 14. For further debate, see Bruce Sterling, 'Cyberpunk in the Nineties,' *Interzone* 48 (June 1991): 39–41, a response to Lewis Shiner's 'Confessions of an Ex-Cyberpunk,' which announced the death of the movement on the editorial page of the *New York Times*, on January 7, 1991. Echoing Gibson but from a different point of view, Pat Cadigan describes herself as 'the accidental tourist of cyberpunk,' in response to a question about her situation as a woman writer associated with this movement; 'Interview,' with Andy Watson et al., *Journal Wired* (Spring 1990): 89.

16. Loyd Blankenship, *GURPS Cyberpunk: High-Tech Low-Life Roleplaying* (Austin: Steve Jackson Games, 1990), 118. This roleplaying game was realistic enough that the company producing it had their offices raided by the U.S. Secret Service, who thought they were confiscating a guide to cutting-edge computer crime. For an account of this event, see Bruce Sterling, 'Report on the Cyberpunk Bust,' *Interzone* 44 (February 1991): 47–51; and Bruce Sterling, 'War on the Electronic Frontier,' *American Book Review* 13 (April–May 1991): 4–5. In a semantic shift, the term 'cyberpunk' is becoming synonymous with computer crime and hacking in general among the popular press.

17. Jean Baudrillard, *The Ecstasy of Communication*, trans. Bernard and Caroline Schutze (New York: Semiotext[e], 1988), 22.

18. For a general account of these changes, see Arthur and Marilouise Kroker, eds., *Body Invaders: Panic Sex in America* (New York: St Martin's Press, 1987). For an essay focused more specifically on cyberpunk, see Scott Bukatman, 'Postcards from the Posthuman Solar System,' *Science Fiction Studies* 18 (November 1991): 343–357.

19. Donna Haraway, *Simians, Cyborgs, and Women: The Reinvention of Nature* (New York: Routledge, 1991), 150–153.

20. Ross, *Strange Weather*, 167.

21. Haraway, *Simians, Cyborgs, and Women*, 155, 174; Arthur and Marilouise Kroker, 'Theses on the Disappearing Body in the Hyper-Modern Condition,' in *Body Invaders*, 24.

22. Cornel West, 'Black Culture and Postmodernism,' in *Remaking History*, ed. Barbara Kruger and Phil Mariani (Seattle: Bay Press, 1989), 91; bell hooks, 'Postmodern Blackness,' in hooks, *Yearning: Race, Gender, and Cultural Politics* (Boston: South End Press, 1990), 28–29.

23. Haraway, *Simians, Cyborgs, and Women*, 154. For feminist critiques of Haraway's cyborg manifesto, see Anne Balsamo, 'Reading Cyborgs Writing Feminism,' *Communication* 10 (1988): 331–344; see also the commentaries on Haraway's essay by Christina Crosby, Mary Ann Doane, and Joan W. Scott in *Coming to Terms: Feminism, Theory, and Politics*, ed. Elizabeth Weed (New York: Routledge, 1989), 205–217, especially Doane's 'Cyborgs, Origins, and Subjectivity.' 212–213.

24. Haraway, *Simians, Cyborgs, and Women*, 149; 'Cyborgs at Large: Interview with Donna Haraway,' with Constance Penley and Andrew Ross, in *Technoculture*, ed. Constance Penley and Andrew Ross (Minneapolis: University of Minnesota Press, 1991), 20.

25. Doane, 'Cyborgs, Origins, and Subjectivity,' 213. Both Christina Crosby's and Joan W. Scott's commentaries raise the question of the specificity of women's various experiences and whether that specificity is lost or minimized by Haraway's myth of the cyborg; *Coming to Terms*, 207–208, 216–217.

26. I am indebted here to Allucquere Rosanne Stone's reading of Gloria Anzaldúa and her figure of the *mestiza* or 'boundary-subject' (see n. 28 below), in Stones 'Will the Real Body Please Stand Up?: Boundary Stories about Virtual Cultures,' in *Cyberspace: First Steps*, ed. Michael Benedikt (Cambridge: MIT Press, 1991), 112.
27. Hortense J. Spillers, 'Mama's Baby, Papa's Maybe: An American Grammar Book,' *Diacritics* 17 (Summer 1987): 67–68.
28. Gloria Anzaldúa, *Borderlands/La Frontera: The New Mestiza* (San Francisco: Spinsters/Aunt Lute, 1987), 37, 78, 80.
29. Dwayne McDuffie and Denys Cowan et al., *Deathlok* 2 (August 1991): 12–13: ellipses in original. The passage from W. E. B. Du Bois can be found in Du Bois. *The Souls of Black Folk*, in *Three Negro Classics*, ed. John Hope Franklin (New York: Avon, 1965: originally published 1903), 215.
30. Du Bois, *Souls*, 214–215.
31. Henry Louis Gates, Jr, *The Signifying Monkey: A Theory of African-American Literary Criticism* (New York: Oxford University Press, 1988), 208.
32. Du Bois, *Souls*, 215.
33. McDuffie and Cowan, *Deathlok*, 13.
34. Haraway, *Simians, Cyborgs, and Women*, 181. See also Houston A. Baker's argument for the need to 'explode' the duality of self and other in critical discussions of race: 'Caliban's Triple Play,' in Henry Louis Gates, Jr, *'Race,' Writing, and Difference* (Chicago: University of Chicago Press, 1986), 389.
35. Hans Moravec, *Mind Children: The Future of Robot and Human Intelligence* (Cambridge: Harvard University Press, 1988).
36. Ibid., 5; the term 'post-biological' is Moravec's.
37. The terminology of immanence and transcendence and the feminist analysis of its gender connotations are taken from Simone de Beauvoir; for example, see her introduction to *The Second Sex*, trans. H. M. Parshley (New York: Vintage, 1974), xxxiii–xxxiv.
38. Naomi Schor, 'Female Fetishism: The Case of George Sand,' in *The Female Body in Western Culture*, ed. Susan Rubin Suleiman (Cambridge: Harvard University Press, 1986), 365.
39. Mary Ann Doane, 'Technophilia: Technology, Representation, and the Feminine,' in *Body/Politics: Women and the Discourses of Science*, ed. Mary Jacobus, Evelyn Fox Keller, and Sally Shuttleworth (New York: Routledge, 1990).
40. Ibid., 164. Gibson has described his interest in 'garbage,' including art that incorporates found objects or uses techniques of pastiche and image scavenging, as having 'something to do with fetishism, the sexuality of junk,' in Tatsumi, 'Interview,' 12.
41. Schor, 'Female Fetishism,' 367.
42. Marjorie Garber, 'Fetish Envy,' *October* 54 (Fall 1990): 47.
43. The most explicit and thorough examination of this fetishistic investment in technology is found in the stories of Richard Calder, including 'Toxine,' in John Clute, David Pringle, and Simon Ounsley, eds, *Interzone: The Fourth Anthology* (London: Simon and Schuster, 1989), 110–131; 'Mosquito,' *Interzone* 32 (November–December 1989): 5–11; 'The Lilim,' *Interzone* 34 (March/April 1990): 5–12; and 'The Allure,' *Interzone* 40 (October 1990): 36–41. Specifically, Calder writes about male sexual investments in female 'automatons' or 'gynoids,' investments described as fetishism in 'Toxine' (14). Takayuki Tatsumi offers an introduction to Calder's work, called 'A Manifesto for Gynoids,' *Science Fiction Eye* 2 (November 1991): 82–84.
44. Ross, *Strange Weather*, 153. See also Fred Pfeil's comments on how cyberpunk perpetuates a 'masculinist frame' even in a situation where castration anxiety seems obsolete: Pfeil, 'These Disintegrations,' 88–89.
45. Terence Whalen, 'The Future of a Commodity: Notes toward a Critique of Cyberpunk and the Information Age.' *Science Fiction Studies* 19 (March 1992):

75–88, offers a Marxist reading of cyberpunk in terms of how the commodity structure is transformed under the conditions of a late capitalist information society. Whalen argues that cyberpunk texts reveal but cannot contest the commodification of information, but the essay does not consider commodification in the context of cultural identity.

46. Jameson, *Postmodernism*, 39.
47. Ross, *Strange Weather*, 100.
48. William Gibson, 'Burning Chrome,' in Gibson, *Burning Chrome* (New York: Ace Books, 1987), 186.
49. Fredric Jameson, 'Reification and Utopia in Mass Culture,' *Social Text* 1 (1979): 148.
50. Jameson, *Postmodernism*, 36.
51. For a critique of the tendency among white critics, including feminists, to treat the experience of African-American women as a 'historicizing presence,' see Valerie Smith's statement that 'when black women operate in oppositional discourse as a sign for the author's awareness of materialist concerns, then they seem to be fetishized in much the same way as they are in mass culture'; Smith, 'Black Feminist Theory and the Representation of the "Other,"' in *Changing Our Own Words: Essays on Criticism, Theory, and Writing by Black Women* (New Brunswick: Rutgers University Press, 1989), 45–46.
52. Michael Warner, 'Fear of a Queer Planet,' *Social Text* 29 (1991): 17, n. 21. On the relation between gay identity and commodification, see also Wayne Koestenbaum's claim that 'mechanical reproduction is *not* second-rate: there is nothing wrong with becoming a clone, wanting to be famous for fifteen minutes, striving to be sexy through mimicry, or commodifying one's life, body, and work. To consider replication degrading is, literally, homophobic: *afraid of the same*'; Koestenbaum, 'Wilde's Hard Labor and the Birth of Gay Reading,' in *Engendering Men: The Question of Male Feminist Criticism*, ed. Joseph A. Boone and Michael Cadden (New York: Routledge, 1990), 182–183.
53. Haraway, *Simians, Cyborgs, and Women*, 177.
54. Ross, *Strange Weather*, 145.
55. Pat Cadigan, *Synners* (New York: Bantam, 1991) and *Patterns* (Kansas City: Ursus Imprints, 1989); Lisa Mason, *Arachne* (New York: William Morrow, 1990); Misha, *Red Spider, White Web* (Scotforth: Morrigan Books, 1990) and *Prayers of Steel* (Union: Wordcraft, 1988); and Laura J. Mixon, *Glass Houses* (New York: Tor Books, 1992).
56. Samuel R. Delany, 'Some *Real* Mothers,' interview with Takayuki Tatsumi, *Science Fiction Eye* 1 (March 1988): 9–10.
57. William Burroughs. *Naked Lunch* (New York: Grove Press, 1959), xi.
58. Langdon Winner argues that Western culture has tended to conceptualize its relation to technology through a typical master-slave dialectic: we are masters of our technology, but by that very token dependent on it, so that the concept of mastery contains within itself the possibility for transformation into enslavement; see Winner, *Autonomous Technology: Technics-out-of-Control as a Theme in Political Thought* (Cambridge: MIT Press, 1977), 187–191. Fred Pfeil associates this dialectic with the traditional science fictional polarity of the utopian and the dystopian. For Pfeil, cyberpunk rejects both the vision of a future in which control of technology necessarily creates a more perfect world and the vision of a dehumanized future in which machines dominate and organize human society; Pfeil. 'These Disintegrations,' 84–85.
59. Mark Poster. *The Mode of Information: Poststructuralism and Social Context* (Chicago: University of Chicago Press, 1990), 19.
60. For examples of such feminist readers, see Joan Gordon's 'Yin and Yang Duke It Out,' *Science Fiction Eye* 2 (February 1990): 37–39: and Hollinger. 'Cybernetic Deconstructions,' 32–33.

12

THE POSTMODERN ROMANCES OF FEMINIST SCIENCE FICTION

Jenny Wolmark

Science fiction has become a key source of metaphors through which to examine the uncertainties of postmodern culture and society. Its narrative strategy of representing the present as the past of some imagined future is particularly appropriate in the context both of the shifting spatial and temporal relations that are characteristic of postmodernism, and of the 'incredulity toward metanarratives' in western culture proposed by Lyotard (Lyotard, 1984: xxiv). It is within feminist science fiction, however, that the erosion of critical and cultural boundaries is most convincingly and enthusiastically explored, boundaries such as those between high and popular culture, nature and culture, self and other. Recent feminist science fiction transforms postmodern anxieties about definitions of the subject into utopian possibilities for the redefinition of gender identity and gender relations. In this chapter I argue that feminist science fiction crosses the boundaries of both gender and genre in two ways: firstly, by drawing on the narrative fantasies of popular romance fiction to offer fantasies of female pleasure and power, and secondly by using the 'hard science' metaphor of the cyborg to redefine definitions of female subjectivity.

Feminist science fiction uses the codes and conventions of both science fiction and romance fiction, and as such it acquires that postmodern ambivalence described by Linda Hutcheon, whereby it is 'doubly coded' because it is 'both complicitous with and contesting of the cultural dominants within which it operates' (Hutcheon, 1989: 142). The feminist science fiction that has been written since the 1970s has negotiated the inherent contradictions of this double coding placing the desiring 'I' of romance fiction at the centre of SF narratives, and in so doing feminist SF has profoundly influenced this most masculinist of genres. The postmodern romances of feminist SF provided an opportunity for women writers to foreground gender relations and to explore the possibilities for redefining them. This was the case in Marge Piercy's *Woman on the Edge of Time* (1976), and Ursula Le Guin's *The Left Hand of Darkness* (1969) and *The Dispossessed* (1974). Women-only future societies were created by a number of feminist SF writers such as Joanna Russ with *The Female Man* (1975), Suzy McKee Charnas with *Walk to the End of the World* (1974) and *Motherlines* (1978), Sally Miller Gearhart with *The Wanderground* (1979), and Alice Sheldon, who used the male pseudonym of James Tiptree Jr for her short story 'Houston, Houston, Do You Read?' (1976). Although none of the societies created by these writers were entirely utopian, they provided the fictional environment for the reworking of both gender identity and gender relations. Despite their ambiguous and sometimes embattled position within a genre that still appears to have a preponderance of white male authors and readers, these narratives have not only been able to make significant inroads into the dominant representations of gender, but they have also stretched the limits and definitions of the genre.

The women-only futures envisioned within feminist science fiction have been criticised, however, on the grounds that the social relations of science and technology which continue to inform the genre have been scrupulously avoided. Joan Gordon has argued that, not only are all feminist SF utopias dominated by images of a pastoral, organic world, but that most feminist SF 'incorporates a longing to go forward into the idealized past of earth's earlier matriarchal nature religions' (Gordon, 1991: 199). It is certainly the case that in the narratives of Sally Miller Gearhart and Suzy McKee Charnas, technology and the urban environment are associated with the repressive structures of patriarchy in a way that comes perilously close to a re-enactment of the duality of nature and culture. Sally Miller Gearhart's *The Wanderground*, for example, describes a community of women living outside the city in a complete and mystical harmony with nature. Such a communion is shown to be intrinsically impossible for men, for once they are outside the city, both they and 'their' technology become impotent and the earth itself rejects them. The emphasis on the essential qualities and differences of masculinity and femininity expressed in *The Wanderground* is also present in the two novels by Suzy McKee Charnas mentioned earlier: in the post-holocaust world of *Walk to the End of the World*, men fear and hate women for their gender, and keep them as dehumanised slaves in their cities. Only when women escape from the cities into the wilderness can they become free, and *Motherlines* focuses entirely on the separatist communities of women that have been established in the wilderness. The binarisms of masculine and feminine, nature and culture, are reproduced in these narratives so that they are as marked by essentialism as any romance text, despite the explicitly feminist intentions of the writers.

A shift of emphasis, however, can be discerned in feminist SF written from the 1980s on, as it confronts the question of gendered subjectivity more explicitly within the context of the masculinist hegemony of technology. Where romance has yet to find a set of metaphors that enables it to fully explore the socially and culturally constructed nature of gender, feminist science fiction has available to it the image of the cyborg, invoked by Donna Haraway in 'A Manifesto for Cyborgs' (Haraway, 1985). In this seminal discussion of the necessity to develop a new and non-essentialist politics of gender based on recognition of difference, Haraway argues that the cyborg, or cybernetic organism, is a metaphor that makes particular sense in the context of the postmodern breakdown of boundaries. The cyborg is a hybrid creature that transgresses and problematises binary oppositions and argues for the need for non-totalising and partial identities. Within feminist SF, the contradictory identities embodied by the cyborg provide a means of questioning existing definitions of gendered subjectivity. As Haraway suggests, 'The cyborgs populating feminist science fiction make very problematic the statuses of man or woman, human, artefact, member of a race, individual identity, or body' (Haraway, 1985: 97). The ambiguous postmodern romances of feminist SF are transgressive because they foreground female desire while evading closure around the binarisms of masculinity and femininity. The ironic 'cyborg monsters' (Haraway, 1985: 99) who inhabit these narratives move between and across the subject positions of masculinity and femininity and allow the relationship between identity and desire to be explored outside the confines of fixed subject positions. The cyborg can be regarded as a disruptive metaphor on number of counts: by problematising the relationship between human and machine, the cyborg also problematises the relationship between self and other, nature and culture. By providing an opportunity for feminist SF to explore possibilities for the redefinition of gender identity in the context of cybernetic systems, the cyborg disrupts the gendered power relations of technology. Finally, by challenging the masculinist hegemony over technology, the cyborg disrupts the generic stability of science fiction itself, since the genre has largely been built around the unquestioned assumption of that hegemony.

The cyborg in feminist SF challenges the power relations that are embedded in postmodern cybernetic systems, but which are usually suppressed. The cyborg therefore stands in opposition to the kind of cultural pessimism that is a noticeable feature of the accounts of postmodern culture provided by both Fredric Jameson and Jean Baudrillard. Jameson has asserted that postmodern culture no longer has the capacity to imagine the future (Jameson, 1982), and from this perspective the utopian longings that infuse both romance fiction and science fiction are absent because the utopian imagination itself has atrophied. Baudrillard demonstrates a similar cultural pessimism in his definition of hyperreality as the point where the contradiction between the real and the imaginary is effaced, the result of which, he suggests, is 'the end of metaphysics, the end of fantasy, the end of SF' (Baudrillard, 1991: 311). Both Jameson and Baudrillard regard the penetration of social and cultural structures by information technology and cybernetic systems as the key to the fragmentation and depthlessness of the postmodern condition. Electronic imaging and the development of Virtual Reality are perhaps the clearest contemporary expressions of this, because of the way in which they give 'virtual' embodiment to an imagined space. The phrase that has been coined to describe the technologies of fantasy at Disneyland, for example, is 'imagineering'; Baudrillard has described it in terms of the hyperreal, that which is more real than the real. For Baudrillard, Disneyland is 'presented as imaginary in order to make us believe that the rest is real' (Baudrillard, 1988: 172). While the boundaries between the real and the imagined are becoming increasingly unstable, so too are the boundaries between the cultural categories of high and popular culture, and the cultural pessimism of both Jameson and Baudrillard is tied to some extent to a model in which those categories remain fixed. In his discussion of the relations between mass culture and modernism, in *After the Great Divide* (Huyssen, 1986), Andreas Huyssen has pointed out that the distinction between high and mass culture is a gendered one, and that historically mass culture has been feminized as modernism's Other. Anxieties about the way in which the boundaries between high and popular culture are being rendered unstable, both by the impact of technology and by the eclecticism of postmodern parody and pastiche, can also, therefore, be seen as gendered.

Far from being constrained by such anxieties, feminist science fiction exploits postmodern instabilities. Within the shifting parameters of cybernetic systems, the boundaries between body and machine, self and other, become increasingly uncertain. As a result, definitions of difference and otherness, and the discourses of power within which those differences are articulated, can be questioned. In the accounts of postmodernism already discussed, this uncertainty is regarded as a threat, both to the unitary categories of culture and, by extension, to unitary definitions of the self. This is certainly the case in science fiction's other notable encounter with the postmodern, 'cyberpunk', in which the virtual realities of cybernetic systems are constantly threatening to undermine the imagined stabilities of gender. The street-wise computer hackers or 'console cowboys' of William Gibson's cyberpunk novels, *Neuromancer* (1986), *Count Zero* (1987) and *Mona Lisa Overdrive* (1989), for example, are driven by the desire to abandon the body and to interface with the matrix, a situation that is surely replete with possibilities for the reconstruction of gender identity and of gender relations. Yet these cyberpunk texts draw back from the possibilities of the interface, in which both self and other could be redefined in non-essentialist terms. Not only are the masculine identities of the console cowboys carefully preserved, even in the virtual realities of cyberspace, but because the artificial intelligences that inhabit the matrix are also defined as masculine, the interface itself is made masculine.

Feminist SF, on the other hand, uses the imagery and metaphors of cybernetic systems to challenge unitary definitions of the self and to offer an alternative and oppositional account of gender identity in which provisionality and multiplicity are emphasised. Because cyborg identity is always in process, it thrives on ambiguity and destabilizes sexual difference. Cyborg imagery is present in the work of feminist SF writers such as Octavia Butler, Vonda McIntyre, Joanna Russ and C.J. Cherryh, all of whom have produced narratives in which the politics of female desire are negotiated through the problematising of female identity. Popular romance narratives also explore female desire through ambivalence and ambiguity, but because that desire is achieved only within the existing relations of dominance and subordination characterising contemporary gender relations, identity remains fixed rather than fluid. In the remainder of this chapter I shall be considering the ambiguity and fluidity of gender identity within the postmodern romances of feminist SF, concentrating on the work of two American writers, Emma Bull and Elizabeth Hand, and an English writer, Gwyneth Jones.

Emma Bull's novel *Bone Dance* (1991) is a post-catastrophe narrative in cyberpunk mode, in which street-wise hustlers do 'biz' in a decaying urban environment. Bull uses several motifs drawn from cyperbunk, such as genetic engineering and mind control, and reuses them to focus on gender identity. It is the cyborgs in the narrative, with their incomplete and partial identities, that call the whole framework of binary divisions into question. All the cyborg bodies in the narrative are genetically engineered, or constructed, as are their identities, and both bodies and identities are subject to change during the course of the novel. This produces a gender incoherence in the narrative, which is explored through the central character of Sparrow, a gender neutral cyborg who takes on the 'look' of masculinity or femininity according to the needs of the moment. Because it is gender neutral, and always other, Sparrow's body becomes unintelligible in terms of the binary definitions of gender. This unintelligibility is also present amongst the second group of cyborgs in the narrative, who are able to take over and control the minds and bodies of whichever 'host' they desire, in a literal embodiment of desire for the Other. The fantasies of power and submission that are central to popular romance are partially lived out in the narrative through the powers of these cyborgs. However, although the cyborgs began as gendered subjects, they also take on both the gender and desires of whichever host body they inhabit, and this multiplicity of sexual identities and desires fatally disrupts the unequal power relations that are embedded in popular romance fiction. The fluid identities of the cyborgs enable them to move between the subject positions of masculinity and femininity, and since gender identity can no longer be resolved in binary terms, both fantasy and desire can be reconstituted as heterogeneous. Sparrow's cyborg body is the literal expression of this fluidity, since it refuses categorization in terms of gender altogether. The cyborgs in *Bone Dance*, then, transform the binarisms which structure popular romance fiction into a transgressive fantasy of multiplicity.

The baroque science fiction novels of Elizabeth Hand, *Winterlong* (1990) and *Aestival Tide* (1992), describe a post-catastrophe future world which includes artificial intelligence, genetic engineering, neurological control and germ warfare. The cyborgs within this environment are contradictory: they embody the dominant and hierarchical relations of control within cybernetic systems but also subvert them. There is an ambivalent relation between power and powerlessness in these novels which parallels that found in popular romance fiction, but in popular romance the subversive potential of this ambivalence is ultimately contained because the fantasies of female desire are structured around the fixed polarities of gender. In Hand's novels, the cyborg signifies an ambivalence and irresolution which calls those polarities into question, and as the boundary between self and other becomes increasingly ambiguous, desire can be differently configured. The novels are peopled with what Donna Haraway has called 'boundary creatures' (Haraway, 1991: 2), those transgressive cyborgs who exist disruptively outside the limits prescribed by the binarisms of gender, of nature and culture, of self and other. One such is Miss Scarlet, a talking ape that has been surgically altered to enable her to speak and to enhance her intellectual capacity, but whose physical embodiment has not changed so that she is neither wholly ape nor wholly human. The potential dissolution of boundaries between human and animal produces a cyborg identity in which the dualisms of nature and culture are thrown into disarray. The identity of a boundary creature like Miss Scarlet is enigmatic and unresolved, and as such it has the capacity to disrupt the binary order which cybernetic systems seek to impose.

In *Aestival Tide* the technologies of genetic engineering have advanced so far that the dead can be regenerated, to become a kind of mindless underclass, subject to the unmediated power relations of cybernetic systems. These are cyborgs without hope, incapable of offering any challenge to the gendered nature of those power relations. The significant challenge is offered instead through the potent combination of a doubly gendered hermaphrodite and a gendered android. These cyborg identities are neither unified nor coherent, and cannot be contained within the familiar constructions of gender identity. It is that incoherence which poses a threat to the already precarious stability of the hierarchical society described in the narrative, and both the android and the hermaphrodite are instrumental in its eventual collapse. The contradictory desires of these cyborgs are neither resolved within nor contained by the narrative, and the significance of their hybrid identities is that they disrupt what Judith Butler calls the 'regulatory fictions' (Butler, 1990: 33) of gender. The partial identities of 'boundary creatures' such as the hermaphrodite, the gendered android, and Miss Scarlet represent transgression against the limits of identity and subjectivity that are imposed by binary oppositions.

The novels of Gwyneth Jones present a more complex discussion of the transgressive potential of the cyborg's hybrid and partial identity. Her most recent novel, *White Queen* (1992), uses the familiar SF motif of an alien-human encounter to explore the collapse of stable boundaries between self and other, human and alien. The cultural construction of 'otherness' is a central concern of the novel, and the mutual incomprehension of gender relations and identities on the part of both humans and aliens is used to ironic effect to suggest that the binary framework is wholly inadequate for an understanding of difference. It is in the earlier novel, *Escape Plans*, however, that Jones engages more directly with the cyborg identities that emerge through the interface between human and machine. The penetration of social and cultural structures by cybernetic information systems noted by Jameson and Baudrillard provides the framework for the narrative, but it is the ability of such cyborg identities to transgress against the limits imposed by those systems that is of interest to Jones.

In *Escape Plans* (1986) Jones takes another familiar SF motif, that of the existence of an infinite universe, or universes, and ironically reverses it. It has been discovered that humanity is actually trapped in a 'bubble universe' that is possibly at the centre of the universe, if it had a centre, and has been since the beginning of the universe, if the universe had a beginning. Jones undermines the romanticised notion of an expansive future that is embedded in many science fiction narratives, and in the process she reveals the way in which the endless longing for an 'elsewhere' in such narratives reinforces an essentially conservative desire for transcendental otherness. In science fiction narratives which celebrate the transcendent otherness of infinitely expanding universes, the existing inequalities of power relations are never questioned. In contrast, Jones' self-reflexive narrative of containment makes the questioning of such relations central by refusing to subscribe to a definition of SF as the romance of the future. In describing a future that is about containment rather than expansion, the doubly coded narrative questions the power relations that produce, and are produced by, the binarisms of space and earth, inner and outer, nature and culture. It is a future in which the human–machine interface is both liberatory and repressive, and the cyborgs of this future both control and are controlled by the technology. Jones suggests that the interface contains not only possibilities for radical change but also possibilities for repression and exploitation, and the narrative refuses closure around either option.

The patterns of control that are embedded in cybernetic systems are scrutinised in the narrative, as are the relations of power that such systems embody. In the past of this future, the seemingly limitless possibilities of space exploration and total automation produced revolutionary changes. Women proved to be better suited to the interface than men and so, in space, they finally achieved equality with men, and as a result of these newly configured cyborg relations, heterosexuality no longer dominated social and sexual relations. However, the eventual discovery that space was finite undermined the utopian aspects of these dvelopments and cybernetic systems became instead a means of reproducing other unequal power relations based primarily on the distinction between the spacefaring elite, which controls the systems, and the 'Subs', those who have remained planet bound and who are ultimately subject to that control. In the present time of the narrative, cybernetic systems have become so ubiquitous that everyone has become 'plugged-in', but not with equal levels of access to the systems. Ironically, the most powerless of all the Subs are those who inter-face most completely with the central computer system: by means of brains sockets burned through the skull they process unimaginable amounts of information, but they are unable to access any of it. The Subs, however, regard the interface in almost mystical terms and the skull holes are valued as an indicator of both difference and status. Since the interface means one thing to the power elite and another to those who are subject to the needs of that elite, it is clear that the cyborg identities in this future are neither innocent nor free. The narrative indicates that, although women achieved social and political equality with men through the interface, unequal structures of power and control have continued to be embedded in the cybernetic systems: gender is, therefore, fully implicated in the perpetuation of those inequalities.

The 'pleasure of the interface' (Springer, 1991) is thus revealed as being somewhat ambiguous, and to emphasise this point, the self referential language of information systems is parodied throughout the narrative. Acronyms are used so extensively that they require a glossary, and personal names and acronyms are conflated as an ironic indication of the universality of the interface: the central character, for example, is known as both Alice and ALIC, depending on the formality of the address. The main focus of the narrative is ALIC's gradual realisation of the inequalities built into what seems to be a perfectly balanced cybernetic system. Her realisation, that it is precisely those inequalities that define definitions of identity and difference, enables the narrative to explore the pleasures and the pains, as well as the possibilities and the limitations, of cyborg identity.

The postmodern romances discussed in this chapter are concerned with the transgression of limits, and with the potential of the human-machine interface to define otherness outside the confines of binary oppositions. The cyborgs in these novels are made but not born, just as they are different but not 'other': it is not transcendence that they seek but rather the dissolution of the boundaries between inner and outer, self and other, nature and culture. The metaphor of the cyborg is a potent one because it is drawn from both the politics of information systems and the politics of gender. The binarisms inherent in the masculine hegemony of technology are also present in constructions of gender, and both sets of inequalities come together, in a very precise way, within cybernetic systems. Cyborg identities, however, are both imprecise and hybrid and as such they have the potential to re-order boundaries and demolish polarities. The doubly coded narratives of feminist science fiction do not pro-pose the magical resolutions to the problem of female desire that are common to popular romance fiction, since the object of desire is differently defined within them. Equally, the idealisation of gender identity that occurs in popular romance is unsustainable within the postmodern romance of feminist SF, because the metaphor of the cyborg undermines the fixed polarities of gender on which such idealisations depend. The cyborg identities of feminist SF emphasise the ambiguities and contradictions of subjectivity, and in the context of such a redefined subjectivity, femininity and female desire are subjected to extensive interrogation.

BIBLIOGRAPHY

Baudrillard, J. (1988) *Selected Writings*, M. Poster (ed.), Cambridge: Polity Press.
Baudrillard, J. (1991) 'Simulacra and Science Fiction', trans. A. B. Evans, *Science Fiction Studies*, 18, Part 3, 309–13.
Bull, E. (1991) *Bone Dance*, New York: Ace.
Butler, J. (1990) *Gender Trouble*, New York and London: Routledge.
Charnas, S. M. (1974) *Walk to the End of the World*, New York: Ballantine.
Charnas, M. (1978) *Motherlines*, New York: Berkley.
Gearhart, S. M. (1979) *The Wanderground*, Watertown: Persephone Press.
Gibson, W. (1986) *Neuromancer*, London: Grafton.
Gibson, W. (1987) *Count Zero*, London: Grafton.
Gibson, W. (1989) *Mona Lisa Overdrive*, London: Grafton.
Gordon, J. (1991) 'Yin and Yang Duke it Out: is Cyberpunk Feminism's New Age?', in L. McCaffery (ed.) *Storming The Reality Studio*, Durham: Duke University Press.
Hand, E. (1990) *Winterlong*, New York: Bantam.
Hand, E. (1992) *Aestival Tide*, New York: Bantam.
Haraway, D. (1985) 'A Manifesto for Cyborgs: Science, Technology, and Socialist Feminism in the 1980s', *Socialist Review*, 80, 65–107.
Haraway, D. (1991) *Simians, Cyborgs, and Women*, London: Free Press.
Hutcheon, L. (1989) *The Politics of Postmodernism*, London: Methuen.
Huyssen, A. (1986) *After the Great Divide: Modernism, Mass Culture, Postmodernism*, Bloomington: Indiana University Press.
Jameson, F. (1982) 'Progress versus Utopia; or, Can We Imagine the Future?', *Science Fiction Studies*, 9, Part 2, 147–58.
Jameson, F. (1991) *Postmodernism, or, The Cultural Logic of Late Capitalism*, London and New York: Verso.
Jones, G. (1986) *Escape Plans*, London: Unwin.
Jones, G. (1992) *White Queen*, London: VGSF.
Le Guin, U. (1969) *The Left Hand of Darkness*, New York: Ace.
Le Guin, U. (1974) *The Dispossessed*, New York: Harper & Row.
Lyotard, J.-F. (1984) *The Postmodern Condition: A Report on Knowledge*, trans. G. Bennington and B. Massumi, Manchester: Manchester University Press.
Piercy, M. (1976) *Woman on the Edge of Time*, New York: Knopf.
Russ, J. (1975) *The Female Man*, New York: Bantam.
Springer, C. (1991) 'The Pleasure of the Interface', *Screen*, 32, no. 3, 303–23.
Tiptree, J. Jr (1978) 'Houston, Houston, Do You Read?', in *Star Songs of an Old Primate*, New York: Ballantine.

PART 3
CYBORG FUTURES

INTRODUCTION

The third part of the book contains essays in which the metaphors of the cyborg and cyberspace appear more obliquely, and science fiction no longer provides the defining narrative context for discussion of the reconstituted subject. Rather, it is the politics of identity and subjectivity that are paramount in the following chapters, and reflections on the posthuman subject are set in a broad historical and materialist framework within which the positive and negative features of hybrid cyber-identities are scrutinised. The chapters in this section are authored by 'inappropriate/d others' who argue for a radical politics of difference in which the dominant orthodoxies surrounding representations of age, race and sexuality are challenged. They are critical of the continuing exclusions that appear in contemporary critical theory and argue the need for a more creative account of difference in postmodern, cybernetic theory.

Chela Sandoval draws on the text of Donna Haraway's 'Manifesto for Cyborgs' to develop a postcolonial, feminist critique of the political and ideological structures of first world postmodernity. She asks 'what constitutes "resistance" and oppositional politics under the imperatives of political, economic, and cultural transnationalization?', and argues for a cyborg consciousness that can usefully be understood as the embodiment of an oppositional consciousness that she describes as 'U.S. third world feminism'. Sandoval suggests that cyborg consciousness develops out of the 'methodology of the oppressed' which can 'provide the guides for survival and resistance under first world transnational cultural conditions', and as such can be thought of as constituting, at the very least, a '"cyber" form of resistance.' She argues that there is a close correlation between the politics of Haraway's 'Manifesto' and those of 'US third world feminist criticism', and that Haraway herself draws on US third world feminism 'for help in modeling the "cyborg" body' that can be capable of challenging what she calls the '"networks and informatics" of contemporary social reality.' Sandoval's engagement with both Haraway's text and the cyborg metaphor produces a complex model of the network of shifting alignments and associated consciousnesses that are necessary for an oppositional and differential politics; as she puts it, 'differential consciousness can be thus thought of as a constant reapportionment of space, boundaries, of horizontal and vertical realignments of oppositional powers.' Differential consciousness exists 'rhyzomatically and parasitically', and as such it functions as a tactical device for deconstructing fixed categories, in the process of which it reorders existing ways of thinking about the relationship between dominant and subordinate cultures and ideologies. For Sandoval, a crucial development for cyborg feminism is that it should become aligned to 'other movements of thought and politics for egalitarian social change.' This is a point that is taken up by Haraway in her own chapter, in the course of a discussion of both the differential consciousness invoked by Sandoval and what might be described as an ensuing set of differential politics.

In her chapter, Jennifer Gonzales is concerned with visual representations of cyborg bodies and 'cyborg consciousness' and their significance for the development of a 'cyborg body politics'. She argues that because the image of the cyborg has often appeared at moments of radical social and cultural change, then the cyborg body 'becomes the historical record of changes in human perception.' Images of the cyborg body signify 'the multiple fears and desires of a culture caught in the process of transformation' and thus represent 'that which cannot otherwise be represented.' Gonzales analyses a disparate group of visual representations of the cyborg body, and suggests that they are linked because they are all informed by a cyborg consciousness which is 'reflected in the spatial and political agency implied by a given cyborg body.' She points out that, historically, visual representations of the 'interface between automaton and autonomy' have reiterated, more or less uncritically, traditional gender roles as they appear in Euro-American culture; visual representations of cyborgs are thus compromised because they inevitably carry the traces of their historical origins. This point is reminiscent of Anne Balsamo's argument that it is important to be aware of the way in which the cyborg metaphor reproduces orthodox gender stereotypes because of the way in which it is embedded in contemporary culture. Gonzales points out that the original focus for her work was an exploration of 'the relationship between representations of cyborgs and representations of race and racial mixtures', and that her research has led her to the conclusion that 'the question of race is decidedly fraught'. She suggests that, despite the utopian assumption that cyborgs can eradicate difference, visual representations of cyborgs do not erase questions of racial identity. She concludes that, notwithstanding the progressive meanings of contemporary cyborgs, 'its "racial" body politics have a long way to go.'

Kathleen Woodward draws attention to two significant omissions from the contemporary technocriticism that is based around the new communication technologies, firstly, recognition of the equal importance of biotechnology, and secondly, the aging process. If technology can be regarded as a 'prosthesis of the human body', then communications technology provides an extension of the body in space, even as it simultaneously displaces the materiality of the body through the capacity to transmit and preserve its image. Developments in biotechnology, however, entail more than the 'mere extension of the body and its images': genetic engineering and reproductive technology are examples of the 'saturation, replication, alteration, and creation of the organic processes of the body – if not the body itself – by technoscience.' Key issues raised by biotechnology concerning gender and the technology of reproduction have been have taken up by feminists, and Woodward points out the importance of Haraway's 'Manifesto' in 'articulating the importance of biotechnology in contemporary cultural criticism': after all, the figure of the cyborg is a convergence of biotechnology and communications technology. Woodward is critical of the fact that, despite the attention paid within technocriticism and cultural theory to the body as a marker of the ideological construction of difference, the importance of age as a marker of difference has remained largely unconsidered, and the binary of 'youth' and 'age' has therefore gone largely unchallenged. She points out the ease with which technology and its artefacts are discarded once they are defined as 'old', which lends credence to her suggestion that technological culture could otherwise be defined as a 'youth culture'. In the rhetoric of technocriticism, the limits of the body are repudiated, but this becomes a more complex issue when age is considered, as in the case of postmenopausal mothers, for example, when 'biological barriers are lifted' by means of technology. Woodward conjectures that feminist support for reproductive rights has persistently conceptualised technology as 'an issue for young and midlife women – not one for women of all ages.' She contends that the possibilities offered by the intersection of biotechnology with concerns relating to the aging body should be explored in way that is positive, but which also ensures that no credence is given to the ultimate technological fantasy of eliminating the process and consequences of aging, since that in itself 'reveals a prejudice against aging that is harmful to us all.'

Donald Morton's discussion of queer theory engages with debates within postmodern theory concerning the universalising narratives of the enlightenment, the deconstruction of the subject, difference and location, and he argues against the abandonment of an epistemology and a politics explicitly based on concepts of equality and social justice. In his chapter, Morton aligns queer theory with what he calls 'ludic (post)modernism', and is equally critical of both. For Morton, queer studies, as opposed to gay and lesbian studies, is a contemporary form of idealism which is divorced from historical materialism, and thus, the 'return of the queer' and other notions such as 'virtual realities, cyberpunk, cybersex, teletheory', can be considered to be manifestations of forms of idealism emerging in late capitalism. For this reason, it is possible to characterise the return of the queer as the '(techno)birth of the cyberqueer.' Morton contends that the return of the queer 'has to be understood as the result, in the domain of sexuality, of the (post)modern encounter with – and rejection of – Enlightenment views concerning the role of the conceptual, rational, systematic, structural, normative, progressive, liberatory, revolutionary, and so forth, in social change.' He argues that the way in which queer theory 'fetishizes desire by rendering it autonomous', and thus ahistorical, is symptomatic of its rejection of enlightenment views, and contrasts this to a historical materialist analysis in which desire is fundamentally related to and structured by global social relations. Morton suggests that queer theory has endorsed Baudrillard's critique of the distinction made in Marxism between use value and exchange value, a critique that ultimately enables the signifier to be elevated over the signified, and desire to displace need; as a result, queer theory 'excludes the Enlightenment project of social progress envisioned by gay studies and renounces ... commonality in the name of uncapturable differences.' In an acknowledgment of contemporary technocriticism in which he also draws on Fredric Jameson's account of postmodernism as the 'cultural logic of late capitalism', Morton suggests that queer theory, and 'the hyperspace, cyberspace, and cyberpunk of technoculture', are all features of late, multinational capitalism. Further, he describes cyberspace as a 'bourgeois designer space' which is symptomatic of the displacement of need by desire, and this displacement ensures that the 'decentred, Interneted, normless society' envisioned by cyberized queer theory can be regarded as having little in common with the theoretically grounded politics and social idealism of the gay movement. Indeed, from the point of view of what Morton describes as the 'classical Marxist perspective', both 'postgay queerity and postleft political cyberpunk are no more than the latest forms of bourgeois idealism.' In a summary of his view of, and differences with, both postmodern and queer theory, Morton describes them as being 'caught in interminable Derridean self-dismantling, in the specificities of Foucauldian closeness, disconnectedness, and localism, or in a Lyotardian libidinal economy'.

In the final chapter of the book, it is clear that although the cyborg metaphor continues to inform Donna Haraway's theoretical practice, it does so in a more diffuse way, appearing here as a 'subject position' into which 'inappropriate/d others' are interpellated. This enables Haraway to avoid the lack of specificity for which the original cyborg metaphor was criticised, and the somewhat imprecise nature of the cyborg politics for which she argued in 'The Cyborg Manifesto' are set out more clearly as a politics of alliances based on the formation of diverse and active 'collectivities'. There are a number of interwoven arguments in the chapter which focus on the necessity to rethink not only definitions of nature and the human, but also by implication, that which is 'unnatural' and 'unhuman'; this then becomes part of the enabling process by which collective political action can be undertaken across a wide range of issues. Describing the essay as a contribution to the 'discourse of science studies *as* cultural studies', she examines the way in which the concept of 'nature' has been constructed in both science and culture, arguing that although it is a shifting and problematic concept, nevertheless the world exists for us as 'nature'. Thus, we cannot not desire it, but we need to rethink our relationship to it. Allucquere Rosanne Stone makes a similar point in her chapter, arguing that the category of 'nature' is as much a cultural construct as is 'technology': by redefining both nature and technology, it becomes possible to think of nature as technology rather than as merely technologised, which provides an opportunity to establish more challenging ways of interacting with it. Haraway suggests that our relationship to nature has been based on 'reification and possession' because of the way in which nature has been constituted as outside and 'other' in 'the histories of colonialism, racism, sexism, and class domination of many kinds'. If, however, nature was seen as 'artifactual', this would constitute both an oppositional and differential view because it recognises that nature is '*made* for us, as both fact and fiction', in both science and in culture.

Haraway declares that the aim of her chapter is to 'provide the roughest sketch for travel, by means of moving within and through a relentless artifactualism, which forbids any direct si(gh)tings of nature, to a science fictional, speculative factual, SF place called, simply, elsewhere.' This is a complex and multi-faceted 'mapping exercise' in which theory 'is *anything* but disembodied', because it is not about 'matters distant from the lived body'. She stresses that the actors who construct the world as nature are 'not all of them human, not all of them organic, not all of them technological', and that nature is just one of the 'cacophonous agencies' that must be admitted to the 'narrative of collective life'. From this point of view, nature is a 'commonplace and a powerful discursive construction, effected in the interactions among material-semiotic actors, human or not.' Haraway draws on the work of Vietnamese-American film maker Trinh Minh-ha to expand on the kind of 'differential artifactualism' that is necessary if a radical vision of a contemporary 'elsewhere' is to be achieved, and suggests that the idea of 'inappropriate/d others' can be particularly useful. As she explains it, 'to be an "inappropriate/d other" means to be in critical, deconstructive relationality, in a diffracting rather than reflecting (ratio)nality – as the means of making potent connection that exceeds domination. To be inappropriate/d is not to fit in the *taxon*, to be dislocated from the available maps specifying kinds of actors and kinds of narratives, not to be originally fixed by difference.' Haraway turns once again to science fiction to explain the way in which such a metaphor can work, because science fiction is generically concerned both with 'the interpenetration of boundaries between problematic selves and unexpected others', and also with 'the exploration of possible worlds in a context structured by transnational technoscience.' Not only do the 'emergent social subjects called "inappropriate/d others" inhabit such worlds' but they are also interpellated into a 'cyborg subject position' which is 'dangerous and replete with the promises of monsters.'

As part of mapping a way through to a new and radical 'elsewhere', Haraway adapts A. J. Greimas's semiotic square to demonstrate, not so much the progress from 'Artifactualism to Elsewhere', which she describes as 'A regenerative politics for inappropriate/d others', as the complex sets of inter-actions which go towards the construction and re-construction of the worlds that we inhabit. It enables her to raise key questions about whether there is a 'consequential difference between a political semiotics of articulation and a political semiotics of representation.' In other words, the question of who speaks, for whom, and in what circumstances, becomes crucial in the con-struction of a new and collective elsewhere. Haraway argues that, where the politics of representation are concerned, 'authorship rests with the representor, even as he claims independent object status for the represented'; the repre-sented are thus disempowered by being 'reduced to the permanent status of the recipient of action, never to be a co-actor in an articulated practice among unlike, but joined, social partners.' In contrast, the political semiotics of articulation can enable the world to be articulated through 'situated knowl-edges', as both human and unhumans interact within a social relationship and in a way that is 'crucial to the generativity of the collective.' A good example of the kind of collectivities and alliances that Haraway is thinking of can be found in her discussion of the human immune system and the AIDS Coalition to Unleash Power (ACT UP). She describes ACT UP as 'a collective built from many articulations among unlike kinds of actors – for example, activists, biomedical machines, government bureaucracies, gay and lesbian worlds, communities of color, scientific conferences, experimental organisms, mayors, international information and action networks, condoms and dental dams, computers, doctors, IV drug-users, pharmaceutical companies, publishers, virus components, counselors, innovative sexual practices, dancers, media technologies, buying clubs, graphic artists, scientists, lovers, lawyers, and more.' Collectivities of this sort can construct a politics of articulation which challenges, and has the potential to change, existing structures and relations of power; because articulation is based on the contingent and the strategic, the alliances that it constructs have constantly to be renewed, thus it is a 'contest-able practice' which has no absolute guarantee of success. For Haraway, however, it remains the most appropriate and potentially powerful form of political and theoretical praxis for the contemporary world.

13

NEW SCIENCES: CYBORG FEMINISM AND THE METHODOLOGY OF THE OPPRESSED

Chela Sandoval

We didn't cross the border, the border crossed us.
(Chicana/o Slogan)

If life is just a highway
Then the soul is just a car
And objects in the rear view mirror
May appear closer than they are
  (Meat Loaf)

If we are imprisioned by language,
then escape from that prison-house requires
language poets.
  (Donna Haraway)

What constitutes 'resistance' and oppositional politics under the imperatives of political, economic, and cultural transnationalization?[1] Current global restructuring is effecting the organizational formations not only of business, but of cultural economies, consciousness, and knowledge. Social activists and theorists throughout the twentieth century have been attempting to construct theories of opposition that are capable of comprehending, responding to, and acting back upon these globalizing forces in ways that renegotiate power on behalf of those Marx called the 'proletariet,' Barthes called the 'colonized classes.' Hartsock called 'women,' and Lorde called the 'outsiders.' If transnational corporations are generating 'business strategy and its relation to political initiatives at regional, national, and local levels,'[2] then, what are the concurrent forms of strategy being developed by the subaltern – by the marginalized – that focus on defining the forms of oppositional consciousness and praxis that can be effective under first world transnationalizing forces?

Let me begin by invoking Silicon Valley – that great land of Lockheed, IBM, Macintosh, Hewlett Packard, where over 30,000 workers have been laid off in the last two years, and another 30,000 more await a similar fate over the year to come: the fate of workers without jobs, those who fear for their livelihood. I begin here to honor the muscles and sinews of workers who grow tired in the required repetitions, in the warehouses, assembly lines, administrative cells, and computer networks that run the great electronic firms of the late twentieth century. These workers know the pain of the union of machine and bodily tissue, the robotic conditions, and in the late twentieth century, the cyborg conditions under which the notion of human agency must take on new meanings. A large percentage of these workers who are not in the administrative sector but in labor-grade sectors are U.S. people of color, indigenous to the Americas, or those whose ancestors were brought here as slaves or indentured servants; they include those who immigrated to the U.S. in the hopes of a better life, while being integrated into a society hierarchized by race, gender, sex, class, language, and social position. Cyborg life: life as a worker who flips burgers, who speaks the cyborg speech of McDonalds, is a life that the workers of the future must prepare themselves for in small, everyday ways.

My argument has been that colonized peoples of the Americas have already developed the cyborg skills required for survival under techno-human conditions as a requisite for survival under domination over the last three hundred years. Interestingly, however, theorists of globalization engage with the introduction of an oppositional 'cyborg' politics as if these politics have emerged with the advent of electronic technology alone, and not as a requirement of consciousness in opposition developed under previous forms of domination.

In this essay I propose another vision, wrought out of the work of cultural theorist and philosopher of science Donna Haraway, who in 1985 wrote the groundbreaking work on 'Cyborg Feminism,' in order to re-demonstrate what is overlooked in current cyborg theory, namely, that cyborg consciousness can be understood as the technological embodiment of a particular and specific form of oppositional consciousness that I have elsewhere described as 'U.S. third world feminism.'[3] And indeed, if cyborg consciousness is to be considered as anything other than that which replicates the now dominant global world order, then cyborg consciousness must be developed out of a set of technologies that together comprise the methodology of the oppressed, a methodology that can provide the guides for survival and resistance under first world transnational cultural conditions. This oppositional 'cyborg' consciousness has also been identified by terms such as 'mestiza' consciousness, 'situated subjectivities,' 'womanism,' and 'differential consciousness.' In the interests of furthering Haraway's own unstated but obvious project of challenging the racialization and apartheid of theoretical domains in the academy, and in the interests of translation, of transcoding from one academic idiom to another, from 'cyborgology' to 'feminism,' from 'U.S. third world feminism' to 'cultural' and to 'subaltern' theory, I trace the routes traveled by the methodology of the oppressed as encoded by Haraway in 'Cyborg Feminism.'

Haraway's research represents an example of scholarly work that attempts to bridge the current apartheid of theoretical domains: 'white male poststructuralism,' 'hegemonic feminism,' 'postcolonial theory,' and 'U.S. third world feminism.' Among her many contributions, Haraway provides new metaphoric grounds of resistance for the alienated white male subject under first world conditions of transnationalization, and thus the metaphor 'cyborg' represents profound possibilities for the twenty-first century (implications of hope, for example, for Jameson's lost subject which 'can no longer extend its protensions and retensions across the temporal manifold.'[4] Under cyborg theory, computer 'travel' can be understood as 'displacing' the 'self' in a similar fashion as the self was displaced under modernist dominations). An oppositional cyborg politics, then, could very well bring the politics of the alienated white male subject into alliance with the subaltern politics of U.S. third world feminism. Haraway's metaphor, however, in its travels through the academy, has been utilized and appropriated in a fashion that ironically represses the very work that it also fundamentally relies upon, and this continuing repression then serves to reconstitute the apartheid of theoretical domains once again. If scholarship in the humanities thrives under the regime of this apartheid, Haraway represents a boundary crosser, and her work arises from a place that is often overlooked or misapprehended under hegemonic understandings.

I have argued elsewhere that the methodology of the oppressed consists of five different technologies developed in order to ensure survival under previous first world conditions.[5] The technologies which together comprise the methodology of the oppressed generate the forms of agency and consciousness that can create effective forms of resistance under postmodern cultural conditions, and can be thought of as constituting a 'cyborg,' if you like, or at least a 'cyber' form of resistance.[6] The practice of this CyberConsciousness that is U.S. third world feminism, or what I refer to as a 'differential postmodern form of oppositional consciousness' has also been described in terms that stress its motion; it is 'flexible,' 'mobile,' 'diasporic,' 'schizophrenic,' 'nomadic' in nature. These forms of mobility, however, align around a field of force (other from motion itself) which inspires, focuses and drives them as oppositional forms of praxis. Indeed, this form of consciousness-in-opposition is best thought of as the particular field of force that makes possible the practices and procedures of the 'methodology of the oppressed.' Conversely, this methodology is best thought of as comprised of techniques-for-moving energy – or better, as *oppositional technologies of power*: both 'inner' or psychic technologies, and 'outer' technologies of social praxis.

These technologies can be summarized as follows: 1) What Anzaldua calls 'la facultad,' Barthes calls semiology, the 'science of signs in culture,' or what Henry Louis Gates calls 'signifyin' and Audre Lorde calls 'deep seeing' are all forms of 'sign-reading' that comprise the first of what are five fundamental technologies of this methodology. 2) The second, and well-recognized technology of the subaltern is the process of challenging dominant ideological signs through their 'de-construction': the act of separating a form from its dominant meaning. 3) The third technology is what I call 'meta-ideologizing' in honor of its activity: the operation of appropriating dominant ideological forms and using them whole in order to trans-form their meanings into a new, imposed, and revolutionary concept. 4) The fourth technology of the oppressed that I call 'democratics' is a process of locating: that is, a 'zeroing in' that gathers, drives, and orients the previous three technologies, semiotics, deconstruction, and meta-ideologizing, with the intent of bringing about not simply survival or justice, as in earlier times, but egalitarian social relations, or, as third world writers from Fanon through Wong, Lugones, or Collins have put it,[7] with the aim of producing 'love' in a de-colonizing, postmodern, post-empire world. 5) Differential movement is the fifth technology, the one through which, however, the others harmonically maneuver. In order to better understand the operation of differential movement, one must understand that it is a polyform upon which the previous technologies depend for their own operation. Only through differential movement can they be transferred toward their destinations, even the fourth, 'democratics,' which always tends toward the centering of identity in the interest of egalitarian social justice. These five technologies together comprise the methodology of the oppressed, which enables the enactment of what I have called the differential mode of oppositional social movement as in the example of U.S. third world feminism.

Under U.S. third world feminism, differential consciousness has been encoded as 'la facultad' (a semiotic vector), the 'outsider/within' (a de-constructive vector), strategic essentialism, (a meta-ideologizing vector), 'womanism' (a moral vector), and as 'la conciencia de la mestiza,' 'world traveling' and 'loving cross-cultures' (differential vectors).[8] Unlike westerners such as Patrick Moynihan who argue that 'the collapse of Communism' in 1991 proves how 'racial, ethnic, and national ties of difference can only ultimately divide any society,'[9] a differential form of oppositional consciousness, as utilized and theorized by a racially diverse U.S. coalition of women of color, is the form love takes in the postmodern world.[10] It generates grounds for coalition, making possible community across difference, permitting the generation of a new kind of citizenship, countrywomen and men of the same psychic terrain whose lives are made meaningful through the enactment of the methodology of the oppressed.

Whether interfaces with technology keep cyborg politics in re-newed contestation with differential (U.S. third world feminist and subaltern) politics is a question only the political and theoretical strategies of undoing apartheid – of all kinds – will resolve. The differential form of social movement and its technologies – the methodology of the oppressed – provide the links capable of bridging the divided minds of the first world academy, and of creating grounds for what must be considered a new form of transdisciplinary work that centers the methodology of the oppressed – of the subaltern – as a new form of post-western empire knowledge formation that can transform current formations and disciplinizations of knowledge in the academy. As we shall see in the following analysis of Haraway's theoretical work, the networking required to imagine and theorize 'cyborgian' consciousness can be considered, in part, a technologized metaphorization of the forms of resistance and oppositional consciousness articulated during the 1970s under the rubric of U.S. third world feminism. However, terms such as 'difference,' the 'middle voice,' the 'third meaning,' 'rasquache,' 'la conciencia de la mestiza,' 'hybridity,' 'schizophrenia,' and processes such as 'minor literature' and 'strategic essentialism' also call up and represent forms of that cyberspace, that other zone for consciousness and behaviour that is being proposed from many locations and from across disciplines as that praxis most able to both confront and homeopathically resist postmodern cultural conditions.

DONNA HARAWAY: FEMINIST CYBORG THEORY AND U.S. THIRD WORLD FEMINISM

Haraway's essay 'Manifesto for Cyborgs' can be defined in its own terms as 'theorized and fabricated hybrid,' a textual 'machine,' or as a 'fiction mapping our social and bodily reality,' phrases which Haraway also calls upon in order to re-define the term 'cyborg,' which, she continues, is a 'cybernetic organism,' a mixture of technology and biology, a 'creature' of both 'social reality' and 'fiction.'[11] This vision that stands at the center of her imaginary is a 'monstrous' image, for Haraway's cyborg is the 'illegitimate' child of dominant society and oppositional social movement, of science and technology, of the human and the machine, of 'first' and 'third' worlds, of male and female, indeed, of every binary. The hybridity of this creature is situated in relation to each side of these binary positions, and to every desire for wholeness, she writes, as 'blasphemy' (149) stands to the body of religion. Haraway's blasphemy is the cyborg, that which reproaches, challenges, transforms, and shocks. But perhaps the greatest shock in her feminist theory of cyborg politics takes place in the corridors of feminist theory, where Haraway's model has acted as a transcoding device, a technology that insists on translating the fundamental precepts of U.S. third world feminist criticism into categories that are comprehensible under the jurisdictions of Women's Studies.

Haraway has been very clear about these intellectual lineages and alliances. Indeed, she writes in her introduction to *Simians, Cyborgs and Women* that one primary aim of her work is similar to that of U.S. third world feminist theory and methods, which is, in Haraway's words, to bring about 'the break-up of versions of Euro-American feminist humanisms in their devastating assumptions of master narratives deeply indebted to racism and colonialism.' (It might be noted there that this same challenge, when uttered through the lips of a feminist theorist of color, can be indicted and even dismissed as 'undermining the movement' or as 'an example of separatist politics.') Haraway's second and connected aim is to propose a new grounds for theoretical and political alliances, a 'cyborg feminism' that will be 'more able' than the feminisms of earlier times, she writes, to 'remain attuned to specific historical and political positionings and permanent partialities without abandoning the search for potent connections.'[12] Haraway's cyborg feminism was thus conceived, at least in part, to recognize and join the contributions of U.S. third world feminist theorists who have challenged, throughout the 1960s, 1970s and 1980s what Haraway identifies as hegemonic feminism's 'unreflective participation in the logics, languages, and practices of white humanism.' White feminism, Haraway points out, tends to search 'for a single ground of domination to secure our revolutionary voice' (160).

These are thus strong ideological alliances, and so it makes sense that Haraway should turn to U.S. third world feminism for help in modeling the 'cyborg' body that can be capable of challenging what she calls the 'networks and informatics' of contemporary social reality. For, she affirms, it has been 'feminist theory produced by women of color' which has developed 'alternative discourses of womanhood,' and these have disrupted 'the humanisms of many Western discursive traditions.'[13] Drawing from these and other alternative discourses, Haraway was able to lay the foundations for her theory of cyborg feminism, yet she remains clear on the issue of that theory's intellectual lineages and alliances:

> White women, including socialist feminists, discovered (that is were forced kicking and screaming to notice) the non-innocence of the category 'woman.' That consciousness changes the geography of all previous categories; it denatures them as heat denatures a fragile protein. Cyborg feminists have to argue that 'we' do not want any more natural matrix of unity and that no construction is whole. (157)[14]

The recognition 'that no construction is whole,' however – though it helps – is not enough to end the forms of domination that have historically impaired the ability of U.S. liberation movements to effectively organize for equality. And for that reason, much of Haraway's ongoing work has been to identify the additional technical *skills* that are necessary for producing this different kind of coalitional, and what she calls 'cyborg,' feminism.

To understand Haraway's contribution, I want to point out and emphasize her correlation of these necessary skills with what I earlier identified as the methodology of the oppressed. It is no accident that Haraway defines, names and weaves the skills necessary to cyborgology through the techniques and terminologies of U.S. third world cultural forms, from Native American concepts of 'trickster' and 'coyote' being (199), to 'mestizaje,' or the category 'women of color,' until the body of the feminist cyborg becomes clearly articulated with the material and psychic positionings of U.S. third world feminism.[15] Like the 'mestiza consciousness' described and defined under U.S. third world feminism which, as Anzaldua explains, arises 'on borders and in margins' where feminists of color keep 'intact shifting and multiple identities' and with 'integrity' and love, the cyborg of Haraway's feminist manifesto must also be 'resolutely committed to partiality, irony, intimacy and perversity' (151). In this equivalent alignment, Haraway writes, feminist cyborgs can be recognized (like agents of U.S. third world feminism) to be the 'illegitimate offspring,' of 'patriarchal capitalism' (151). Feminist cyborg weapons and the weapons of U.S. third world feminism are also similar with 'transgressed boundaries, potent fusions and dangerous possibilities' (154). Indeed, Haraway's cyborg textual machine represents a politics that runs parallel to those of U.S. third world feminist criticism. Thus, insofar as Haraway's work is influential in feminist studies, her cyborg feminism is capable of insisting on an alignment between what was once hegemonic feminist theory with theories of what are locally apprehended as indigenous resistance, 'mestizaje,' U.S. third world feminism, or the differential mode of oppositional consciousness.[16]

This attempted alignment between U.S. feminist third world cultural and theoretical forms and U.S. feminist theoretical forms is further reflected in Haraway's doubled vision of a 'cyborg world,' which might be defined, she believes, as either the culmination of Euro-American 'white' society in its drive-for-mastery, on the one hand or, on the other, as the emergence of resistant 'indigenous' world views of mestizaje, U.S. third world feminism, or cyborg feminism. She writes:

> A cyborg world is about the final imposition of a grid of control on the planet, about the final abstraction embodied in Star Wars apocalypse waged in the name of defense, about the final appropriation of women's bodies in a masculinist orgy of war. From another perspective a cyborg world might be about lived social and bodily realities in which people are not afraid of their joint kinship with animals and machines, not afraid of permanently partial identities and contradictory standpoints. (154)[17]

The important notion of 'joint kinship' Haraway calls up here is analogous to that called for in contemporary indigenous writings where tribes or lineages are identified out of those who share, not blood lines, but rather lines of affinity. Such lines of affinity occur through attraction, combination, and relation carved out of and in spite of difference, and they are what comprise the notion of mestizaje in the writings of people of color, as in the 1982 example of Alice Walker asking U.S. black liberationists to recognize themselves as mestizos. Walker writes:

> We are the African and the trader. We are the Indian and the Settler. We are oppressor and oppressed ... we are the mestizos of North America. We are black, yes, but we are 'white,' too, and we are red. To attempt to function as only one, when you are really two or three, leads, I believe, to psychic illness: 'white' people have shown us the madness of that.[18]

Mestizaje in this passage, and in general, can be understood as a complex kind of love in the postmodern world where love is understood as affinity – alliance and affection across lines of difference which intersect both in and out of the body. Walker understands psychic illness as the attempt to be 'one,' like the singularity of Roland Barthes' narrative love that controls all meanings through the medium of the couple-in-love. The function of mestizaje in Walker's vision is more like that of Barthes' prophetic love, where subjectivity becomes freed from ideology as it ties and binds reality. Prophetic love undoes the 'one' that gathers the narrative, the couple, the race into a singularity. Instead, prophetic love gathers up the the mexcla, the mixture-that-lives through differential movement between possibilities of being. This is the kind of 'love' that motivates U.S. third world feminist mestizaje, and its theory and method of oppositional and differential consciousness, what Anzaldua theorizes as *la conciencia de la mestiza*, or 'the consciousness of the Borderlands.'[19]

Haraway weaves such U.S. third world feminist commitments to affinity-through-difference into her theory of cyborg feminism, and in doing so, begins to identify those skills that comprise the methodology of the oppressed, as indicated in her idea that the recognition of differences and their corresponding 'pictures of the world' (190) must not be understood as relativistic 'allegories of infinite mobility and interchangeability.' Simple mobility without purpose is not enough, as Gayatri Spivak posits in her example of 'strategic essentialism,' which argues both for mobility *and* for identity consolidation at the same time. Differences, Haraway writes, should be seen as examples of 'elaborate specificity' and as an opportunity for 'the loving care people might take to learn how to see faithfully from another point of view' (190). The power and eloquence of writings by certain U.S. feminists of color, Haraway continues, derives from their insistence on the 'power to survive' not on the basis of original innocence, (the imagination of a 'once-upon-a-time wholeness' or oneness), but rather on the insistence of the possibilities of affinity-through-difference. This mestizaje or differential consciousness allows the use of any tool at one's disposal (as long as its use is guided by the methodology of the oppressed) in order to both ensure survival and to remake the world. According to Haraway, the task of cyborg feminism must similarly be to 'recode' all tools of 'communication and intelligence,' with one's aim being the subversion of 'command and control' (175).

In the following quotation, Haraway analyzes Chicana intellectual Cherrie Moraga's literary work by applying a 'cyborg feminist' approach that is clearly in strong alliance with U.S. third world feminist methods. She writes:

> Moraga's language is not 'whole'; it is self-consciously spliced, a chimera of English and Spanish, both conqueror's languages. But it is this chimeric monster, without claim to an original language before violation, that crafts the erotic, competent potent identities of *women of color. Sister Outsider* hints at the possibility of world survival not because of her innocence, but because of her ability to live *on the boundaries*, to write without the founding myth of original wholeness, with its inescapable apocalypse of final return to a deathly oneness . . . Stripped of identity, the bastard race teaches about the power of the margins and the importance

of a mother like Malinche. Women of color have transformed her from the evil mother of masculinist fear into the originally literate mother who teaches survival. (175–76)

Ironically, U.S. third world feminist criticism, which is a set of theoretical and methodological strategies, is often understood by readers, even of Haraway, as a demographic constituency only ('women of color', a category which can be used, ironically, as an 'example' to advance new theories of what are now being identified in the academy as 'postmodern feminisms'), and not as itself a theoretical and methodological approach that clears the way for new modes of conceptualizing social movement, identity, and difference. The textual problem that becomes a philosophical problem, indeed, a political problem, is the conflation of U.S. third world feminism as a theory and method of oppositional consciousness with the demographic or 'descriptive' and generalized category 'women of color,' thus depoliticizing and repressing the specificity of the politics and forms of consciousness developed by U.S. women of color, feminists of color, and erasing the specificity of what is a particular *form* of these: U.S. third world feminism.

By 1991 Haraway herself recognizes these forms of elision, and how by gathering up the category 'women of color' and identifying it as a 'cyborg identity, a potent subjectivity synthesized from fusions of outsider identities' (i.e. Sister Outsider), her work inadvertently contributed to this tendency to elide the specific theoretical contributions of U.S. third world feminist criticism by turning many of its approaches, methods, forms and skills into examples of cyborg feminism (174). Haraway, recognizing the political and intellectual implications of such shifts in meaning, proceeded to revise her position, and six years after the publication of 'Cyborg Feminism' she explains that today, 'I would be much more careful about describing who counts as a "we" in the statement "we are all cyborgs"'. Instead, she asks, why not find a name or concept that can signify 'more of a family of displaced figures, of which the cyborg' is only one, 'and then to ask how the cyborg makes connections' with other non-original people who are also 'multiply displaced'?[20] Should we not be imagining, she continues, 'a family of figures' who could 'populate our imaginations' of 'postcolonial, postmodern worlds that would not be quite as imperializing in terms of a single figuration of identity'?[21] These are important questions for theorists across disciplines who are interested in effective new modes of understanding social movements and consciousness in opposition under postmodern cultural conditions. Haraway's questions remain unanswered across the terrain of oppositional discourse, however, or rather, they remain multiply answered and divided by academic terrain. And even within feminist theory, Haraway's own cyborg feminism and her later development of the technology of 'situated knowledges,' though they come close, have not been able to effectively bridge the gaps across the apartheid of theoretical domains described earlier.

For example, if Haraway's category 'women of color' might best be understood, as Haraway had earlier posited, 'as a cyborg identity, a potent subjectivity synthesized from fusions of outsider identities and in the complex political-historical layerings of her biomythography' (174), then why has feminist theory been unable to recognize U.S. third world feminist criticism itself as a mode of cultural theory which is also capable of unifying oppositional agents across ideological, racial, gender, sex or class differences, even if that alliance and identification would take place under the gendered, 'raced' and transnational sign 'U.S. third world feminism'? Might this elision be understood as yet another symptom of an active apartheid of theoretical domains? For, as I have argued, the nonessentializing identity demanded by U.S. third world feminism in its differential mode creates what Haraway is also calling for, a mestiza, indigenous, even cyborg identity.[22]

We can see Haraway making a very similar argument for the recognition of U.S. third world feminist criticism in her essay in *Feminists Theorize the Political*. Haraway's essay begins by stating that women who were 'subjected to the conquest of the new world faced a broader social field of reproductive unfreedom, in which their children did not inherit the status of human in the founding hegemonic discourses of U.S. society.'[23] For this reason, she asserts, 'feminist theory produced by women of color' in the U.S. continues to generate discourses that confute or confound traditional western standpoints. What this means, Haraway points out, is that if feminist theory is ever to be able to incorporate the visions of U.S. third world feminist theory and criticism, then the major focus of feminist theory and politics must make a fundamental shift to that of making 'a place for the different social subject.'[24]

This challenge to feminist theory – indeed, we can read it as a challenge to all social movement theory – represents a powerful theoretical and political shift, and if answered, has the potential to bring feminism, into affinity with such theoretical terrains as post-colonial discourse theory, U.S. third world feminism, postmodernism, and Queer Theory.

How might this shift be accomplished in the domain of feminist theory? Through the willingness of feminists, Haraway proposes, to become 'less interested in joining the ranks of gendered femaleness,' to instead become focused on 'gaining the INSURGENT ground as female social subject (95).'[25] This challenge to Women's Studies means that a shift must occur to an arena of resistance that functions outside the binary divide male/female, for it is only in this way, Haraway asserts, that 'feminist theories of gendered *racial* subjectivities' can 'take affirmative AND critical account of emergent, differentiating, self-representing, contradictory social subjectivities, with their claims on action, knowledge, and belief.'[26] Under this new form of what Haraway calls an 'anti-racist,' indeed, even an *anti-gender* feminism, Haraway asserts, 'there is no place for women,' only 'geometrics of difference and contradiction crucial to women's cyborg identities' (171).

It is at this point that Haraway's work begins to identify the specific technologies that fully align her theoretical apparatus with what I have called the methodology of the oppressed. How, then, might this new form of feminism, or what I would call this new form of oppositional consciousness, be brought into being? By identifying a set of skills that are capable of dis-alienating and realigning what Haraway calls the human 'join' that connects our 'technics' (material and technical details, rules, machines and methods), with our 'erotics' (the sensuous apprehension and expression of 'love'-as-affinity).[27] Such a joining, Haraway asserts, will require what is a savvy kind of 'politics of articulation,' and these are the primary politics that lay at 'the heart of an anti-racist feminist practice'[28] that is capable of making 'more powerful collectives in dangerously unpromising times.'[29] This powerful politics articulation, this new 'anti-racist' politics that is also capable of making new kinds of coalitions, can be recognized, argues Haraway, by identifying the 'skilled practices' that are utilized and developed within subaltern classes. Such skills, or technologies, what Haraway calls 'the standpoints of the subjugated' are preferred, she writes, because

> In principle they are least likely to allow denial of the critical and interpretive core of all knowledge. They are savvy to modes of denial through repression, forgetting, and disappearing acts – ways of being nowhere while claiming to see comprehensively. The subjugated have a decent chance to be on to the god-trick and all its dazzling – and therefore, blinding – illuminations. 'Subjugated' standpoints are pre-ferred because they seem to promise more adequate, sustained, objective, transforming accounts of the world. *But HOW to see from below is a problem requiring at least as much skill with bodies and language, with the mediations of vision, as the 'highest' technoscientific visualizations.* (191, emphasis mine)

Haraway's theoretical work outlines the forms taken by the subjugated knowl-edges she identifies. These forms required, as she writes, 'to see from below,' are particular skills that effect 'bodies,' 'language' and the 'mediations of vision.' Haraway's understanding of the nature of these skills cleaves closely to those same skills that comprise the methodology of the oppressed, which including the technologies of 'semiotics,' 'deconstruction,' 'meta-ideologizing,' 'democratics,' and 'differential movement.' It is these technologies that permit the constant, differential repositioning necessary for perception and action from what Haraway identifies as 'the standpoints of the subjugated.' Indeed, Haraway's essay on cyborg feminism identifies all five of these technologies (if only in passing) as ways to bring about what she hopes will become a new feminist methodology.

Of the first 'semiotic' technology, for example, Haraway writes that 'self knowledge requires a semiotic-material technology linking meanings and bodies ... the opening of non-isomorphic subjects, agents, and territories to stories unimaginable from the vantage point of the cyclopian, self-satiated eye of the master subject' (192). Though Haraway does not identify the technologies of 'deconstruction,' or 'meta-ideologizing' separately, these two interventionary vectors are implied when she writes that this new contribution to social movement theory, cyborg feminism, must find many 'means of understanding and intervening in the patterns of objectification in the world.' This means 'decoding and transcoding plus translation and criticism: all are necessary.' 'Democratics' is the technology of the methodology of the oppressed that guides all the others, and the moral force of this technology is indicated in Haraway's assertion that in all oppositional activity, agents for change 'must be accountable' for the 'patterns of objectification in the world' that have now become 'reality.' In this effort to take responsibility for the systems of domination that now exist, Haraway emphasizes that the practitioner of cyborg feminism cannot be 'about fixed locations in a reified body.' This new oppositional actor must be 'about nodes in fields' and 'inflections in orientation.' Through such mobilities, an oppositional cyborg feminism must create and enact its own version of, 'responsibility for difference in material-semiotic fields of meaning' (195). As for the last technology of the methodology of the oppressed, called 'differential movement,' Haraway's own version is that cyborg feminism must understand 'the impossibility of innocent "identity" politics and epistemologies as strategies for seeing from the standpoints of the subjugated.' Rather, oppositional agents must be 'committed' in the enactment of all forms-of-being and all skills, whether those 'skills' are semiotic, 'decoding,' 'recoding' or 'moral' in function, to what Haraway calls 'mobile positioning and passionate detachment' (192).

I have argued that the 'cyborg skills' necessary for developing a feminism for the twentieth century are those I have identified as the methodology of the oppressed. Their use has the power to forge what Haraway asserts can be a potentially 'earth-wide network of connections' including the ability to make new coalitions across new kinds of alliances by translating 'knowledges among very different – and power-differentiated – communities' (187). The feminism that applies these technologies as 'skills' will develop into another kind of science, Haraway asserts, a science of 'interpretation, translation, stuttering, and the partly understood.' Like the 'science' proposed under the differential mode of consciousness and opposition – U.S. third world feminism – cyborg feminism can become the science of those Haraway describes as the 'multiple subject with at least double vision.' Scientific 'objectivity' under this new kind of science, writes Haraway, will mean an overriding commitment to a practice capable of facing down bureaucratic and administrative sciences, a practice of 'objectivity' that Haraway calls 'situated knowledges' (188). For, she writes, with the advent of U.S. third world feminism and other forms of feminisms, it has become clear that 'even the simplest matters in feminist analysis require contradictory moments and a wariness of their resolution.' A scholarly and feminist consciousness-of-science, then, of objectivity as 'situated knowledges' means, according to Haraway, the development of a different kind of human relation to perception, objectivity, understanding, and production, that is akin to Hayden White and Jacques Derrida's use of the 'middle voice,' for it will demand the scholar's situatedness 'in an ungraspable middle space' (111). And like the mechanism of the middle voice of the verb, Haraway's 'situated knowledges' require that what is an 'object of knowledge' also be 'pictured as an actor and agent' (198), transformative of itself and its own situation while also being acted upon.

In other words, Haraway's situated knowledges demands a form of differential consciousness. Indeed, Haraway names the third part of her book *Simians, Cyborgs and Women* 'differential politics for inappropriated/d others.' This chapter defines a coalescing and ever more articulated form of social movement from which 'feminist embodiment' can resist 'fixation' in order to better ride what she calls the 'webs of differential positioning' (196). Feminist theorists who subscribe to this new postmodern form of oppositional consciousness must learn, she writes, to be 'more generous and more suspicious – both generous and suspicious, exactly the receptive posture I seek in political semiosis generally. It is a strategy closely aligned with the oppositional and differential consciousness'[30] of U.S. third world feminism.

It was previously assumed that the behaviors of oppressed classes depend upon no methodology at all, or rather, that they consist of whatever acts one must commit in order to survive, both physically and psychically. But this is exactly why the methodology of the oppressed can now be recognized as that mode-of-being best suited to life under postmodern and highly technologized conditions in the first world. For to enter a world where any activity is possible in order to ensure survival is also to enter a cyberspace-of-being and consciousness. This space is now accessible to all human beings through technology, (though this was once a zone only accessible to those forced into its terrain), a space of boundless possibilities where meanings are only cursorily attached and thus capable of reattaching to others depending upon the situation to be confronted. This cyberspace is Barthes' zero degree of meaning and prophetic love, Fanon's 'open door of every consciousness,' Anzaldua's 'Coatlique' state, and its processes are linked closely with those of differential consciousness.

To reiterate, the differential mode of oppositional consciousness finds its expression through the methodology of the oppressed. The technologies of semiotic reading, de-construction of signs, meta-ideologizing, and moral commitment-to-equality are its vectors, its expressions of influence. These vectors meet in the differential mode of consciousness, carrying it through to the level of the 'real' where it can guide and impress dominant powers. Differential consciousness is itself a force which rhyzomatically and parasitically inhabits each of these five vectors, linking them in movement, while the pull of each of the vectors creates on-going tension and re-formation. Differential consciousness can be thus thought of as a constant reapportionment of space, boundaries, of horizontal and vertical realignments of oppositional powers. Since each vector occurs at different velocities, one of them can realign all the others, creating different kinds of patterns, and permitting entry at different points. These energies revolve around each other, aligning and realigning in a field of force that is the differential mode of oppositional consciousness, a Cyber-Consciousness.

Each technology of the methodology of the oppressed thus creates new conjunctural possibilities, produced by ongoing and transforming regimes of exclusion and inclusion. Differential consciousness is a crossing network of consciousness, a transconsciousness that occurs in a register permitting the networks themselves to be appropriated as ideological weaponry. This cyber-space-of-being is analogous to the cyberspace of computer and even social life in Haraway's vision, but her understanding of cyberspace is more pessimistic: 'Cyberspace seems to be the consensual hallucination of too much complexity, too much articulation ... In virtual space, the virtue of articulation, the power to produce connection threatens to overwhelm and finally engulf all possibility of effective action to change the world.'[31] Under the influence of a differential oppositional consciousness understood as a form of 'cyberspace,' the technologies developed by subjugated populations to negotiate this realm of shifting meanings are recognized as the very technologies necessary for all first world citizens who are interested in re-negotiating contemporary first world cultures with what we might call a sense of their own 'power' and 'integrity' intact. But power and integrity, as Gloria Anzaldua suggests, will be based on entirely different terms then those identified in the past, when, as Jameson writes, individuals could glean a sense of self in opposition to the centralizing dominant power that oppressed them, and then determine how to act. Under postmodern disobediencies the self blurs around the edges, shifts 'in order to ensure survival,' transforms according to the requisites of power, all the while, under the guiding force of the methodology of the oppressed carrying with it the integrity of a self-conscious awareness of the transformations desired, and above all, a sense of the impending ethical and political changes that those transformations will enact.

Haraway's theory weds machines and a vision of first world politics on a transnational, global scale together with the apparatus for survival I call the methodology of the oppressed in U.S. third world feminism, and it is in these couplings, where race, gender, and capital, according to Haraway, 'require a cyborg theory of wholes and parts' (181), that Haraway's vision contributes to bridging the gaps that are creating the apartheid of theoretical domains. Indeed, the coding necessary to re-map the kind of 'disassembled and reassembled postmodern collective and personal self' (163) of cyborg feminism must take place according to a guide capable of placing feminism in alignment with other movements of thought and politics for egalitarian social change. This can happen when being and action, knowledge and science, become self-consciously encoded through what Haraway calls 'subjugated' and 'situated' knowledges, and what I call the methodology of the oppressed, a methodology arising from varying locations and in a multiplicity of forms across the first world, and indominably from the minds, bodies, and spirits of U.S. third world feminists who have demanded the recognition of 'mestizaje,' indigenous resistance, and identification with the colonized. When feminist theory becomes capable of self-consciously recognizing and applying this methodology, then feminist politics can become fully synonymous with anti-racism, and the feminist 'subject' will dissolve.

In the late twentieth century, oppositional actors are inventing a new name and new languages for what the methodology of the oppressed and the 'Coatlicue,' differential consciousness it demands. Its technologies, from 'signifyin'' to 'la facultad,' from 'cyborg feminism' to 'situated knowledges,' from the 'abyss' to 'differance' have been variously identified from numerous theoretical locations. The methodology of the oppressed provides the schema for the cognitive map of power-laden social reality for which oppositional actors and theorists across disciplines, from Fanon to Jameson, from Anzaldua to Lorde, from Barthes to Haraway, are longing.

NOTES

1. Dedicated to those who move in resistance to the 'proper,' and especially to Chicana feministas Yolanda Broyles-Gonzalez, Antonia Castaneda, Deena Gonzalez, Emma Perez, Gloria Anzaldua, Shirley Munoz, Norma Alarcon, Ellie Hernandez, Pearl Sandoval, and Tish Sainz.
2. Richard P. Appelbaum, 'New Journal for Global Studies Center,' CORI: Center for Global Studies Newsletter, Vol 1 No 2, May 1994.
3. See 'U.S. Third World Feminism: The Theory and Method of Oppositional Consciousness in the Postmodern World,' which lays the groundwork for articulating the methodology of the oppressed. Genders 10, University of Texas Press, Spring 1991.
4. Fredric Jameson, 'Postmodernism: The Cultural Logic of Late Capitalism,' New Left Review, No 146, July–August, pp. 53–92.
5. The Methodology of the Oppressed, forthcoming, Duke University Press.
6. The term 'cybernetics' was coined by Norbert Wiener from the Greek word 'Kubernetics,' meaning to steer, guide, govern. In 1989 the term was split in two, and its first half 'cyber' (which is a neologism with no earlier root) was broken off from its 'control' and 'govern' meanings to represent the possibilities of travel and existence in the new space of computer networks, a space, it is argued, that must be negotiated by the human mind in new kinds of ways. This cyberspace is imagined in virtual reality films like Freejack, The Lawnmower Man and Tron. But it was first termed 'cyberspace' and explored by the science fiction writer William Gibson in his 1987 book Neuromancer. Gibson's own history, however, passes through and makes invisible 1970's feminist science fiction and theory, including the works of Russ, Butler, Delany, Piercy, Haraway, Sofoulis, and Sandoval. In all cases, it is this 'Cyberspace' that can also adequately describe the new kind of movement and location of differential consciousness.
7. For example, Nellie Wong, 'Letter to Ma,' This Bridge Called My Back; Maria Lugones, 'World Traveling;' Patricia Hill Collins, Black Feminist Thought; June Jordan, 'Where is the Love?,' Haciendo Caras.
8. It is through these figures and technologies that narrative becomes capable of transforming the moment, of changing the world with new stories, of meta-ideologizing. Utilized together, these technologies create trickster stories, stratagems of magic, deception, and truth for healing the world, like Rap and CyberCinema,

which work through the reapportionment of dominant powers.

9. *MacNeil/Lehrer NewsHour*, November 1991.
10. See writings by U.S. feminists of color on the matter of love, including June Jordan, 'Where is the Love?,' Merle Woo, 'Letter to Ma,' Patricia Hill Collins, *Black Feminist Thought*, Maria Lugones, 'Playfulness, "World-Traveling", and Loving Perception,' and Audre Lorde, *Sister Outsider*.
11. Donna Haraway, *Simians, Cyborgs, and Women: The Reinvention of Nature*, (Routledge: New York, 1991), 150. All quotations in this section are from this text (especially chapters eight and nine 'A Cyborg Manifesto: Science, Technology, and Socialist-Feminism in the Late Twentieth Century' and 'Situated Knowledges: The Science Question in Feminism and the Privilege of Partial Perspective') unless otherwise noted. Further references to this work will be found in the text.
12. Ibid., p. 1.
13. Donna Haraway, 'Ecce Homo, Ain't (Ar'n't) I a Woman, and Inappropriate/d Others: The Human in a Post-Humanist Landscape,' *Feminists Theorize the Political*, ed. Judith Butler and Joan Scott, (New York: Routledge, 1992), 95.
14. This quotation historically refers its readers to the impact of the 1970's U.S. third world feminist propositions which significantly revised the women's liberation movement by, among other things, renaming it with the ironic emphasis 'the *white* women's movement.' And perhaps all uncomplicated belief in the righteous benevolence of U.S. liberation movements can never return after Audre Lorde summarized seventies' women's liberation by saying that 'when white feminists call for "unity"' among women 'they are only naming a deeper and real need for homogeneity.' By the 1980's the central political problem on the table was how to go about imagining and constructing a feminist liberation movement that might bring women together across and through their differences. Haraway's first principle for action in 1985 was to call for and then teach a new hoped-for constituency, 'cyborg feminists,' that '"we" do not want any more natural matrix of unity and that no construction is whole.'
15. See Haraway's 'The Promises of Monsters,' *Cultural Studies*, (New York: Routledge, 1992), 328, where the woman of color becomes the emblematic figure, a 'disturbing guide figure,' writes Haraway, for the feminist cyborg, 'who promises information about psychic, historical and bodily formations that issue, perhaps from some other semiotic processes than the psychoanalytic in modern and postmodern guise' (306).
16. U.S. third world feminism recognizes an alliance named 'indigenous mestizaje,' a term which insists upon the kinship between peoples of color similarly subjugated by race in U.S. colonial history (including but not limited to Native peoples, colonized Chicano/as, Blacks, and Asians), and viewing them, in spite of their differences, as 'one people.'
17. Haraway's contribution here is to extend the motion of 'mestizaje' to include the mixture, or 'affinity,' not only between human, animal, physical, spiritual, emotional and intellectual being as it is currently understood under U.S. third world feminism, but between all these and the machines of dominant culture too.
18. Alice Walker, 'In the Closet of the Soul: A Letter to an African-American Friend,' *Ms. Magazine* 15 (November 1986): 32–35. Emphasis mine.
19. *Borderlands*, p. 77.
20. Constance Penley and Andrew Ross, 'Cyborgs at Large: Interview with Donna Haraway,' *Technoculture*, Minneapolis: University of Minnesota Press, 1991), 12.
21. Ibid., p. 13.
22. We might ask why dominant theoretical forms have proven incapable of incorporating and extending theories of black liberation, or third world feminism. Would not the revolutionary turn be that theorists become capable of this kind of 'strategic essentialism?' If we believe in 'situated knowledges,' then people of any racial, gender, sexual categories can enact U.S. third world feminist practice. Or do such

practices have to be transcoded into a 'neutral' language that is acceptable to all separate categories, 'differential consciousness,' for example, or 'cyborgology'?

23. Haraway, 'Ecce Homo,' 95.
24. Ibid. Emphasis mine.
25. The new theoretical grounds necessary for understanding current cultural conditions in the first world and the nature of resistance is not limited to feminist theory, according to Haraway. She writes, 'we lack sufficiently subtle connections for collectively building effective theories of experience. Present efforts – Marxist, psychoanalytic, feminist, anthropolitical – to clarify even "our" experience are rudimentary' (173).
26. Ibid., p. 96. Emphasis mine.
27. Haraway, 'The Promises of Monsters,' 329.
28. Ibid.
29. Ibid., p. 319.
30. Ibid., p. 326.
31. Ibid., p. 325.

14

ENVISIONING CYBORG BODIES: NOTES FROM CURRENT RESEARCH

Jennifer González

> The truth of art lies in this: that the world really is as it appears in the work of art. (Herbert Marcuse)

The cyborg body is the body of an imagined cyberspatial existence. It is the site of possible being. In this sense it exists in excess of the real. But it is also imbedded within the real. The cyborg body is that which is already inhabited and through which the interface to a contemporary world is already made. Visual representations of cyborgs are thus not only utopian or dystopian prophesies, but are rather reflections of a contemporary state of being. The image of the cyborg body functions as a site of condensation and displacement. It contains on its surface and in its fundamental structure the multiple fears and desires of a culture caught in the process of transformation. Donna Haraway has written,

> A cyborg exists when two kinds of boundaries are simultaneously problematic: 1) that between animals (or other organisms) and humans, and 2) that between self-controlled, self-governing machines (automatons) and organisms, especially humans (models of autonomy). The cyborg is the figure born of the interface of automaton and autonomy.[1]

Taking this as a working definition, one can consider any body a cyborg body that is both its own agent and subject to the power of other agencies. To keep to the spirit of this definition but to make it more specific, an *organic cyborg* can be defined as a monster of multiple species, whereas a *mechanical cyborg* can be considered a techno-human amalgamation (there are also conceivable overlaps of these domains). While images of *mechanical cyborgs* will be the focus of the short essays that follow, both types of cyborgs, which appear frequently in Western visual culture, are metaphors for a third kind of cyborg – a cyborg consciousness.[2] This last, is both manifest in all the images included here, and is the invisible force driving their production, what Michel Foucault might call a 'positive unconscious.'[3] This unconscious is reflected in the spatial and political agency implied by a given cyborg body. Unlike some of my contemporaries, I do not see the cyborg body as primarily a surface or simulacrum which signifies only itself; rather the cyborg is like a symptom – it represents that which cannot otherwise be represented.

The following group of eclectic commentaries is meant to address a sample of the issues which may arise in the consideration of representing cyborg bodies. The images here were chosen not because they are the most beautiful, the most frightening or the most hopeful of the cyborg visions I have found, but because they incarnate what seem to be important features for any consideration of a 'cyborg body politics.'[4] This is only a beginning.

MECHANICAL MISTRESS

Flanked by rows of cypress, demurely poised with a hand on one hip, the other hand raised with a pendulous object hanging from plump and delicate fingers stands 'L'Horlogère' (The Mistress of Horology) (Figure 14.1). From above her head stares the circular face of time, supported by a decorative frame through which her own face complacently gazes. Soft feminine shoulders descend into a tightly sculpted bust of metal. Cinched at the waist, her skirts flounce into a stiffly ornamental 'montre emboeté' which rests on dainty feet, toes curled up to create the base. An eighteenth-century engraving, this image by an unknown printer depicts what we today might call a cyborg. The body of the woman is not merely hidden inside the machine (despite the two tiny

Figure 14.1 *L'Horlogère*

human feet that peek out from below), nor is the organic body itself a mechanical replica, rather the body and the machine are a singular entity. In contrast to, but within the context of, the popular depictions of entirely mechanical automata – the predecessors of our modern-day robots – this image represents an early conception of an ontological merging of 'cultural' and 'natural' artifacts.

Taken as a form of evidence, the representation of an amalgam such as this can be read as a symptom of the pre-industrial unconscious. *L'Horlogère* substantiates an ideology of order, precision, and mechanization. French philosopher Julien Offray de La Mettrie, in his essay 'L'Homme Machine' (1748) wrote 'The human body is a machine which winds its own springs. It is the living image of perpetual movement.'[5] The beating hearts of many Europeans at this time no doubt sounded an apprehensive ticking. Wound up to serve the industrial impulse, the human model of perfection culminated in a mechanized identity. Variations of this ideal continued into the nineteenth century with the expansion of large-scale industrial production. Scholar Julie Wosk writes,

> ... artists' images of automatons became central metaphors for the dreams and nightmares of societies under-going rapid technological change. In a world where new labor-saving inventions were expanding human capabilities and where a growing number of people were employed in factory systems calling for rote actions and impersonal efficiency, nineteenth-century artists confronted one of the most profound issues raised by new technologies: the possibility that people's identities and emotional lives would take on the properties of machines.[6]

But was this not exactly what was desired by one part of the population – that another part of the population become mechanical? Was this not also exactly what was feared? The artists, depicting an experience already lived by a large portion of the population, were reflecting a situation in which the relation – and the distinction – between the machine and the human became a question of gender and class. Those who had access to certain machines were privileged, those who were expected to behave like certain machines were subjugated. The same is true today.

The pre-industrial representation of *L'Horlogère* thus functions as an early prototype of later conceptual models of the cyborg. The woman is a clock, the clock is a woman – complex, mechanical, serviceable, decorative. Her history can be traced to automatic dolls with clockwork parts dating back to at least the fifteenth century in Europe[7] and much earlier in China, Egypt and Greece.[8] Of the examples which I have found of such automata, a decided majority represent female bodies providing some form of entertainment.

> The idea of automatons as useful servants and amusing toys continued in the designs of medieval and Renaissance clockmakers, whose figures, deriving their movements from clock mechanisms, struck the hours ... [9]

The history of the automaton is thus imbedded in the mechanical innovations of keeping time, and *L'Horlogère* undoubtedly derives much of her status from this social context. She is clearly an embodied mechanism, but she has the privilege of her class. The imaginative engraver who produced this image undoubtedly wished to portray *L'Horlogère* as aristocratic; as one who could acquire, and therefore represent, the height of technological development. As machine, she displays the skill and artistry of the best engineers of her epoch.

The fact that she represents a female body is indicative of the role she is meant to play as the objectification of cultural sophistication and sexuality. Her gender is consistent with the property status of an eighteenth-century decorative artifact.

L'Horlogère is not merely an automaton. As part human, she should have human agency, or some form of human being. Her implied space of agency is, nevertheless, tightly circumscribed. This cyborg appears more trapped by her mechanical parts than liberated through them. If a cyborg is 'the figure born of the interface of automaton and autonomy,' then to what degree can this cyborg be read as a servant and toy, and to what degree an autonomous social agent? In order to determine the character of any given cyborg identity and the range of its power, one must be able to examine the *form* and not merely the fact of this interface between automaton and autonomy. For, despite the potentially progressive implications of a cyborg subject position,[10] the cyborg is not necessarily more likely to exist free of the social constraints which apply to humans and machines already. 'The machine is us, our processes, an aspect of our embodiment,'[11] writes Donna Haraway. It should therefore come as no surprise that the traditional, gendered roles of Euro-American culture are rarely challenged in the visual representations of cyborgs – a concept which itself arises from an industrially 'privileged' Euro-American perspective. Even the conceptual predecessors of the cyborg are firmly grounded in everyday social politics; 'Tradition has it that the golem first did housework but then became unmanageable.'[12] The image of *L'Horlogère* thus provides a useful ground and a visual tradition from which to explore and compare more contemporary examples of cyborg bodies.

SIGNS OF CHANGING CONSCIOUSNESS

The image of the cyborg has historically recurred at moments of radical social and cultural change. From bestial monstrosities, to unlikely montages of body and machine parts, to electronic implants, imaginary representations of cyborgs take over when traditional bodies fail. In other words, when the current ontological model of human being does not fit a new paradigm, a hybrid model of existence is required to encompass a new, complex and contradictory lived experience. The cyborg body thus becomes the historical record of changes in human perception. One such change may be reflected in the implied redefinition of the space the cyborg body inhabits.

Taking, for example, the 1920 photomontage by Hannah Höch entitled *Das schöne Mädchen* (The Beautiful Girl) (Figure 14.2) it is possible to read its dynamic assemblage of images as an allegory of modernization. Allied with the Dadaists of the Weimar Republic, Höch provided a chaotic vision of the rapid social and cultural change that followed in the wake of World War I. In *Das schöne Mädchen* the figure of a woman is set in the midst of a disjointed space of automobile and body parts. BMW logos, a severed hand holding a watch, a flying wig, a parasol, a hidden feminine face and a faceless boxer leaping

Figure 14.2 Hannah Höch, *Das schöne Mädchen* (The Beautiful Girl), 1919–20, 35 × 29 cm, photomontage, private collection

through the tire of a car surround the central figure whose head has been replaced with an incandescent light bulb – perhaps as the result or condition of her experience.

Many of Höch's early photomontages focus on what might be called the 'New Woman' in Weimar Germany. These images are not simply a celebration of new, 'emancipated' roles for women in a period of industrial and economic growth, they are also critical of the contradictory nature of this experience as depicted in the mass media. 'Mass culture became a site for the expression of anxieties, desires, fears and hopes about women's rapidly transforming identities,' writes Maude Lavin in her recent book *Cut With the Kitchen Knife: The Weimar Photomontages of Hannah Höch.* 'Stereotypes of the New Woman generated by the media could be complex and contradictory: messages of female empowerment and liberation were mixed with others of dependence, and the new consumer culture positioned women as both commodities and customers.'[13] Existing across several domains, the New Woman was forced to experience space and presence in new and ambiguous ways.

Traveling through and across this space could therefore be both physically and psychologically disorienting. The experience of a disjointed modern space takes form in Höch's collage as a cyborg body suspended in chaotic perspective (the hand-held parasol in the image is reduced to one-tenth the size of the figure's floating hair), with body parts chopped off, and with new mechanical/electric parts added in their place (the missing hand severed at the figure's wrist reappears holding the watch in the foreground). It is impossible to tell exactly which spatial plane is occupied by the body and of what sort of perception this body is capable. Yet, despite her Dadaist affiliation, Höch's work tends not to be random. Her images produce a discordant but strikingly accurate appraisal of an early twentieth-century experience of modernism. Here, existence as a self-contained humanist subject is overcome by an experience of the body in pieces – a visual representation of an unconscious state of being that exceeds the space of the human body. Perception is aligned to coincide with the machine. The effects of such an alignment are made alarmingly transparent by Virginia Woolf in her 1928 novel *Orlando*. In her description of the uncanny event of experiencing the world from the perspective of the automobile, her metaphor of torn scraps of paper is particularly appropriate to Höch's use of photomontage.

After twenty minutes the body and mind were like scraps of torn paper tumbling from a sack, and indeed, the process of motoring fast out of London so much resembles the chopping up small of body and mind, which precedes unconsciousness and perhaps death itself that it is an open question in what sense Orlando can be said to have existed at the present moment.[14]

The questionable existence of Orlando, Virginia Woolf's protagonist who changes gender and who adapts to social and cultural changes across many centuries and continents, is not unlike the questionable existence of the cyborg. It is the existence of a shifting consciousness that is made concrete only in moments of contradictory experience. The attempt to represent and reassemble – but not to repair – the multiple scraps of body and mind that are scattered at such historical junctures has, in fact, been a central activity of modernism.

Photomontage has served as a particularly appropriate medium for the visual exploration of cyborgs. It allows apparently 'real' or at least indexically grounded representations of body parts, objects and spaces to be rearranged and to function as fantastic environments or corporal mutations. Photographs seduce the viewer into an imaginary space of visually believable events, objects and characters. The same can be said of assemblage – the use of found or manufactured (often commonplace) objects to create a three-dimensional representational artifact. The common contemporary practice of representing a cyborg through photomontage or assemblage resembles the poetic use of everyday words: the discrete elements are familiar, though the total result is a new conceptual and ontological domain.

Figure 14.3 Raoul Hausmann, *Tête méchanique. L'Esprit de notre temps*, ca. 1921, assemblage of wood, metal, cardboard, leather and other materials, 12.5 × 7.5 × 7.5 inches (Musée National d'Art Moderne, Centre Georges Pompidou, Paris. Photo: Musée National d'Art Moderne, Paris)

One of Höch's compatriots, Dadaist Raoul Hausmann, pictured this new ontological domain in several of his own photomontages and found-object assemblages. Unlike Höch's representations, however, Hausmann's images represent a more cerebral concept of the modern experience. Rather than a body in pieces, he depicts a mechanical mind. His assemblage *Tête Méchanique* (Mechanical Head) (Figure 14.3) is a particularly appropriate example of a cyborg mind. Also called 'The Spirit of Our Times,' this assemblage consists of the wooden head of a mannequin to which are attached diverse cultural artifacts. A wallet is fixed to the back, a typographic cylinder in a small jewel box is on one side of the head, a ruler attached with old camera parts is on the other side, the forehead is adorned with the interior of a watch, random numbers and a measuring tape, and a collapsible metal cup crowns the entire ensemble. Timothy O. Benson writes that this assemblage depicts a man imprisoned in an unsettling and enigmatic space, 'perceiving the world through a mask of arbitrary symbols.'[15] At the same time it functions as a hyper-historical[16] object collection; a testament to Hausmann's own contemporary material culture. The cyborg in this case is not without origins, though it is without origin myths. Donna Haraway contends that cyborgs have no natural history, no origin story, no Garden of Eden and thus no hope of, nor interest in, simplistic unity or purity. Nevertheless, given their multiple parts, and multiple identities, they will always be read in relation to a specific historical context. According to scholar Matthew Biro,

> ... by fashioning his cyborgs out of fragments of the new mass culture which he found all around him, Hausmann also believed he was fulfilling the primary positive or constructive function he could still ascribe to dada: namely, the material investigation of the signs and symbols bestowed on him by his historical present.[17]

Until the desire to define identities and the power to do so is lost or relinquished, even the most spontaneous cyborgs cannot float above the lingering, clinging past of differences, histories, stories, bodies, places. They will always function as evidence.

The new spatial relations of the human body are thus traced onto the cyborg body. Höch's figure is fragmentary and dispersed, floating in an untethered perspective. Hausmann's figure implies a calculable context that is linear, a cerebral space of measurement and control. Each cyborg implies a new spatial configuration or territory – a habitat. For Höch's *Beautiful Girl*, the world is a space of multiple perspectives, consumer goods, lost identities and fleeting time. The world in which Hausmann's *Mechanical Head* operates is one that links identity to material objects, and is simultaneously an environment in which knowledge is the result of a random encounter with the world of things. Hausmann himself described this assemblage as representing an everyday man who 'has nothing but the capacities which chance has glued to his skull.'[18] Sense perception itself is accounted for only to the degree that appears in the image of the cyborg body. Sight, hearing, and tactile senses can only be implied by the body's exterior devices. The *Mechanical Head*, for example, has no ears to 'hear' with, only a mechanical ruler and jewel case. *The Beautiful Girl* has neither eyes nor ears, nor mouth to speak with, only a light bulb for illumination. The human head which gazes from the corner of this image is at best the memory of what has been displaced. A new social space requires a new social being. A visual representation of this new being through an imaginary body provides a map of the layers and contradictions that make up a hyper-historical 'positive unconscious.' In other words, the cyborg body marks the boundaries of that which is the underlying but unrecognized structure of a given historical consciousness. It turns the inside out.

WHITE COLLAR EPISTEMOLOGY

Turning the outside in, Phoenix Technologies Ltd produced an advertisement for their new Eclipse Fax in 1993. The advertisement's lead-in text reads: 'Eclipse Fax: if it were any faster, you'd have to send and receive your faxes internally.' The text floats over an image of a pale woman's head and shoulders. It is clear from the image that the woman is on her back as her long hair is splayed out around her head. Her shoulders are bare, implying that she is unclothed. Mechanical devises comprised of tubes, metal plugs, cables, hoses and canisters appear to be inserted into her ears, eye sockets and mouth. Two electrodes appear to be attached to the woman's forehead, with wires extending out to the sides, almost like the antennae of an insect. A futuristic Medusa's head of wires, blinded with technology, strapped to the ground with cables and hoses, penetrated at every orifice with the flow of information technologies, this is a subjugated cyborg. Her monstrous head is merely a crossroads. All human parts of the image are passive and receptive. Indeed it seems clear that the blind silence of this clearly female creature is the very condition for the possibility of information flow. This is not a cyborg of possibilities, it is a cyborg of slavery. The advertisement promises that the consumer will be able to send and receive faxes 'without being interrupted,' and concludes that '... to fax any faster, you'd have to break a few laws. Of physics.' Here the new technology is not only seen as always available, but also as somehow pushing the boundaries of legality – even if only metaphorically – in the use of the body. This is the bad-boy fantasy prevalent in so many images of feminized cyborgs. The textual emphasis on speed in this advertisement is but a thin veil through which the underlying visual metaphor of information flow as sexual penetration bursts forth. For many, this is already an apt metaphor for cyborg body politics: knowledge as force-fed data.

What *are* the consequences of a montage of organic bodies and machines? Where do the unused parts go? What are the relations of power? Is power conserved? Is the loss of power in one physical domain the necessary gain of power for another? Who writes the laws for a cyborg bill of rights? Does everyone have the 'right' to become any kind of cyborg body? Or are these 'rights' economically determined? These are questions that arise in the attempt to figure a politics of cyborg bodies. A visualization of this hypothetical existence is all the more important for its reflection of an already current state of affairs.

THE POWER OF PLENITUDE

Robert Longo's sculpture/installation entitled *All You Zombies: Truth before God*, (Figure 14.4) stages the extreme manifestation of the body at war in the theater of politics. The glow of many painted lights hang within the frame of a

Figure 14.4 Robert Longo, *All You Zombies: Truth Before God*, L.A. County Museum of Art, Rizzoli, NY, 1990

it semi-circular canvas – an opera house or concert hall – that surrounds a monstrous cyborg soldier who takes center stage on a revolving platform. The chasm between the implied context of cultural refinement and the uncanny violence of a body that defies any and all such spaces, visually enunciates the collaboration that is always found between so-called civilization and its barbarous effects. The central figure is a cultural and semiotic nightmare of possibilities; an inhabitant of what Hal Foster has described as 'the war zone between schizoid obscenity and utopian hope.'[19] In a helmet adorned with diverse historical signs (Japanese armor, Viking horns, Mohawk-like fringe and electronic network antenna), the cyborg's double face with two vicious mouths snarls through a mask of metal bars and plastic hoses that penetrate the surface of the skin. One eye is blindly human, the other is a mechanical void. A feminine hand with razor sharp nails reaches out from the center of the chest, as if to escape from within. With arms and legs covered in one-cent scales, clawed feet, legs with fins, knee joints like gaping jaws, serpents hanging from the neck, insects swarming at the genitals, hundreds of toy soldiers clinging to the entrails and ammunition slung across the body, this beast is a contemporary monster – what Longo has called 'American machismo.'[20] The cyborg might be what Robert Hughes saw in the Dadaist obsession with war cripples, 'the body re-formed by politics: part flesh, part machine.'[21] In this light, it may seem to embody the very 'illegitimate offspring of militarism and patriarchal capitalism' that Haraway problematizes in her 'Cyborg Manifesto.'[22] But in fact Longo's sculpture describes a rebellion against these institutions, who are better represented by the familiar corporate or government-owned, sterile, fantasy figures such as RoboCop.

Instead of an asexual automaton, Longo's creature represents a wild manifestation of human, animal and mechanical sexual potency and violence. With one artificially-rounded bare breast, and one arm raised, holding a torn flag to a broken pole, the figure is remarkably reminiscent of Delacroix's *Liberty Leading the People*. Her incongruous presence has the power to capture the imagination of the viewer through an embodiment of a maternal wrath and revolutionary zeal. At the same time the creature is not without his penis, protruding but protected in a sheath of its own armor (the wings of a powerful and no doubt stinging insect). Whether the cyborg is bisexual or not, it certainly has attributes of both human sexes. Interestingly, none of the three critics writing about this work in the exhibition catalogue mention this fact. Indeed, they all fail to acknowledge that the creature has any female attributes at all. Although the body overall has a masculine feel of weight and muscular bulk, this is clearly not a single-sex being. It storms across several thresholds; that between male and female, life and death, human and beast, organic and inorganic, individual and collective. To a certain degree then, this might be considered a hybrid body; a body which 'rejoices' in 'the illegitimate fusions of animal and machine.'[23]

Historically, genetic engineering and cyborg bodies have produced similar fears about loss of human control – if there ever was such a thing – over the products of human creation. Barbara Stafford in her book *Body Criticism: Imagining the Unseen in Enlightenment Art and Medicine*, writes that in the eighteenth century,

> The hybrid posed a special problem for those who worried about purity of forms, interfertility, and unnatural mixtures. Both the plant and animal kingdoms were the site of forced breeding between species that did not amalgamate in the wild. The metaphysical and physical dangers thought to inhere in artificial grafts surfaced in threatening metaphors of infection, contamination, rape, and bastardy.[24]

Robert Longo's sculpture functions as an iconography of this metaphysics. It appears to be the very amalgamation of organic and inorganic elements that is the result of a dangerous and threatening mutation. But what makes this 'hybrid' fusion 'illegitimate'?

Fraught with many contradictory cultural connotations, the term 'hybrid' itself demands some explanation before it can be used in any casual way – as it has been – to describe a cyborg body. The term appears to have evolved out of an early seventeenth-century Latin usage of *hybridia* – a crossbred animal.[25] Now the word has several meanings, among them: a person or group of persons reflecting the interaction of two unlike cultures, traditions, etc.; anything derived from heterogeneous sources or composed of elements of different or incongruous kinds; bred from two distinct races, breeds, varieties, species or genera. These definitions reveal a wide range of meaning, allowing for easy application, but little semantic substance. What makes the term controversial, of course, is that it appears to assume by definition the existence of a non-hybrid state – pure state, a pure species, a pure race – with which it is contrasted. It is this notion of purity that must, in fact, be problematized. For if any progress is to be made in a politics of human or cyborg existence, heterogeneity must be taken as a given. It is therefore necessary to imagine a world of composite elements without the notion of purity. This, it seems, is the only useful way to employ the concept of the hybrid: as a combination of elements

273

that, while not in themselves 'pure' nonetheless have characteristics that distinguish them from the other elements with which they are combined. Hybridity must not be tied to questions of legitimacy or the patriarchal lineage and system of property which it implies. Rather, it must be recognized that the world is comprised of hybrid encounters that refuse origin. Hybrid beings are what we have always been – regardless of our 'breeding.' The visual representation of a hybrid cyborg thus becomes a test site for possible ways of being in the world. Raging involuntarily even against its own existence, the hybrid figure in *All You Zombies: Truth before God*, stands as its own terrible witness of a militarized capitalist state. As a body of power, active within its multiple selves, though mercenary in its politics, this cyborg is as legitimate as any other.

PASSING

When I began to explore visual representations of cyborg bodies, I was originally motivated by a desire to unravel the relationship between representations of cyborgs and representations of race and racial mixtures.[26] I was brought to this point by the observation that in many of the texts written about and around the concept of cyborgs the term 'miscegenation' was employed. Not only that, there seemed to be a general tendency to link the 'otherness' of machines with the otherness of racial and sexual difference. I encountered statements such as the following:

> But by 1889 [the machine's] 'otherness' had waned, and the World's Fair audience tended to think of the machine as unqualifiedly good, strong, stupid and obedient. They thought of it as a giant slave, an untiring steel Negro, controlled by Reason in a world of infinite resources.[27]

> Hence neither the identification of the feminine with the natural nor the identification of the feminine with the cultural, but instead, their uncertain mixture – the miscegenation of the natural and the cultural – is what incites, at once, panic and interest.[28]

The history of a word is significant. While the word 'hybrid' has come to have ambiguous cultural connotations, words such as 'illegitimate' and 'miscegenation' are much more problematic. The latter is believed to have been 'coined by U.S. journalist David Goodman Croly (1829–89) in a pamphlet published anonymously in 1864.'[29] 'Miscegenated' unlike 'hybrid' was originally conceived as a pejorative description, and I would agree with scholar Stephanie A. Smith that 'This term not only trails a violent political history in the United States but is also dependent on a eugenicist, genocidal concept of illegitimate matings.'[30] At the same time, this may be the very reason that certain writers have employed the term – to point out the 'forbidden' nature of the 'coupling' of human and machine. (But, as others have made quite clear, this dependency upon metaphors of sexual reproduction is a problem of, rather than the solution to, conceptions of cyborg embodiment.) Lingering in the connotations of this usage are, of course, references to racial difference.

Figure 14.5 *Silent Möbius*

While there are several images I have encountered of cyborgs that appear to be racially 'marked' as not 'white' (*Cyborg* by Lynn Randolf being among the better known and more optimistic of these images) none struck me as so emblematic of the issues with which I was concerned as those found by my colleague Elena Tajima Creef in the 1991 Japanese comic book *Silent Möbius* (Figures 14.5 and 14.6).[31] In part one, issue six, there is the story of a young woman of color. (Her hair is green, and her face is structured along the lines of a typical Euro-American comic-book beauty with big eyes [that not insignificantly fluctuate between blue and brown] and a disappearing nose and mouth. Yet, her skin is a lovely chocolate brown. She is the only character 'of color' in the entire issue and she is clearly a 'hybrid.') When we encounter her at the beginning of the story she is identified as a member of a futuristic feminine police force who is rehabilitating in the hospital. She is then seen racing off in a sporty jet vehicle after some dreaded foe. After pages of combat with an enemy called 'Wire,' it appears that she has won by default. She then declares that she must expose something to her love interest, a white young man with red hair. She disrobes, pulls out a weapon of some kind and proceeds to melt off her beautiful skin (looking more and more like chocolate as it drips away from her body), revealing to her incredulous and aroused audience that she is in fact a cyborg. Her gray body underneath looks almost white. She says, 'This is my body Ralph. Seventy percent of my body is bionic, covered with synth-flesh. Three years ago, after being cut to pieces, I was barely saved by a cyber-graft operation. But I had it changed to a combat graft.' 'Why?' Ralph asks. 'So I could become as strong as Wire, the thing that destroyed my life.' But she goes on to say, 'Eventually I started to hate this body. I wasn't feminine anymore. I was a super-human thing. I hated this body even though I wanted it. I didn't want to accept it. I kept feeling it wasn't the way I was supposed to be.' As she speaks the reader is given more views of her naked body with gray, cyborg parts laid bare beneath disappearing brown skin. In the end she says, 'I think I can finally live with what I have become.'[32]

'Kiddy,' for that is her diminutive name, is typical of contemporary (mostly male-produced) cyborg fantasies: a powerful, yet vulnerable, combination of sex toy and techno-sophisticate – in many ways not unlike *L'Horlogère*. But she is not an awkward machine with tubes and prosthetics extending from joints and limbs. She is not wearing her power on the outside of her body, as does Robert Longo's sculpture, nor is she broken apart and reassembled into disproportionate pieces, as in the case of Hannah Höch's photomontage. Rather she is an 'exotic' and vindictive cyborg who passes – as simply human. It is when she removes her skin that she becomes the quintessential cyborg body. For in the Western imaginary, this body is all abut revealing its internal mechanism. And Kiddy is all about the seduction of the strip tease, the revelation of the truth, of her internal coherence; which, ultimately, is produced by

Figure 14.6 *Silent Möbius*

the super-technicians of her time. Her 'real' identity lies beneath the camouflage of her dark skin – rather than on its surface.

'Passing' in this case has multiple and ominous meanings. Initially this cyborg body must pass for the human body it has been designed to replicate. At the same time, Kiddy must pass for the feminine self she felt she lost in the process of transformation. Staging the performance of 'true' identity, Kiddy raises certain questions of agency. Must she reveal the composition of her cyborg identity? Which seventy percent of her original self was lost? Which thirty percent was kept? Who is keeping track of percentages? (The historical shadow of blood-quantum measurement and its contemporary manifestations looms over these numerical designations. Still serious and painful, especially in many communities of color, are the intersecting meanings of percentage, passing and privilege.)[33] What are the consequences of this kind of 'passing,' especially when the body is so clearly marked with a sign – skin color – of historical oppression in the West? Why does Kiddy feel her body is not the way it is 'supposed to be'? Has it been allowed unusual access to technological freedoms? Is the woman of color necessarily a cyborg? Or is Kiddy also only 'passing' as a woman of color? What are the possible consequences of this reading? I leave these questions open-ended.

From my encounter with the world of cyborgs, and any cyberspace that these bodies may inhabit, I have seen that the question of race is decidedly fraught. Some see cyborgs and cyberspace as a convenient site for the erasure of questions of racial identity – if signs of difference divide us, the logic goes, then the lack of these signs might create a utopian social-scape of equal representation. However, the problem with this kind of e-race-sure is that it assumes differences between individuals or groups to be primarily superficial – literally skin deep. It also assumes that the status quo is an adequate form of representation. Thus the question over which so much debate arises asks: are there important differences between people (and cyborgs), or are people (and cyborgs) in some necessary way the same? The answers to this two-part question must be yes, and yes. It is the frustration of living with this apparent contradiction that drives people to look for convenient alternatives. As in-dustrialized Western cultures become less homogeneous, this search intensifies. Cyborg bodies will not resolve this contradiction, nor do they – as yet – function as radical alternatives. It may be that the cyborg is now in a new and progressive phase, but its 'racial' body politics have a long way to go. At best, the configuration of the cyborg, which changes over time, will virtually chart human encounters with a contradictory, lived experience and continue to provide a vision of new ontological exploration.

ACKNOWLEDGEMENTS

I would like to thank all of the friends and colleagues who have helped, and continue to help, in my search for images of cyborg bodies, and who have pointed me in the direction of useful literature on the topic; especially Elena

Tajima Creef, Joe Dumit, Douglas Fogle, Chris Hables Gray, Donna Haraway, Vivian Sobchack and the Narrative Intelligence reading group at the MIT Media Lab.

NOTES

1. Donna Haraway. *Primate Visions: Gender, Race and Nature in the World of Modern Science*. New York: Routledge. 1989. p. 139.
2. Donna Haraway alludes to such a consciousness in her essay 'A Cyborg Manifesto,' in *Simians, Cyborgs and Women: The Reinvention of Nature*. New York: Routledge. 1991.
3. Michel Foucault. *The Order of Things: An Archeology of the Human Sciences*. New York: Vintage Books. 1973. Foucault explains his project in this text as significantly different from earlier histories and epistemologies of science by suggesting that he wishes to reveal a 'positive unconscious' of knowledge – 'a level that eludes the consciousness of the scientist and yet is part of scientific discourse.' p. xi. This 'positive unconscious' can be thought of as 'rules of formation, which were never formulated in their own right, but are to be found only in widely differing theories, concepts, and objects of study.' p. xi. One might also think of this notion as akin to certain definitions of ideology. It is useful for my purposes to the degree that it implies an unconscious but simultaneously proactive and wide-ranging discourse, in this case, of cyborg bodies.
4. See 'The Cyborg Body Politic Meets the New World Order' by Chris Hables Gray and Steven Mentor. In this volume.
5. Julie Wosk. *Breaking Frame: Technology and the Visual Arts in the Nineteenth Century*. New Brunswick: Rutgers University Press. p. 81.
6. Ibid. p. 79.
7. Ernst Von Bassermann-Jordan. *The Book of Old Clocks and Watches*. Trans. H. Alan Lloyd. London: George Allen & Unwin Ltd. 1964. p. 15.
8. 'The delight in automatons extended back even earlier to ancient Egyptian moving statuettes with articulated arms, and to the automatons of ancient Chinese, ancient Greek and medieval Arab artisans, as well as to European clockmakers of the medieval and Renaissance periods. [...] Homer in *The Illiad* described two female automatons who aided the god Haphaestus, the craftsman of the gods. The women were "golden maidservants" who "looked like real girls and could not only speak and use their limbs but were endowed with intelligence and trained in handiwork by the immortal gods."' See Julie Wosk, pp. 81–82.
9. Wosk. p. 82.
10. 'From another perspective, a cyborg world might be about lived social and bodily realities in which people are not afraid of their joint kinship with animals and machines, not afraid of permanently partial identities and contradictory standpoints. [...] Cyborg unities are monstrous and illegitimate; in our present political circumstances, we could hardly hope for more potent myths for resistance and recoupling.' Donna Haraway. 'A Cyborg Manifesto,' p. 154.
11. Ibid. p. 180.
12. Patricia S. Warrick, *The Cybernetic Imagination in Science Fiction*. Cambridge, MA: MIT Press. 1980. p. 32.
13. Maude Lavin. *Cut With a Kitchen Knife*. New Haven: Yale University Press. 1993. p. 2.
14. Virginia Woolf. *Orlando*. New York: Harcourt Brace Jovanovich Publishers. 1928. p. 307.
15. Timothy O. Benson. *Raoul Hausmann and Berlin Dada*. Ann Arbor: University of Michigan Research Press. 1987. p. 161.
16. I use the term hyper-historical to connote an object or existence that appears to be

synchronous with a very specific moment in history, but that seems to have had no coherent evolutionary past, nor developing future. The concept was originally conceived in order to salvage the seemingly a-historical status of the cyborg in Donna Haraway's description of this term (see *Simians, Cyborgs and Women: The Reinvention of Nature*). She writes that the cyborg has no myth of origin, but I do not think she means to imply that it has no historical presence. Rather it has a presence that is so entirely wrapped up in a state of contemporary and multiple being that it is an exceptionally clear marker of any given historical moment – even when it makes references to the past.

17. Matthew Biro. 'The Cyborg as New Man: Figures of Technology in Weimar and the Third Reich.' Unpublished manuscript delivered at the College Art Association Conference, New York 1994.
18. Benson. p. 161.
19. Hal Foster. 'Atrocity Exhibition,' in *Robert Longo*, ed. Howard N. Fox. Los Angeles: Los Angeles County Museum of Art and New York: Rizzoli. 1989. 'It is in the war zone between schizoid obscenity and utopian hope that the art of Robert Longo is now to be found,' p. 61
20. Howard N. Fox. 'In Civil War,' in *Robert Longo*. Los Angeles: Los Angeles County Museum of Art and New York: Rizzoli. 1989. 'Longo describes the monster as an image of "American machismo," a confusion of vitality and vigor with warlike destructiveness,' p. 43.
21. Robert Hughes. *The Shock of the New*. New York: Alfred A. Knopf Inc. 1980. p. 73.
22. Haraway. 'A Cyborg Manifesto,' p. 151.
23. Haraway. 'A Cyborg Manifesto,' p. 154.
24. Barbara Maria Stafford. *Body Criticism: Imagining the Unseen in Enlightenment Art and Medicine*. Cambridge, MA: The MIT Press. 1991. p. 264.
25. Random House Dictionary. Unabridged edition. 1993.
26. The term 'race' is itself problematic and relies upon a history of scientific and intellectual bias. I use it here because the term has come to have a common-usage definition referring to genetic phenotype. Regardless of my distaste for this usage, I nevertheless recognize the real relations of power that are structured around it. It would be naive to ignore the ways in which this concept is employed and deployed. Here I try to point out the role of the idea of 'race' in the conception of cyborg bodies.
27. Hughes p. 11.
28. Mark Seltzer. *Bodies and Machines*. New York: Routledge. 1992. p. 66.
29. Random House Dictionary. Unabridged edition. 1993.
30. Stephanie A. Smith. 'Morphing, Materialism and the Marketing of *Xenogenesis*,' in *Genders*, No. 18, winter 1993. p. 75.
31. Elena Tajima Creef is Assistant Professor of Women's Studies at Wellesley College. She discusses *Silent Möbius* in relation to issues of race and representation in, 'Towards a Genealogy of Staging Asian Difference,' a chapter of her dissertation 'Re/orientations: The Politics of Japanese American Representation.'
32. Kia Asamiya. 'Silent Möbius,' Part 1, No. 6. Trans. James D. Hudnall and Matt Thorn. Viz Communications Japan, Inc. 1991. pp. 34–36.
33. For a good discussion of passing see 'Passing for White, Passing for Black,' by Adrian Piper in the British journal *Transition*.

15

FROM VIRTUAL CYBORGS TO BIOLOGICAL TIME BOMBS: TECHNOCRITICISM AND THE MATERIAL BODY

Kathleen Woodward

> Technology discloses man's mode of dealing with nature, the process of production by which he sustains his life, and thereby also lays bare the mode of formation of his social relations and of the mental conceptions that flow from them. (Karl Marx 1887)

> Ageism is usually regarded as being something that affects the lives of older people. Like ageing, however, it affects every individual from birth onwards – at every stage putting limits and constraints on experience, expectations, relationships and opportunities. Its divisions are as arbitrary as those of race, gender, class and religion. (Catherine Itzin 1986)

The dominant narrative of the historical development of technology that emerged from the technocriticism of the 1960s, '70s, and early '80s was grounded in the cluster of technologies associated with the so-called 'revolution' in communications – the electronic media, the computerization of the workplace and other spaces, and the cybernetics of war. The postindustrial prophets, as they were called by William Kuhns in 1971 – Marshall McLuhan, Norbert Wiener, Buckminster Fuller, and Harold Adams Innis, among them – were all noted for their fervent utopian and dystopian visions of technological change (today we might also note that these prophets were all male).[1] Information was singled out as the primary commodity of what was variously called the 'technological society,' the 'postindustrial society,' or simply, the 'information society.' Since the 1970s these terms have waned considerably in their effectiveness. Nevertheless, the power of this story persists.

Witness in the popular press, for example, *Time* magazine's decision to make CNN's Ted Turner the Man of the Year in 1991. Turner himself was, of course, merely a surrogate for the corporate institution that was being lauded as an unprecedented agent of social and cultural change – the network of global communications satellites. The tradition of American individualism requires a human hero, hence the human 'head' of CNN was featured on *Time*'s cover, not a graphic representing the technology itself. Or consider an April 1992 *Newsweek* cover story trumpeting 'Computers to Go' as the 'Next Electronic Revolution' in its headlines. It portrayed two young male tech-nerds as representative heroes – a ludicrously debased notion of revolution if ever there were one.

In recent years we have also seen the 'return' of communication technologies in the writing of what I refer to as the second generation of academic technocritics, most prevalently in cultural studies. They include, for example, the prominent French intellectuals Paul Virilio and Jean Baudrillard; the former has trenchantly critiqued the cybernetic techniques of visualization that are used in contemporary warfare (specifically, the Gulf War) while the latter has extolled the power of the mass media to produce simulacra – as well as a powerful and passively aggressive silent majority. In the United States a plethora of books have recently appeared that focus on what we might broadly call cyberculture, including Avital Ronell's *The Telephone Book: Technology – Schizophrenia – Electric Speech* (1989), Mark Poster's *The Mode of Information: Poststructuralism and Social Context* (1990), Constance Penley and Andrew Ross's edited volume *Technoculture* (1991), George Landow's *Hypertext: The Convergence of Contemporary Critical Theory and Technology* (1992), Scott Bukatman's *Terminal Identity: The Virtual Subject in Postmodern Science Fiction* (1993), and Andrew Ross's *Strange Weather: Culture, Science, and Technology in the Age of Limits* (1991). Echoing the technocritics of the 1970s, Ross designates the information revolution as 'the chief capitalist revolution of our times,' observing in a post-McLuhan tone that 'the cybernetic countercultures of the nineties are already being formed around the *folklore of technology* – mythical feats of survivalism and resistance in a data-rich world of virtual environments and posthuman bodies.'[2]

There was, however, another story that was told in the formative years of technocriticism in the United States, a narrative rooted not in communications technology but rather in biotechnology. Why did the major technological character in this story not receive equal attention? What versions of this narrative are being told today? What role does gender play in the technocriticism grounded in biotechnology? How do these two narratives of technological change – the dominant narrative based on communications and cybernetics, the recessive narrative based on biotechnology – figure the human body? These questions are central to what follows.

But first I want to briefly note several of the overlapping assumptions that underlie my thinking about technology: one, that technology is not 'neutral'; two, that it is absurd to think of technology as 'out of control' or 'autonomous' because it is in fact thoroughly embedded in social processes; three, that it is preferable to remain as concrete as possible when we think about technology, referring to particular technologies in specific contexts (solar energy for a home in a suburban setting, for instance, or laser brain surgery for a child in a high-tech medical complex) rather than to Technology as a monolithic demonic or liberating historical force; and four, that it is important to recognize when the term 'technology' is being used primarily metaphorically to refer to something other than itself (as when, for example, Michel Foucault uses 'technology' to refer to the *discursive workings* of a particular cultural formation).[3] We must be careful, in other words, to tease out the ideological implications of technocritical thought and rhetoric. Specifically, I will argue here that, in general, discourse about technology in Western culture is fundamentally bound up in a rhetoric about age, one that has negative – if for the most part subliminal – consequences for all of us.

One way of reading the narrative of technological development based on communications that has been elaborated over these two generations of technocritics is as a story about the human body. It is a story of an inevitable evolutionary process mapped on the anatomy and physiology of the body, one that has a surprising ending. It goes like this: Over hundreds of thousand of years the body, with the aid of various tools and technologies, has multiplied its strength and increased its capacities to extend itself in space and over time. According to this logic, the process culminates in the very immateriality of the body itself. In this view technology serves fundamentally as a prosthesis of the human body, one that ultimately displaces the material body, transmitting instead its image around the globe and preserving that image over time. (Today that image of the body is itself capable of being altered with the help of any number of software programs.) What began slowly has proceeded at an ever-accelerating pace. If we understand the agricultural revolution in terms of the extension of the arm by the tool, and the industrial revolution in terms of the augmentation of the power and dexterity of the human body as a whole with complex machines, then it follows that the postindustrial revolution is defined in terms of the extension of our minds by information technologies. In McLuhan's ecstatic phrase, the media serve to extend 'our central nervous system itself in a global embrace, abolishing both space and time as far as our planet is concerned.'[4]

It is paradoxical – seductively so – that while the new communications and cybernetic networks permit increased visual access to far-flung parts of the world as well as to the inner recesses of the human body, they are based on technologies that are 'unseen.' There is a beguiling, almost mesmerizing relationship between the progressive vanishing of the body, as it were, and the hypervisuality of both the postmodern society of the spectacle (Virilio) and the psychic world of cyberspace. As the philosopher Hans Jona has insisted in his work on technology, electricity is 'disembodied, immaterial, unseen'; electronics 'creates a range of objects imitating nothing' (the satellite is a perfect example of this).[5] How can we account for this fascination with the simulacra of postmodernism, with copies without originals, with the paradoxical insubstantiality and invincibility of the body in the space of virtual reality?

In 1930 Freud, in a passage rare for him, suggested a compelling and time-honored answer to this question. With 'every tool man is perfecting his own organs, whether motor or sensory, or is removing the limits to their functioning,' he wrote in *Civilization and Its Discontents*. For Freud, technological 'progress' represents the fulfillment of a deeply-held psychic wish – that 'Man ... become a kind of prosthetic God.'[6] From a psychological, if not psychoanalytic perspective, then, the possibility of an invulnerable and thus immortal body is our greatest technological illusion – that is to say, *delusion*. This desire for the impregnability of the body is expressed unmistakably in the scenarios composed for countless video games. It is played out, over and over, in virtual reality – and in the stories that are written about it.

A key text that illustrates this with consummate if not bizarre clarity is the cult science-fiction film *The Lawnmower Man*, directed by Brett Leonard and released in 1992. *The Lawnmower Man* is a contemporary version of Daniel Keyes's 1966 novel *Flowers for Algernon* (a narrative perhaps better known in its film incarnation as the 1968 *Charly*, starring Cliff Robertson). It betrays, however, none of the moving poignancy of the conclusion of *Flowers for Algernon*. In *The Lawnmower Man* we find instead a demonic 1990s version of McLuhan's utopian prediction of a global village united through communications technology.

In *The Lawnmower Man* a good-looking youth named Joe (Jeff Fahey) becomes part of an experiment to stimulate the growth of learning, an experiment designed by an ambitious but misguided scientist named Dr Angelo (Pierce Brosnan). Joe is what we used to call retarded. Amazingly, his capacity for cognitive learning increases exponentially during the course of the experiment. He is injected with drugs. More significantly, he is hooked up to a high-tech virtual-reality trapeze in the scientist's government laboratory, aptly if unimaginatively named Virtual Space Industry. The once sweet and naive Joe rapidly surpasses his teacher Dr Angelo in mental ability. As he does, his appetite for knowledge and power – the two are inextricably linked in *The Lawnmower Man* – grows insatiable. (Interestingly enough, his appetite for sex, which is stimulated at the beginning of the experiment, appears to dwindle as he comes to desire instead domination over the entire globe.) Joe takes over the administration of the experiment himself, displacing his teacher. Enabled by his enhanced mental abilities to *will* the destruction of others, his taste for bodily violence reaches horrific proportions. He uses his hastily acquired and formidable mental skills to ruthlessly murder people. He pulverizes their bodies into pixels. In a hilariously gruesome scene he turns a 'virtual' power lawn mower against a neighbor he has learned to hate, cutting his body to shreds.

Fairly early on in the narrative Dr Angelo says, 'Sometimes I think I've discovered a new planet, one that I'm inventing.' For Joe this 'planet' is the psychic 'universe' of virtual reality. With his hypermental ability he is ultimately capable of interfacing his mind with the global telecommunications network, expanding the definition of the universe of virtual reality to hyperbolic proportions. Joe will allow no one to interfere with his access to this universe – which in *The Lawnmower Man* literally means control of the entire planet. The film concludes with the image of Joe, dressed in his skin-tight, silver-and-grey body suit, as he takes on the demonic persona of his imaginary two-dimensional body in virtual reality – and leaves his organic body behind. He merges triumphantly with the interconnected communications system that circles the globe, never to emerge again in human form.[7] Corresponding to his contempt for the bodily integrity and suffering of others is his overweening desire to 'peel back' the layer of his own organic body, to rid himself of what has become its intolerable limits, the organic body itself. What Freud referred to as a 'prosthetic God' is figured in *The Lawnmower Man* in terms of overpowering evil. We are given to understand that the mind of Joe is now omnipresent, permutated around the globe, that he will exert dictatorial control through telecommunications technology.

I've alluded to Freud again because his work inevitably returns us to the reality principle, which in the context of this essay is represented by the very *materiality of the human body* – the technological possibilities for altering the body itself. As I suggested earlier, in the late 1960s and throughout the '70s many did insist – and I agree with them – that the most significant and far-reaching technological revolution on the horizon is predicated on biotechnology, not communications technology. This revolution entails not the mere extension of the body and its images, but more fundamentally, the saturation, replication, alteration, and creation of the organic processes of the body – if not the very body itself – by technoscience. Genetic engineering and reproductive technologies are two prime examples. (It's important to note here that the way these technologies intervene in the biological processes of the body is radically different from the way discursive formations, as Foucault has so persuasively demonstrated, construct and discipline the body.)

Why was this important story not reproduced in the academic marketplace of technocriticism in the '70s?[8] First, it was almost exclusively a depressingly dystopian discourse; its shrill predictions of the disastrous breakdown of the 'natural' law and order of the body were not counter-balanced by utopian claims that could have led to productive and tempered debate. It is notable that many of the books that dealt with the biotechnological revolution carried what were essentially warning titles – for example, *The Biological Time Bomb* (1968), *The Biocrats* (1970), *Genetic Fix: The Next Technological Revolution* (1973), *Pre-Meditated Man: Bioethics and the Control of Future Human Life* (1975), *Prolongevity* (1976), *No More Dying: The Conquest of Aging and the Extension of Human Life* (1976), and *Bio-Babel: Can We Survive the New Biology?* (1978).

Second, its concerns intersected with those of medicine, which is the U.S. has generally been thought to be a more 'private' or 'personal' matter, not one that requires 'public' debate or policy. (This is, of course, a faulty assumption, as today's debates on national health care demonstrate; moreover, it should be abundantly clear to us that no public policy at all is in fact a policy.) And it is indeed true that in any of a number of these books one will find chapters devoted to issues such as patient consent and the right to die, issues that may have nothing to do with biotechnology explicitly and are generally categorized as questions of medical ethics.

Third, in contrast to the study of communications technology (the media and cybernetics), debate about biotechnology did not find an academic niche in the American university of the 1970s. During this decade many undergraduate interdisciplinary programs were established in the U.S. to develop curricula that would address the study of technology from the perspective of the humanities; the ideal was often cast as a partnership between faculty from engineering (definitely not the sciences) and faculty from the humanities and social sciences (history, philosophy, and literature were the disciplines most widely represented). Known as science, technology, and society studies (STS), these programs devoted virtually no attention to issues of biotechnology. Thus, for example, in the program called Cultural and Technological Studies, in which I taught at the University of Wisconsin, Milwaukee, in the late '70s, only two courses out of some fifty dealt with biotechnological issues (and these courses addressed problems of population control, medical ethics, and ecology as well as genetic engineering).[9]

And fourth, I suspect that biotechnology did not receive the sustained scholarly attention it deserves because many of its most spectacular results, as well as its more mundane concerns, were associated with women – with motherhood, children, and care-giving. I am thinking, for example, of test-tube babies and birth-control devices, of surrogate motherhood and the biotechnological age of 'viability' of 'fetuses,' of the right to die and life-support equipment.

It is altogether fitting, then, that it is feminists who have taken up issues raised by biotechnology, focusing on gender and the technology of *reproduction* (not *production*) – that is, on the politics of biological reproduction. Ranging from Shulamith Firestone to artist Valie Export, many have argued that these new technologies give us the possibility of seizing control of reproduction. As Firestone approvingly put it in her book *The Dialectic of Sex* (1971), women can be freed from the biology that has tyrannized them; the bearing of children can be 'taken over by technology.'[10] Yet many women, not to say men, find this prospect threatening. Here, for example, is Richard Restak on biotechnology in 1975: 'the list of potential horrors – in vitro fertilization, anonymous sperm banking, behavior-control technology in the prisons and ghettos, experiments on unwitting human subjects – is lengthening.'[11] That Restak would conflate in-vitro fertilization and anonymous sperm banking with the horrors of experiments on unwitting subjects is, of course, telling.

In the United States it is, however, feminist biologist Donna Haraway's wide-ranging and brilliant 'A Cyborg Manifesto,' a piece that appeared a decade ago, that has been the most seminal in articulating the importance of biotechnology in contemporary cultural criticism. Haraway achieves this in great part by smuggling, as it were, the subject of biotechnology into an essay that deals with communications technology, ever understood as the hallmark of a postmodern world. In mapping what she calls the current 'movement from an organic, industrial society to a polymorphous, information system,' Haraway points to the shift from organic sex-role specialization to optional genetic strategies as one of its major features.[12] Haraway's essay is theoretical science fiction; she envisions a convergence of biotechnology and communications technology in the figure of the cyborg. A hybrid of organism and machine, the cyborg represents for Haraway the possibility of a world that is postgender – for her, arguably its most important aspect. Like much literary science fiction at its best, Haraway's essay is a critique of the present, one that is markedly clear in its acknowledgment of the social and historical constitution of gender and race.

But what of that other ubiquitous and virtually unanalyzed marker of the body – age? I am concerned here, in other words, primarily with the aging body, not with the sexual geography of the body (which is in any case being altered by surgery – I am referring, of course, to sex-change operations). Much recent research on the body in what is broadly referred to as cultural studies has been concerned with the ideology of difference, with understanding how differences are produced by discursive formations, social practices, and material conditions. And yet in cultural studies the only bodily difference that has received little scholarly attention is the category of 'age.' In the great binary of youth and age, it is an understatement to say that in Western culture the favored of the two terms is youth. What implications does this have for technology? And technology for it? I am also interested in the relations between age and technology in general, or in what we might call the biotechnology of age in the culture of advanced capitalism. Most post-industrial societies are aging societies; that is, they are growing older in demographic terms, their populations are aging. In what follows, I will speculate on these issues, focusing on the body and drawing on literature, film, and video.

As I have already suggested, the notion of technology as a prosthesis of the body underscores the extension of the body in space. But what of time? In a sense, one could say that many technologies – from the technologies of writing and speaking to those of film and photography – extend the body over time as well as in space. But these technologies produce only representations of the body. What about the extension of the organic body over time? If the new reproductive technologies can be said to destabilize power relations in our culture between male and female, can we conclude that new or emergent bio-technologies – I am thinking here primarily of longevity technologies – might destabilize dominant relations in our culture between the young, middle-aged, and old? Are there, in our culture, age-related prejudices associated with technology? Does our discourse about technology conceal an ideology of age?

In Gordon Rattray Taylor's mass-marketed paperback *The Biological Time Bomb*, the chapter bluntly entitled 'Is Death Necessary?' contains section headings that transparently suggest some of our culture's concerns about pro-longevity: 'Eternal youth – allotted span – deep freeze – social consequences – arrested death – a friend in need – severed heads – immortality.'[13] Note that the first phrase reads 'eternal youth,' not 'eternal life' or 'eternal age.' It is clear that youth is to be valued; old age, not. Worse, old age – specifically, a long old age – is to be feared; there is an 'allotted span,' and we should hope not to live beyond it when death is 'arrested,' when death is our proverbial 'friend in need' but does not come, when our bodies have withered away and all that is left of us is 'severed heads.'

Taylor's reference to 'severed heads' recalls science-fiction writer Olaf Stapledon's *Last and First Men* (1930). This ambivalent history of the future presents us with a stunning scene of giant human brains. Immortal but physic-ally immobile, they thrive in beehive-shaped cells, nourished by a system of chemical plants. The implication? The desire for longevity results in the obliteration of the body of the species, and it is disgusting. Or consider an altogether different version of the future of the body extending itself in time – a narrative not of the human species but, rather, of an individual in which the aging process is literally reversed. In a little-known short story by F. Scott Fitzgerald, a grown man reverts day by day to infancy; the dream of eternal youth turns into the nightmare of increasing infantilization. It would seem that when it comes to biological time, we cannot think of it in any way other than as an irreversible variable; that is one of the reasons why in the United States we cannot tolerate the possibility of people in their later years losing some of the abilities they have acquired over their life, reverting, as we say, to a 'second childhood.' (Many people have sex-change operations; can we imagine the real possibility of an age-change operation? I will turn to this later in a discussion of the film *Brazil*.)

In Aldous Huxley's *Brave New World* the ultimate fantasy of technological domination is over the human body, one achieved through biotechnology. Published over sixty years ago, this novel continues to speak to our society's dominant fantasies in relation to age today. In this new world, old age, defined as a problem of hideous proportions, has been conquered. All the physiological and psychological stigmata of old age have been eliminated. 'Characters remain constant throughout a whole lifetime'; the body is programmed bio-logically to maintain youth and vitality over the long stretch of the life span – and then to die rapidly. As Huxley envisioned this 'utopia,' the process of aging and dying is telescoped into a single, short moment: 'senility galloped so hard that it had no time to age the cheeks – only the heart and brain.'[14] Thus, both the individual and society at large are conveniently spared any disquieting knowledge of physical aging. Moreover, the state is spared the troublesome task of having to decide whether to devote any resources to older, presumably 'less productive' members of the population. There is no need for irritating debates on generational equity, for example. The problem has been solved with biotechnology. In consumer culture we have a name for this – planned obsolescence.

The head of state in *Brave New World* says, 'We haven't any use for old things here.'[15] For a society such as ours that is built on the technological values of efficiency, cost-control, and innovation (the 'new'), what is perceived as 'old' is understood as not only antithetical to our dominant values, but dangerous. Our language in everyday life confirms that a technology that is 'getting old' is suspect per se, as we see, for example, in this headline that appeared a few years ago in the *New York Times*: 'Aging Nuclear Reactor Hazardous, Experts Say.'[16] Every nuclear reactor is surely perilous, but setting this aside, what the content of the article makes clear is that the real problem is *not* in fact the age of the reactor but rather inadequate design – and, in particular, a design that is not Western or 'modern.'

Fundamentally, the term 'aging' is derived from the organic realm. Aging is also a social process, and as such it is accorded different cultural values in different settings. The terms 'youth' and 'age' belong to the rhetoric of this continuum of the biological life span, which is given different meanings in different societies. But we also apply these terms to artifacts, as we just saw. And when we do, the negative connotations attached to the term 'age' when it is used to describe, for example, a material object that is falling apart (such as a car or a washing machine), will return to the human domain laden with negative associations. We also use, of course, the terms 'new' and 'old' to describe both artifacts and organic beings – it's a 'new' tree, we say, or a 'new' car. But we do not say that a computer or a blender are 'young.' The problem, then, is that the term 'old' but not the term 'young' is transferable from the organic realm to the technological realm –and then back again. And generally speaking, in our technological culture there is nothing good about an artifact or a technology that is old. The supreme value is to work efficiently (which is understood to be synonymous with being new), not to break down. In short, I would argue that the rhetoric (as well as social practices) of the technological culture of advanced capitalism contributes to widespread ageism against older generations.

Age relations are power relations – with the young generally holding the power over the old (the young only need to hold on; in most cases, the passage of biological time will take care of any power differential in their favor). In Terry Gilliam's shrewd 1985 film *Brazil*, this relation between two generations is momentarily upset by technology; it is further complicated by gender. The setting is a bureaucratically impacted, grey retro world where one of the only vivacious beings is an 'older' woman, a woman who is (of all taboo things) the mother of a grown son. With the aid of one plastic surgery after another, the mother (the beautiful Katherine Helmond) of the bureaucratically beleaguered Sam Lowry (Jonathan Pryce) appears to grow progressively younger. In a final scene she appears to be the youthful age of twenty-five, which unambivalently delights her even as it unsettles her irresolute son. Her plastic surgeon makes a million promises. 'I'll make you twenty years younger,' he says, 'they won't know you when I finish with you.' But another surgeon warns prophetically. 'I'll will never last ... In six months she'll look like Grandma Moses.' The younger generation would prefer the older generation to stay where they are – to stay older than *they* themselves are: it is a matter of power.

In *Brazil* nothing works efficiently. Actually nothing seems to work at all. One of the dominant institutions, Information Retrieval, is more than a frustrating joke, it becomes a bureaucratic site of torture. More importantly for my purposes, tampering with the biological body – particularly an older woman altering her own body – is implied to be not only unnatural, but worse, horribly doomed to fail. Katherine Helmond's temporarily youthful body is a postmodern parable for the injunction against biological engineering that would give an edge to the old. It is understood that she will be punished for her desire to claim the sexually admiring glances of men her son's age.

The disavowal of the limits of the body in technocriticism – or what I have been referring to as one of the ideologies of technology in general – is thus especially complex with regard to age. In *Brazil* what you want (and therefore what is 'right') depends on just how old you are. In *Brazil* both the son and the mother want to escape the limits of the body. Both hope that technology will enable them to do so. The son tries to escape certain torture by fleeing in a car – actually he is only *imagining* the car. The playful, designing mother calls on the aid of surgical technology for her escape – actually, it is an *old-fashioned* technology (clumsy clamps are involved), so we as spectators, smugly ensconced in a technological future relative to the awkward technological culture of *Brazil*, know it won't work. In fact neither the mother nor the son succeed in escaping. In one of the several fascinatingly grotesque scenes of the film, the son witnesses the future of his mother in the fate of one of her friends. This older woman has also had a series of plastic surgeries, and they begin to fail – that is, to reverse themselves. During her funeral Sam accidentally upsets her coffin – only to spill a disgusting gelatinous mass of decomposed matter that was once her body.

We also see ageism at work in the reactions reported recently to several new advances in reproductive technology. Consider the story about cloning that made the headlines in the United States and Europe in the fall of 1993. In a piece in the *New York Times* headed 'Cloning Human Embryos: Furious Debate Erupts Over Ethics,' several reasons were cited as to why the cloning of embryos is unacceptable – including, for example, that the sanctity of the individual would be violated, and that the human body should not be allowed to become a commodity. These reasons we could have easily predicted. But one objection in particular intrigues me. The director of the Center for Bioethics at the University of Minnesota, Arthur Caplan, raised the specter of age; the scenario of twins being born years apart was disturbing to him. '"Twins that become twins separated by years or decades let us see things about our future that we don't want to," Dr Caplan said. "You may not want to know, at 40, what you will look like at 60. And parents should not be looking at a baby and seeing the infant 20 years later in an other sibling"'.[7] Why not? Why should this be so potentially distressing, not to say intolerable? Why should a vision into our biological future be so disturbing? In our technological culture the future is associated with progress, with development, with the new. What is 'aging' and 'old' is considered anathema to it, associated with decline, diminishment, and disease.

Many object to certain high-tech reproductive procedures in principle for all women. But some only object to them for 'old' women. The prejudice against older women is blatantly revealed, for example, in the horrified reactions to the reports of the 'old' women – a fifty-nine-year-old English woman and a sixty-one-year-old Italian woman – who, with the aid of in-vitro fertilization, recently gave birth to children. This was considered sensational news. A January 1994 story in *People* magazine, which was accompanied by full-page photographic spreads, carried the title 'Turning the Clock Back.' Many insisted vehemently that a postmenopausal woman should not have babies because it is 'unnatural,' not in the proper biological order of things. Obviously, however, what is also at stake is a deep-seated prejudice against older women; a pregnant older woman has long been understood as a monstrous contradiction in terms, both biologically and culturally. When the biological barriers are lifted, as they were in this case by technology, what remains so glaringly to be seen is the cultural bias against older women. In France, for instance, the government decided to introduce legislation to outlaw artificial insemination of women who are postmenopausal. As many were quick to point out, our culture does not condemn older men for fathering children; indeed older fathers are often seen as especially virile. So why should older women be vilified? What this story shows is that women are *aged by society* earlier than are men. What it also reveals is that if feminists have turned to the study of technology primarily through examining and championing reproductive rights for women, they have been conceptualizing technology as an issue for young and midlife women – not one for women of all ages.

We also see the ageism – and sexism – embedded in the ideology of technology in the reactions to the recently reported possibility of transplanting ovaries taken from fetuses into the bodies of women who are infertile. Again at stake is the presumed 'natural' order of generations; many find the prospect of disrupting this 'natural' biological sequence 'grotesque,' as a lawyer who is an ethicist at Harvard put it. '"Should we be creating children whose mother is a dead fetus? What do you tell a child? Your mother had to die so that you could exist?" George Annas asks.'[18] The obvious implication of his objection is that a woman has to be a biological mother before she can be a grandmother. But why necessarily so? The proper focus, in my view, is on detaching biological roles and sequences from social roles, such as parenting, when it allows for increased opportunities to fulfill those roles.

Thus, on the one hand, I think we should explore in a positive spirit the possibilities offered by biotechnology as they intersect with issues concerning the aging body. Why should women not extend their biological capabilities with, for example, such reproductive technologies? On the other hand, the grandiose fantasy of eliminating altogether the inevitable process and consequences of aging reveals a prejudice against aging that is harmful to all of us. Indeed, the limits to the extension of the power and control of the human body with the aid of technology can be clearly seen in the context of the body as it ages.

Stanley Kubrick's elegant 1968 science-fiction film *2001: A Space Odyssey* begins, for example, as a technological romance with an ape who discovers in the midst of war that a bone multiplies the strength of his arm; he joyously hurls it into the air. The film then cuts to the future; we see a spaceship soaring aloft accompanied by the waltzing strains of the *Blue Danube*. The bone extends the power of the arm; the spaceship, the power of locomotion, of travel.

Kubrick's superb anthropological fable concludes, however, with a series of sobering scenes that remind us of the limit to this narrative of the technological evolution of the power of the prosthetic body – that limit is the aging body itself. The sole surviving member of the crew (Keir Dullea) has been captured by alien intelligence. Alone in an exquisitely tasteful, spare suite of rooms, he is witness, in a slow relay of looks, to his own future. It is the progressive aging of his body. He ages before his own eyes, soberly accepting this visual knowledge that Arthur Caplan (to whom I referred earlier in reference to the cloning of embryos) suggested we might not be able to acknowledge psychologically. Technology, it is thus implied, will never provide an escape from the upper reaches of the biological life span. (Earlier in *2001* the fantasy of prolongevity is also punctured when the computer HAL unplugs the life-support machines to which the hibernating crewmen are attached.) The sequence ends with the captain, frail and immobile, confined to his bed. But to this sequence Kubrick added a coda. A fetal figure of human life serves as a redemptive image, displacing aging onto the birth of life, recasting the narrative of technology as a mystery play in which the cycle of death and the resurrection of life is endlessly and magically repeated.

If we rewrite this epic narrative as a story of everyday life without the rescue of a coda, it might go like this: as children we take pleasure in progressively acquiring prosthetic expertise by learning how to use a fork, ride a bicycle, and operate a computer; as older adults we may welcome, although possibly ambivalently, the prosthetic innovations of bifocals and aids for hearing; but most of us fear the future prospect of frailty as a cyborg, 'hooked up,' as we commonly say, to a machine.

'The ultimate technological fantasy,' Andreas Huyssen has written, 'is creation without the mother.' This is a widespread male fantasy embedded within patriarchy, the fantasy that women – especially older women – are not necessary. But from the perspective of age, not gender, the ultimate technological fantasy is, as we saw in *Brave New World*, the perpetuation of the power, health, and beauty associated with youth coupled with the psychic assurance of immortality. Another way of making this point is by asking Donna Haraway the age of the cyborg she so optimistically celebrates. It is, I suspect, young, at the most middle-age. As I have been insisting, technological culture is a youth culture.

In video artist Cecelia Condit's magical miniopera *Not a Jealous Bone* (1987), we see a surprisingly enchanting reversal of the dominant values of youth and age. As in *2001*, the bone here too is a symbol of power – technological power, that is to say, cultural power. There are two main figures – the eighty-some-year-old heroine, Sophie, with dentures and an every-which-way black wig, and the conventionally beautiful younger woman, who is in, I would guess, her late twenties. In the struggle between them, it is Sophie who wrests the bone from the younger woman, who is as narcissistically preoccupied with her mirror image as the jealous Queen in *Snow White and the Seven Dwarfs*. It is a 'magic bone,' which represents power for the old in a technological culture fixated on youth.

When the video opens, we see Sophie making a spectacle of herself, her body ungainly and unsteady in its old-fashioned, black-skirted bathing suit. At the narrative's end, Sophie, now the possessor of the bone, dresses up with charming pleasure. No longer culturally reprehensible, she dances lightly and with self-possession on a center stage. Moreover, she tosses the bone into the unseen audience, handing it over to others who would catch it. It is the desire for life, not youth or the appearance of it, that is given value – in the figure of an older woman.

In *The Biological Time Bomb* Taylor warned of the dangers of prolongevity for capitalist culture. Who does he single out as representing a sure threat? Older women. Consider this passage, which is rife with both sexism and ageism: 'Another probable consequence of the prolongation of life would be a further increase in the excess of women over men in the higher age groups of the population ... A society containing a large percentage of such "dragons" would be intolerable.'[20] The eighty-some-year-old Sophie dancing on the beach, her makeup askew and her body unseemly by our youthful standards: Taylor would cast her as a *dragon*, as a *biological time bomb* that is threatening the vitality of our culture. Taylor would want to stop her clock for his own purposes.

In *The Biological Time Bomb* Taylor also worried that a significant increase in the older portion of the population relative to those younger 'might bring about a gerontocracy and an "age-centered" society; it is difficult to see how the latter kind could be as vigorous as the former.'[21] In terms of the rhetoric of age, we see here again the metaphorical use of the term 'age.' But here the rhetoric of biological age – 'young,' 'aging,' and 'old' – is applied not to an individual or to an artifact but rather to an entire society. Implicit in the notion of an aging society is the fear that it – therefore we – might die. I read the marvelous figure of Sophie as a protest against the youth-oriented – and sexist – ethic of our capitalist culture. As our population ages – in part, ironically, because of technological change of many kinds – the United States has grown anxious that it is losing its competitive edge to 'younger' economies in Asia, a fear that only works to further intensify the ageism inherent in an 'advanced' technological society.

NOTES

1. William Kuhns, *The Post-Industrial Prophets: Interpretations of Technology* (New York: Weybright & Talley, 1971).
2. Andrew Ross, *Strange Weather: Culture, Science, and Technology in the Age of Limits* (London: Verso, 1991), pp. 9, 88.
3. See Michel Foucault, *The History of Sexuality*, Vol. 1: *An Introduction*, trans. Robert Hurley (New York: Vintage Books, 1980).
4. Marshall McLuhan, *Understanding Media: The Extensions of Man* (New York: New American Library, 1964), p. 19.
5. Hans Jonas, *Philosophical Essays: From Ancient Creed to Technological Man.* (Englewood Cliffs, N.J.: Prentice Hall, 1974), p. 78.
6. Sigmund Freud, *Civilization and Its Discontents* (1930), *The Standard Edition of the Complete Psychological Works of Sigmund Freud*, trans. and ed. James Strachey, 24 vols. (London: Hogarth and Institute of Psycho-Analysis, 1953–74) 21, pp. 90, 91–92.
7. Allucquére Roseanne Stone, in an essay on the body and virtual reality, astutely comments: 'The quality of mutability that virtual interaction promises is expressed as a sense of dizzying, exciting physical movement occurring within a phantasmic space – again an experimental mode psychoanalytically linked to primal experiences. It is no wonder, then, that inhabitants of virtual systems seem to experience a sense of longing for a space that is simultaneously embodied and imaginary, such as cyberspace suggests. This longing is frequently accompanied by a desire, inarticulately expressed, to penetrate the interface and merge with the system' in 'Virtual Systems,' *Incorporations*, ed. Jonathan Crary and Sanford Kwinter (New York: Zone Books, 1992), p. 619.
8. Much has appeared in the popular press, however. *Time* ran a cover story in November 1993 entitled 'Cloning Humans,' for example, and a major essay entitled 'The Genetic Revolution' and authored by Philip Elmer-Dewitt appeared in the January 17, 1994, issue, summarizing the revolution quite competently as follows: 'The ability to manipulate genes – in animals and plants, as well as humans – could eventually change everything: what we eat, what we wear, how we live, how we die and how we see ourselves in relation to our fate,' p. 42.
9. I am aware, of course, that there is a goodly amount of scholarly attention devoted to many issues in bioethics; in the university, however, courses on bioethics are often taught only in medical or nursing schools.
10. Shulamith Firestone, *The Dialectic of Sex: The Case for Feminist Revolution* (New York: Bantam, 1971), p. 228. See Valie Export, 'The Real and Its Double: The Body,' *Discourse* 11, no. 1 (Fall–Winter 1988–89): pp. 3–27. For a persuasive argument that the point of Firestone's book is 'to stimulate a revolutionary consciousness here and now rather than to dictate how babies should be made,' see Carl Hedman, 'The Artificial Womb – Patriarchal Bone or Technological

Blessing,' *Radical Philosophy* 56 (Autumn 1990), p. 19. Much of the research by academic feminists regarding reproductive technology is conservative. Shelly Minden's point is representative; she writes, 'Although these technologies promise things that many women want – possibilities of healthier babies and of reduced infertility – the price that they exact is no less than that of women's autonomies over their own bodies,' in 'Patriarchal Designs: The Genetic Engineering of Human Embryos,' in *Made to Order: The Myth of Reproductive and Genetic Progress*, ed. Patricia Spallone and Deborah Lynn Steinberg (Oxford: Pergamon Press, 1987), pp. 102–103.

11. Richard M. Restak, *Pre-Meditated Man: Bioethics and the Control of Future Human Life* (New York: Penguin Books, 1977), p. 167.

12. Donna J. Haraway, 'A Cyborg Manifesto: Science, Technology, and Socialist-Feminism in the Late Twentieth Century,' in Haraway, *Simians, Cyborgs, and Women: The Reinvention of Nature* (New York: Routledge, 1991), p. 161.

13. Gordon Rattray Taylor, *The Biological Time Bomb* (New York: North American Library, 1968).

14. Aldous Huxley, *Brave New World* (New York: Harper & Row, 1969), pp. 37, 135.

15. Ibid., p. 148.

16. The story appeared in the *New York Times* 25 March 1992.

17. Gina Kolata, 'Cloning Human Embryos: Furious Debate Erupts Over Ethics,' *New York Times* 26 October 1993.

18. Gina Kolata, 'Fetal Ovary Transplant Is Envisioned,' *New York Times* 6 January 1994.

19. Andreas Huyssen, *After the Great Divide: Modernism, Mass Culture, Postmodernism* (Bloomington: Indiana University Press, 1986), p. 70.

20. Taylor, p. 115.

21. Ibid., p. 115.

16

BIRTH OF THE CYBERQUEER

Donald Morton

In today's dominant, 'post-al' academy, the widely celebrated 'advance' in the understanding of culture and society brought about by ludic (post)modernism has been enabled by a series of displacements: of the signified by the signifier, of use value by exchange value, of the mode of production by the mode of signification, of conceptuality by textuality, of the meaningful by the meaningless, of determination by indeterminacy, of causality by undecidability, of knowing by feeling, of commonality by difference, of political economy by libidinal economy, of need by desire, and so on.[1] In the domain of sexuality, the new space of queer theory is a postgay, postlesbian space. Ludic (post)-modernism, which promotes the localizing of cultural phenomena, discourages any effort to render these developments systematically coherent and intelligible. Hence, the reappearance of *queer* today is given local 'explanation' – for example, as an oppressed minority's positive reunderstanding of a once negative word, as the adoption of an umbrella to encompass the concerns of both female and male homosexuals and bisexuals, or as the embracing of the latest fashion over an older, square style by the hip youth generation.[2]

I argue here that explanations relying on trends, styles, and the sexual subject's 'voluntary' intentions trivialize the issue of queerness for the purpose of occluding the ideological significance of the return of the queer. In other words, queer studies – as a superseder of the older and presumably outmoded Enlightenment-inspired gay and lesbian studies – participates in the contemporary shift brought about by ludic (post)modernism toward a theoretically updated form of idealism and away from historical materialism. This idealism comes to light when the return of the queer is historicized as part of a systematic development connected to the appearance in late capitalism of such notions as virtual realities, cyberpunk, cybersex, teletheory. The return of the queer today is actually the (techno)birth of the cyberqueer.[3]

I

That more is at stake in the return of the queer than mere trends, styles, or fads is suggested by the strongly contestatory, even violent, edge that the reappearance has had. Popular accounts of the queer emphasize its negative relation to the gay: 'The one thing on which everybody was agreed was that whatever else it may or may not be, queer definitely is not gay' (Mitchell and Olafimihan 38). The authors of a queer Canadian 'zine' declare, 'BIMBOX is at war against lesbians and gays. A war in which modern queer boys and queer girls are united against the prehistoric thinking and demented self-serving politics of the above-mentioned scum' (Cooper 31). The conclusion that gay liberation failed to meet its goals, possibly motivating the *Bimbox* outburst, is certainly true. As a historian of the movement puts it, 'The paradox of the 1970's was that gay and lesbian liberation did not produce the gender-free, communitarian world it envisioned, but faced an unprecedented growth of gay capitalism and a new masculinity' (Adam 97). The question remains, Did lesbian and gay liberation fail because of the limitations of particular persons (as *Bimbox* implies) or because of the movement's limited theoretical grasp of its project? Does the very project of emancipation need to be abandoned (along with the notions of reliable knowledge, conceptuality, theory, and, most important, need), as the queer perspective suggests, or must the project of emancipation be retheorized to overcome the weaknesses that helped produce 'an unprecedented growth of gay capitalism'?

Rather than as a local effect, the return of the queer has to be understood as the result, in the domain of sexuality, of the (post)modern encounter with – and rejection of – Enlightenment views concerning the role of the conceptual, rational, systematic, structural, normative, progressive, liberatory, revolutionary, and so forth, in social change. The queer commitment to desire, emblematized in Frank Browning's *The Culture of Desire: Paradox and Perversity in Gay Lives Today*, is a commitment to pleasure that is on 'a journey separate from the path to equity, democracy, and justice' (Browning 104). Behind Browning's popularly articulated 'paradox' lies an urgent philosophical and political issue: while homosexual citizen subjects and their sexual needs must be defended, the defense must be part of an argument for social justice for all citizens and cannot be founded on pleasure (whether pleasure is labeled *desire, sensation*, or *jouissance*). The public sphere – the space of shared citizenship – is not formed by common sensations (individuals do not all have or even desire the same sensations) or by sensation at all: the public sphere is constructed through concepts, the very notion desire theorists reject.

Gay liberation, envisioning a 'gender-free, communitarian world,' did not promote the separation of which Browning speaks. The explanation for the shift from gay and lesbian studies, based on the category gender, to queer theory, which fetishizes desire by rendering it autonomous, is not self-evident. It is commonly assumed that (post)modern queer studies has made a decisive and radical advance over modernism (and its precursors), which assigned questions of sexuality and desire to secondary social and intellectual status. Even while giving sexuality and desire central importance in his theory, Freud, as a modernist thinker still committed to Enlightenment assumptions, stressed that the rational regulation of sexuality and desire was necessary to civilized life, despite the inevitable 'discontents' that accompany civilization as a result. Against such supposedly outmoded modernist assumptions, ludic (post)modern theory produces an atmosphere of sexual deregulation. As a – if not the – leading element in this development, queer theory is seen as opening up a new space for the subject of desire, a space in which sexuality becomes primary. As Eve Sedgwick puts it, '[A]n understanding of virtually any aspect of modern Western culture must be, not merely incomplete, but damaged in its central substance to the degree that it does not incorporate a critical analysis of modern homo/heterosexual definition' (*Epistemology* 1). In this new space, desire is regarded as autonomous – unregulated and unencumbered. The shift is evident in the contrast between the model of necessary sexual regulation promoted by Freud in *Civilization and Its Discontents* and the notion of sexual deregulation proposed by Gilles Deleuze and Félix Guattari. Deleuze and Guattari represent the deregulating process – in which desire becomes a space of 'pure intensities' (*A Thousand Plateaus* 4) as a breakthrough beyond the Oedipus complex (that 'grotesque triangle' [*Anti-Oedipus* 171]), which colonizes the subject and restricts desire. Since the oedipal model is explicitly heterosexual, its supersession appeals particularly to many queer theorists, who take up the call for sexual deregulation.

Contesting this lineage is a countertradition (of which this essay is a part) that acknowledges the significance of desire but insists on relating desire to the historical world (not an ideal one) and to worldly materiality. The materialist tradition ranges from the ancients (Heraclitus, for instance) to the (post)moderns but becomes especially critical with the rise of capitalism and class society and thus finds acute expression in the writings of eighteenth-century European Enlightenment philosophers and its strongest theorization in Marx's historical materialism. Marx, of course, sees history as the transforming of social organization by changes in the mode of production. The encompassing commitments of historical materialism are expressed through its goal, as Marx formulates it, 'of comprehending theoretically the historical movement as a whole' (64). Here Marx defines materialism as an understanding that is theoretical, historical, and global. In this perspective, desire too must be theorized, historicized, and situated in global social relations.

Some will doubtless object that materialism is everywhere in mainstream academic and intellectual writing, including queer theory, and that my idealism-materialism distinction is overdrawn. After all, a broad range of contemporary cultural and social critics and theorists – such as those working within the frames of (post)structuralism, psychoanalysis, feminism, Foucauldianism, and the new historicism – espouse something they call materialism. However, these claims are themselves expressions of idealism since they deploy the notion of the material in such a way as to erase class conflict. Materialism is a structure of conflicts.

Against Freud's more biological- and physiological-sounding terms, (post)-modern theory gives desire linguistic and significatory explanations: 'Desire is the perpetual effect of symbolic articulation' (Lacan viii). While Kristeva's phrase 'desire in language' suggests that desire is not autonomous, being in language does not define desire, since in the (post)modern paradigm everything is ultimately language-based. Desire is distinctive rather because it is the unruly and uncontainable excess that accompanies the production of meaning. Generated in the unbridgeable gap between organic need and linguistic or symbolic demand (the subject's call for its need to be satisfied through language), desire 'is not an appetite: it is essentially excentric and insatiable' (Lacan viii). Desire then is the excess produced at the moment of the human subject's entry into the codes and conventions of culture. As such, desire is an autonomous entity outside history, an uncapturable, inexpressible, and actually meaningless remainder left over when the person becomes a socialized participant in what Lacan calls the symbolic. (Post)structuralist theorists articulate the disruption of conscious, convention-bound, and voluntary meaning making by unconscious desire as the disruption of the signified (the conceptual, meaning-*full* part of the sign) by the signifier (the quasi-sensory, meaning-*less* sound-image). In short, (post)structuralism represents desire as an autonomous, if language-embedded, entity that inexorably disrupts sociality, the domain of collective codes and conventions.

Jean Baudrillard exemplarily stages the displacement of the economic account of need by the linguistic account of desire. He proposes a shift from the (Marxist) analysis of political economy (which focuses on the mode of production and which privileges what Baudrillard calls 'utility,' 'needs,' 'use value,' 'economic rationality' [*For a Critique* 191]) to the analysis of 'the political economy of the sign' (which concerns itself with the mode of signification). Marxism, Baudrillard argues, has made a false distinction between use value and the more 'abstract and general' level of exchange value, a distinction in which 'lard is valued as lard and cotton as cotton' and they 'cannot be substituted for each other, nor thus "exchanged"' (130). Capitalism, which requires exchangeability or equivalence between commodities, is founded on exchange value and is driven by the need to produce from it a surplus value that is responsible for the difference of class. Overcoming capitalism would involve a return to the more fundamental level of addressing human need, represented by use value, and entail the cancellation of all the (needless) desires produced by capitalist commodity fetishism. But, against Marx, Baudrillard argues that '[u]se value is [also] an abstraction. It is an abstraction of the *system of needs* cloaked in the false evidence of a concrete destination and purpose, an intrinsic finality of goods and products,' and thus that 'use value and exchange value' are 'regulated by an identical abstract logic of equivalence' (131). Use value is not distinct from but dependent on exchange value, and both are equally subject to the operations of commodification and fetishization. Hence, no overturning of capitalism could ever return the subject to a 'simple relation to his work and his products' (130).

Need is therefore caught up in the same language-based relations as is desire. Baudrillard expresses this correspondence in a modification of the familiar Saussurean formula:

$$\frac{\text{exchange value}}{\text{use value}} = \frac{\text{signifier}}{\text{signified}}$$

Commenting on this ratio, Baudrillard remarks:

> Absolute preeminence redounds to exchange value and the signifier. Use value and needs are only an effect of exchange value. Signified (and referent) are only an effect of the signifier. ... Neither [signified nor referent] is an autonomous reality that either exchange value or the signifier would express or translate in their code. At bottom, they are only simulation models, produced by the play of exchange value and of signifiers. ... Use value and the signified do not constitute an *elsewhere* with respect to the systems of the other two; they are only their alibis. (137)

Thus, not only is the mode of production (as the basis for accounting for need) subsumed by the mode of signification (as the basis for accounting for desire) but also, in broad terms, the human needs to which materialism wants to respond are merely simulacra of desire. In particular, historical materialism's social goals, founded on a rational, concept-based understanding of the social totality, and any of its possible allied political aims and agendas, such as emancipation from injustice and from exploitation, are rendered chimerical. Baudrillard reverses the relation of concept (signified) to sound-image (signifier) in Saussure, rendering the concept a subordinate mirage of the operations of sound-images. The conventional meanings (concepts) by which collective social life is conducted are merely evanescent mirages thrown up by signification, which is driven by autonomous and socially meaningless desire. In the end, the desire-bound signifier is postulated as autonomous, outside history and its materiality. Furthermore, illusory, concept-based social understandings cannot be the basis of an effective political position. Following this conclusion, deconstructionists propose that pedagogy should be regarded not as the production of concepts but as the unleashing of desire: Barbara Johnson speaks of 'teaching ignorance' (68) and Jane Gallop of 'thinking through the body.' Articulating this conclusion emphatically for queer studies, Ed Cohen offers a slogan: 'We fuck with categories' (174–75).

The return of the queer today was produced in the wake of the historical developments of ludic (post)modern theory. Following Baudrillard's paradigm, which disrupts the conceptual series of Enlightenment thought by elevating desire and displacing need, queer theory creates a similar disruption by rewriting the sexual ratio:

$$\frac{\text{queer}}{\text{gay}} = \frac{\text{signifier}}{\text{signified}} = \frac{\text{desire}}{\text{need}}$$

Gay liberation grew historically out of the lessons of the civil rights and feminist movements of the sixties and depended for its self-understanding and its political agenda on conceptualization, the formulation of a reliable knowledge on which a politics could be grounded. Gay liberation depended on the development of concepts such as exploitation and oppression, on the one hand, and social justice, on the other, which could lead to the collective goal of overcoming the former to achieve the latter. (The movement thus counted also, and not incidentally, on the concept of causality.) But just as ludic (post)-modernism proposes that the conceptual part of the sign is only a 'meaning effect' and that need is a 'need effect,' queer theory sees gayness as nothing more than a gay effect, a mirage of signification.

Sensing the ghostly status of the gay in the current cultural environment, Barbara Smith, a lesbian political worker, observes that 'today's "queer" politicos seem to operate in a historical and ideological vacuum,' so that 'revolution seems like a largely irrelevant concept to the gay movement of the nineties' (13). Smith is correct, but there are deeper theoretical changes at work. It is conceptuality itself, not simply the concepts of history, ideology, and revolution, that has been displaced in (queer) politics: 'rational politics, in the sense of the concept, is over,' declares Jean-François Lyotard; 'that is the swerve of this fin-de-siècle' (Lyotard and Thébaud 75). Embracing what Lyotard designates the libidinal economy and rejecting the conceptual economy, queer theory excludes the Enlightenment project of social progress envisioned by gay studies and renounces (concept-based) commonality in the name of uncapturable difference: 'erotic desire always introduces the phenomenon of difference,' Karl Toepfer remarks, 'which subverts the great unifying ambition of revolution' (136). Today the 'social universe is formed by a plurality of games' for which there is 'no common measure' (Lyotard and Thébaud 58, 50).

The rejection of the signified, of course, has consequences at the level of subjectivities. Queer theory carries further the injunction of difference politics forbidding anyone to speak for the other. As nothing more than insatiable signifiers of insatiable signifiers, the authors of the queer zines are said to 'speak for one – not for an emerging generation and not even for themselves. If anything, they position themselves as savage cheerleaders feeding off and fueling their contemporaries' violent mood swings.' Cheerleading queer subjects, existing in a dimensionless dimension and an immaterial materiality, devote themselves to the intensification of pure intensities. Composed of 'quirky outcasts,' the queer community (if it can be called that) is 'intense' but 'scattered,' unorganized and unorganizable. Although almost instinctively sensitive to exclusion, the queer can have no principles (concepts) of inclusion or exclusion: 'they don't expect anything close to a consensus, and so are able to fight among themselves with a degree of affection' (Cooper 31). If judgment is called for, the queer subject responds like the ludic (post)modern subject, which 'judges without criteria' (Lyotard and Thébaud 16).

The queer subject is deprived of the possibility not only of speaking for (others or even itself) but also of speaking in the name of: it cannot speak in the name of any principle, such as social justice (an up-to-date position articulated in Stanley Fish's declaration 'I don't have any principles' [298]). As a social construct that can only act self-reflexively, by deconstructing itself, the (post)-modern subject can only perform, not practice. In the terms made familiar by Judith Butler, whose work deconstructs the notion of (gender) identity, the subject's actions are 'not expressive but performative' (*Gender Trouble* 141). In other words, they do not express the subject's inner essence (soul, spirit, psyche, etc.), as the modernist tradition proposes, or even some constructed and existing identity, as the (post)modernist position might imply. Just as Baudrillard understands the simulacrum to be a copy that has no original and that renders all representations copy effects (see *Simulations*), Butler under-stands gender as a gender effect, a simulation or mimicry of nothing that is prior to it, a nonreferential repetition. 'There is,' Butler argues, 'no gender identity behind expressions of gender; that identity is performatively consti-tuted by the very "expressions" that are said to be its results' (*Gender Trouble* 25). The subject becomes what Deleuze and Guattari call an 'asignifying particle' (*A Thousand Plateaus* 4). Such a position leads Butler to declare that although she will use 'the sign of lesbian,' she will do so only on condition that it is 'permanently unclear what precisely that sign signifies' ('Imitation' 14). To be gay is to have a mere identity; to be queer is to enter and celebrate the ludic space of textual indeterminacy. As Gregory Bredbeck declares in the queer mode, 'Homosexuality *is* textuality in its most potent and postmodern form' (255).

The shift from gay studies to queer studies has been accompanied by the displacement of gender as too conceptual a notion.[4] Eve Sedgwick insists that the dissociation of gender from sexuality is 'axiomatic' for 'antihomophobic inquiry' today (*Epistemology* 27). In contrast to queer studies, gay studies – which conducted an 'intense critique of gender' (Adam 97) – was grounded on the premise that any effective fight against sexual oppression and exploitation requires a reliable understanding of their causes. Like predeconstructive feminism, gay studies conceptualized gender as a part of a larger determinative system – as a structural and systematic regulation of sexual practices resulting from the binary power arrangements generated by patriarchy. On this logic, gay liberation could only comprehend the struggle against sexual oppression as a struggle to overthrow the gender system and the patriarchy as the causes. The movement had explanatory, causal, determinative, and globalizing impulses, an ensemble unacceptable to queer theory.

Queer theory departs from traditional humanist literary and aesthetic studies (and from gay and lesbian studies) by virtue of its absorption of ludic (post)modern theoretical developments along their two main axes. Aside from the overtly ludic Derridean-Deleuzean axis, in which 'liberated' desire subverts the official relations of signifieds (conceptuality) and signifiers (textuality), there is the historicist Foucauldian strand, which insists that outside the text are material institutions, enabled by discourses but not textualist in the Derridean sense.[5] These institutions (as against historical materialism's global account of them) are disconnected and autonomous, and they can be sites of liberation where marginal groups seize power (which is voluntarily reversible). For these historicists, social inequality is a measure of the inequality of power among groups and is not, as conceived by Marx, produced by exploitation during capitalism's extraction of surplus value. On the political plane, Foucault's work converges finally with Derrida's and diverges from Marx's. It is undoubtedly some seeming agreements between Marx and Foucault (for instance, in the view that desire is not so much repressed as produced) that results in the use of such misleading phrases as 'Foucauldian Marxism' (Kernan 207), an expression that blurs the differences between the forms of materialism in Marx and Foucault and creates the impression that Foucauldian materialism is a better (because more up-to-date) Marxism.

While indeed rejecting Derrida's pantextualism, Foucault's work nevertheless coincides in crucial ways with ludic theory. The desire or sexuality Foucault writes about in *The History of Sexuality* is discursive: sex is 'produced' in those interminable discourses early in church confessionals and later on the psychiatrist's couch. Of course, Foucault extends the notion of materiality (beyond textualism) by tying the generation of discourses to specific historically developed institutions such as the church, the prison, and the asylum. But at the same time, he theorizes these institutions as purely local sites that emerge islandlike on the surface of a culture and, like Lyotard's language games, have no common measure ('Nietzsche' 148–52). While Foucault's localization of the material has provided theoretical support for localist political actions, by groups like Act Up and Queer Nation, it has also blocked the possibility of theorizing, as Marx does, systematic global exploitation in relation to the mode of production. Furthermore, in 'Nietzsche, Genealogy, History,' Foucault reiterates the Lyotardian shift away from the conceptual and toward the libidinal by arguing that the 'effective history' he wishes to write has 'more in common with medicine than philosophy ... since among the philosopher's idiosyncrasies is a complete denial of the body' (156). Effective history will thus shun the kind of abstraction philosophy deals with and embrace bodily concreteness. Foucault's later work, particularly the final published volume of his history of sexuality, *The Care of the Self*, anchors the new split, postindividual subject firmly in the body. Just as Barthes promotes a mode of pleasure-*full* 'reading with the body' defined by attention to what is 'close up' (66–67), Foucault declares that '[e]ffective history studies what is closest' ('Nietzsche' 156). Foucault's notion of 'material' politics is captured in this declaration: 'The essence of being radical is *physical*' (Miller 182: emphasis added). While not rejecting outright concept-based theoretics for body-based erotics, Foucault's work nevertheless moves strongly in that direction. He establishes the distance between his materialism and Marx's by aiming 'to move less toward a "theory" of power than toward an "analytics" of power' (*Introduction* 82). This analytics is a middle ground between a historical materialist, concept-based theoretics and (the destination toward which Foucault finally points) a play-inspired, desire-guided queer erotics.

'New' readings of Marxian concepts and texts are being published today in the dominant academy, and following Derrida's recent pronouncement that '[t]here will be no future ... without Marx' ('Spectres' 132), more will surely come. However, these readings are commonly ludic, libidinalizing, localizing. For instance, commentators on the relation of lesbianism and commodification describe local consumer features and represent commodification as a generalized form of reification of the subject in which desire is oppressively fixed (Clark; Wiegman). Other writers produce queer (fetishistic) readings of Marxian texts (Kipnis; Parker; Keenan). In *Marx: A Video*, Laura Kipnis has a worker remark that Marx 'looked to the material foundations of the moment, he looked to the body' (253). It is now commonplace for up-to-date bourgeois critics – under the influence of ludic (post)modern theory – to reread the concept-based theoretical texts of Marx through the lens of the local, the fetishistic, the libidinal, and the bodily.

II

Forward-looking gay theory had a historical vision of a future more just than the present. Caught in interminable Derridean self-dismantling, in the specificities of Foucauldian closeness, disconnectedness, and localism, or in a Lyotardian libidinal economy, queer theory points not toward a differently ordered utopia but toward a nonconditioned and nonordered atopia. When queer theorists envision a future, they portray an ever-expanding region of sensuous pleasure, ignoring the historical constraints need places on pleasure. Queer theory – like ludic (post)modernism in general – can be critically and historically understood as a part of the confluence of elements that constitute late, multinational capitalism, sharing fundamental features in particular with the hyperspace, cyberspace, and cyberpunk of technoculture. Hints of these shared features appear, for instance, when cyberspace is described in the same nostalgic, ahistorical, and atavistic formulas used to characterize the queer. As the queer is said to be older than gay identity, cyberspace is said to tap into 'desires far older than digital computers' (Davis 11). The ahistoricity of the queer is indicated by its violent separation from the historically minded gay.

Dennis Cooper describes the queer as 'the punky, anti-assimilationist, transgressive movement on the fringe of lesbian and gay culture.' He then yokes the punk and the queer to a common antileft position: both 'spring from an idealism that radicals have abandoned for the pleasures of a compromised but stable Left' (31). In the new queer space, the word *compromised* does not mean 'complicit' in the old political sense; any position whatsoever that is stable is already 'compromised' because it is an illusion. Thus, what Cooper calls queer and punk idealism cannot be the kind of social idealism supported by materialist left theory (the commitment to social progress). In envisioning a society with different norms, the gay movement had a social idealism tied to conceptual understandings of material and historical conditions and believed in a politics that was theoretically grounded (and, for the Left, at least related to the tradition of Marx). Cybercized queer theory, with its roots in the anarchic skepticism of Nietzsche, envisions a decentered, Interneted, normless society (if that is not a contradiction in terms). Indeed, one cyberpunk professes a post-Nietzschean project: 'Nihilism doesn't go far enough for me' (Caniglia 95). The association with cyberspace (entry to which is literally not open to all) allies queer idealism with the self-interested individualistic idealism of the bourgeois subject: another cyberpunk states that his life goal is 'to move up' (Caniglia 92).

Cyberspace is the universe of virtual reality produced by the cybernetic electronic systems of contemporary plugged-in or air-wave technoculture. From 3-D yesterday to VR today, there has been a tremendous advance in the enhancement of illusion: 'VR may ultimately project the user into the midst of a digital space as concrete, chimerical, and manipulable as a lucid dream' (Davis 10). Considered globally, the illusions of cyberspace link with the dreams of its Euro-American inventors: beyond merely practical and aesthetic questions lies the politics of cyberspace. Cyberspace is a bourgeois designer space in which privileged Western or Westernized subjects fantasize that instead of being chosen by history, they choose their own histories. By manipulating the machines, the user-subjects write virtual histories according to their desires and seek to evade present historical conditions. Cyberspace is thus symptomatic of the (post)modern displacement of need by desire (the material by the ideal). Julian Dibbell emphasizes the connection of cyberspace to (post)structuralist notions of textuality: 'cyberspace is a place all right, but it is an insistently textual one' whose purpose is the same as that of all writing – 'to fake presence,' Dibbell notes (13–14), citing Derrida. Part of not only Derrida's textualist world but also Lyotard's libidinal economy, cyberspace is produced when 'the libidinal pleasures of video-game joysticks' are connected 'with dry data banks' (Davis 10).

For my purposes here, one of the most important new zones in the technocultural expansion of the bourgeois subject's desire is cybersex, or '"teledildonics" (simulated sex at a distance)' (Rheingold 19). Like cyberspace, cybersex is often accounted for pragmatically: in the age of AIDS, cybersex, which produces what Susie Bright refers to as 'the virtual orgasm' (60), is the newest form of safe sex. From telephone sex lines that still rely on the user's imagination to erotic messages and pictures transmitted by e-mail and computer bulletin boards to the full-fledged graphic virtual sexual experiences enabled by VR goggles and gloves, cybersex provides 'tactile telepresence' (Rheingold 346) that simulates the conditions and sensations required to produce sexual stimulation and release. One cyberfan predicts, 'Thirty years from now, when portable telediddlers become ubiquitous, most people will use them to have sexual experiences with other *people*, at a distance, in combinations and configurations undreamed of by precybernetic voluptuaries' (345).

Reporting on a telephone-sex operation run by two women, Bright reveals not so much its pragmatics as its political economy. Bright deploys tellingly contradictory discourses: on the one hand, she describes phone sex as 'out-of-body' travel (body-less-ness; 67); on the other hand, such travel is undertaken in search of sensuous experience (body-full-ness). The contradiction is resolvable in materialist terms: cybersex enables travel out of the actual, historical body and into a virtual body of Deleuzean pure intensity. Bright listens as a sex-line hostess with hips that 'haven't seen thirty-six inches in about ten years' describes to a client a virtual body meant to satisfy him: 'I'm a 36C–25–36.' This virtual body is the commodified, ahistorical body fetishized by the imaginary and promoted for profit: 'You can be any BODY you want to be on a fiber optic network' (60–61). While the features fetishized in different representations will of course vary with the target audience (indeed, a wholly ludic self-shifting subject-audience – bisexual, pan-sexual, gay-male woman, male lesbian, etc. – would permit the most efficient market), the same process is at work throughout contemporary culture in the United States – from explicit pornography to crotch grabbing by popular performers.

Bright reveals not only virtual reality's micropolitical uses (the enhancement of erotic consciousness) but also its macro- and geopolitical applications (the salving of Western powers' bad consciences). The United States Defense Department, which has 'the most advanced telerobotic, long distance computer scanning equipment' available, used it to make the Persian Gulf war a '"virtual" war' in which the pilots of cyberbombers strapped on special 'goggles [to] see where they [were] going and where ... to drop bombs.' The advantage was that 'technological imaging' 'get[s] human beings out of the loop.' While Bright sees the use of virtual reality in war as 'the dead opposite of the technoeroticism' she promotes (63), these micro-uses and macro-uses converge on the political plane: virtual reality situates both bomber pilots and cybersex participants in an ahistorical space supposedly disconnected from actuality, putting them beyond social responsibility.

Alongside Bright's popular report (and produced by the same dominant assumptions) are such sophisticated theoretical accounts of contemporary cyberculture as, for instance, Avital Ronell's *The Telephone Book: Technology, Schizophrenia, Electric Speech*. Ronell's work is self-consciously informed by all the ludic (post)modern assumptions that give primacy to the mode of signification. Ronell takes the signifiers of the telephone book to constitute the human community (thus understood as a linguistic entity). Therefore, she conceives of her inquiry into the operation of the telephone not as reflection (that is, traditional philosophy) but as performance, as 'response,' or 'answering a call' (3). For Ronell, the telephone is not so much a technological instrument developed at a particular historical moment as an emblem of an endless pan-historical connectedness: 'The telephone splices a party line stretching through history' (295). The subject of this party line is the schizophrenic: Deleuze and Guattari's schizo 'desiring machine' becomes the subject 'who has distributed telephone receivers along her body' (4). Thus Ronell's contemporary subject exists in a state of generalized perversity (queerness), in which it is difficult to decide or conclude any question. Human intersubjectivity is a system of shifting relays that never reveals who calls and who answers. The same political question of social responsibility haunts Bright's essays and Ronell's book.

If *Bimbox* queers and cyberpunks are similar in mixing 'youthful rage and technological terrorism' (Caniglia 88), the difference between the two is that cyberpunks derive their technology from differential, digital, and binary computer culture while cyberqueers base their technotheory on the differential and binary theories of ludic (post)modernism. From the classical Marxist perspective, both technologies are the consequences of changes in the mode of production under late capitalism. As virtual, post-al spaces supposedly beyond historical consciousness where those who can afford it choose their reality, postgay queerity and postleft political cyberpunk are the latest forms of bourgeois idealism.

The same violent imposition of virtual reality over historical actuality is present in 'more civilized' queer studies. 'White Glasses,' for instance, is an intimate essay by Eve Sedgwick written as a memorial in advance for a friend dying of AIDS (the gay poet, teacher, scholar, and activist Michael Lynch). In the form of a journal or memoir, the essay deals also with the author's sense of her own mortality brought home by the discovery of her breast cancer. Certainly such matters are personal, but what better opportunity to raise transpersonal, collective issues? Instead of inquiring into the global historical conditions that determine health care availability and quality, instead of using AIDS and breast cancer as occasions for a conceptual inquiry into unjust patterns of resource distribution on a global scale, Sedgwick follows the queer mode: she lets desire play by fetishizing a signifier (white eyeglasses) associated metonymically with the dying friend (Lynch was wearing white glasses on the day Sedgwick met and 'fell in love with' him [194]). The problem with Sedgwick's text is not that she organizes it around what she calls 'obsessive imagery' (198) but that the fetishized glasses become the emblem of the current privileged ludic mode of understanding, for questions of health and illness, for queer studies (since Lynch and Sedgwick have played prominent roles in the development of that mode) and 'otherness' itself, and indeed for any kind of inquiry. My critique concerns the broader effects of 'White Glasses' as a text by a committed queer studies critic and pedagogue that promotes fetishes. This modality is an intellectual problem because, as Laura Mulvey puts it, the function of the fetish is not to produce new understanding but 'to guard against the encroachment of knowledge' (12). By wearing white glasses, one produces local (fashion) effects, sets oneself apart through association with a new desire (for the signifier, white glasses); and by appropriating a new difference, one feels in control over difference itself. As Mulvey suggests, the substitution of feeling for knowledge trivializes the understanding of how differences are produced and of the place of particular differences within prevailing exploitative relations. My point is neither to deny nor to erase fetishes but to inquire into their production and use. In Sedgwick's essay, what starts out as a personal fetish ('to me the glasses meant ... nothing but Michael' [195]) is transformed into a universal one, a 'prosthetic device that attaches to, extends, and corrects the faulty limb of our vision' (197). Sedgwick also provides a striking instance of the merger of the fetish as erotic with the fetish as commodity. She reports that her 'first thought' on seeing Lynch was this: 'Within a year, every fashion-conscious person in the United States is going to be wearing white glasses. ... I want white glasses first' (193). In the 'magical' realm (197) made available by the white glasses (which create, Sedgwick says, a 'series of uncanny effects' [197]), reality is queered: a woman who thinks of herself (and believes that others think of her) as 'fat' can escape that historical condition and can moreover slip the moorings of her social construction and become, 'along with Michael, a gay man' (198). In the end, the white glasses produce the same effect as VR goggles: the bourgeois subject, whose desire is (relatively) autonomous, is in a position not only to have the latest commodity first but also, more broadly, to write her own virtual history.[6]

III

Although the ahistorical thinking characteristic of queer theory and cyberspace seems most readily visible in 'avant-garde' texts, it constitutes the understanding of the real in all texts of the dominant culture. A popular example that lies at the intersections of queerity, perversion, psychopathology, cyberspace, and virtual reality is Paul Verhoeven's film *Basic Instinct*. Queer activists vigorously protested the making and distribution of this film, in which the characters Catherine, Roxy, and Beth are involved in lesbian sex and are intending to be, suspected of being, or finally proved to be murderers. The queer activists' criticism is ethical and moral and, therefore, idealistic: it accuses the producers of the film, who know that most lesbians and gay men are not murderers, of circulating a false and negative stereotype. Such criticism, however, merely mirrors the religious fundamentalists' moral account of homosexuality, replacing their negative associations with positive ones. Furthermore, it is pointless to issue moral protests from the queer frame of reference, for by definition it sets aside questions of (humanist and modernist) morality as irrelevant. By contrast, materialist critique investigates how associations themselves (positive or negative) are produced and connects questions of sexual practices not to morality but to the politics of class and of other grounds of oppression and to ideology, of which morality is one expression.

One of the most urgent aspects of *Basic Instinct* is its demand that the viewer separate sexuality from humanist, modernist, and moralist frames of reference. The central character, Catherine Trammell (Sharon Stone), is an exemplary ludic (post)modern subject: wealthy, she is a writer who graduated from Berkeley with a double major in literature and psychology. She has written a mystery story that anticipates in its details the murder of one of her sex partners (this sensational ice-pick murder of a man by a woman, occurring at the point of orgasm during sexual intercourse, is the opening scene of the film). Catherine works hard to separate her understanding of sexuality in general and of her relationship with the murdered man in particular from normative social views. It was not a loving relationship but a sexual one of pure desire. She corrects a detective on this point: 'I wasn't dating him. I was fucking him.' The film is an investigation of the principles that history is nothing more than a text and that the text's meaning is undecidable. On the one hand, Catherine becomes the prime suspect because the murder follows the plot of her fictional account; on the other hand, she has what Nick Curran (Michael Douglas), a detective investigating the case, calls the perfect alibi (she would not be so stupid as to commit a murder that matches her own fictional plot). As Catherine becomes involved sexually with Nick, she is writing (on a computer) a novel about a detective who, after falling for the 'wrong woman,' is killed by her. Part of the film's tension derives from the fear that fiction may come true again. Eventually, the relationship of the rich and free woman and the employed and thus constrained man takes on the aspect of an allegory not just of class relations but also of the contrast between ludic (post)modernist and modernist ways of living. Nick is represented through familiar discourses: he is an ordinary joe and an ex-alcoholic and ex-cocaine user who is trying to resist falling off the wagon. For Nick, who must work for a living ('I don't have any money,' he says), desire is pitted against need. For Catherine, wealth sustains pleasure-seeking adventures (with drugs, sex, etc.) and produces a state of consciousness in which desire seems to have escaped mere need; the regulation of desire – resisting temptation – seems as irrelevant to her as oedipal regulation is to Deleuze and Guattari. Like the subject articulated in ludic theory, Catherine writes playfully (the plots of more than one of her published novels parallel events in her life) and strives to live playfully too.

From the materialist perspective, to protest *Basic Instinct* in behalf of the queer is less useful than to critique the kind of queered reality the film explores and to relate the film's 'message' to the queer political project at large. The central character is not just queer and a murder suspect but, more important, a bourgeois subject free to write her virtual reality. In her novel, Catherine composes her own *jouissance*, her pure intensity (Nick calls her 'the fuck of the century,' evoking the desire to escape from history), and lives out her desire in a virtual space. Not everyone can let desire play as Catherine can: Gus, a middle-aged detective who is Nick's best friend and a voice of normative reality, remarks: 'I can get laid by blue-haired women, but I don't want 'em.' This reminder that desire is (relatively) autonomous only for some calls in question the occasional claims that the queer agenda includes free sexual expression for all. For a ludic (post)modern subject like Catherine, desire *is* relatively auton-omous. Although the film's sex scenes do not immediately recall the technically engineered pleasures of cybersex, sex in *Basic Instinct* is nevertheless techni-cally (pharamacologically and otherwise) enhanced. Furthermore, virtual reality is by no means unfamiliar to the film's director, who also made the technoculture hit *Robocop*. Catherine, for whom historical reality is only virtual, uses her novel to unsettle her identity as murderer for those trying to solve the crime. Her manipulation of representations according to her desire recalls the manipulation of gender identity by the queer theorist Judith Butler: both Catherine and Butler relate gender to desire and occlude the connections of gender and sexuality to need (class).

Right-wing criticism of the film would no doubt focus on its sexual images and dialogue, placing them in a 'moral' frame. Assuming the existence of timeless and universal truths (and seeking a 'proper,' 'pure,' and thus 'moral' expression of those truths), conservatives would find the sexual content a sign of moral decay and an affront to 'family values.' The observation that the film offends bourgeois morality, however, overlooks the way in which bourgeois morality alters to keep up with historical changes in the mode of production: the morality needed by the bourgeoisie under earlier stages of capitalism and the morality needed by the same class under late, multinational capitalism are different. Thus, the film's sexual content is not so much a calculated transgres-sion of bourgeois values for shock effect as it is the mark of a transition in bourgeois consciousness; the frank sexuality results from the (post)modern erasure of the distinction between the public and the private, the outside and the inside. It is often remarked that modernist depth has disappeared and that nothing but (post)modern 'surfaces' remain (Jameson 6). On this view, nothing is hidden any longer; the outside is the inside and vice versa. The human subject no longer has an 'interiority': consciousness and unconsciousness are written on the surface of the body, which is a text. In the new world of the cyberqueer, modesty or reticence in speech or behavior is irrelevant. The film's content reflects the collapse of the space of civil behavior into the private space of instinctual reaction.

One of the most telling aspects of the queer response to *Basic Instinct* is that while rejecting the negative image of homosexuals, the queer perspective would support the sexual frankness – and would justify both positions on the moral or ethical grounds of accuracy of representation. This contradictory response raises another question that I have yet to see asked in queer studies: Does not cyberqueerity (which promotes the supposed autonomy of desire and suppresses the issue of need) play a pivotal role today – while seeming to oppose this development – in the transition to that new bourgeois morality and state of consciousness which is desperately needed by late, multinational capitalism to maintain its exploitative and oppressive regime of class relations?

NOTES

1. On the 'post-al,' see *Transformation* 1, especially Mas'ud Zavarzadeh's lead essay, 'Post-ality.' As I understand (post)modernism, the term denotes not a positive entity 'out there' but a political construct – the collectivity of reading strategies deployed to make sense of a particular historical series. Following Teresa L. Ebert, I use *ludic (post)modernism* to designate the understanding that sees (post)modernity as a problematics of representation and conceives representation as a rhetorical issue, a matter of signification, in which the process of signification itself articulates the signified. Under the law of *différance*, representation is always incommensurate with the represented. Ludic (post)modernism, therefore, posits the real as an instance of 'simulation' and in no sense the origin of the truth that can provide a ground for a political project. *Différance* is regarded as the effect of the unending playfulness (thus the term *ludic*) of the signifier in signifying processes that can no longer acquire representational authority by anchoring themselves in what Derrida calls the 'transcendental signified' (*Of Grammatology* 20). For more sustained discussions of the different understandings of (post)modernity in relation to current competing forms of cultural studies, see Zavarzadeh and Morton 152–54.
2. On the queer, see de Lauretis; Berlant and Freeman; Warner; Smyth.
3. For other theorizations of mutations in subjectivity (besides cyberqueerity) in technoculture, see Haraway; Penley and Ross.
4. For a sustained inquiry into this shift from gender to sexuality, see Morton, 'Politics.'
5. Exemplary instances of the first mode are Hocquenghem; Fuss (*Essentially Speaking and Inside/Out*); Bristow; Munt; Meese; Edelman; Warner; Wolfe and Penelope; Abelove, Barale, and Halperin; Doan; and Butler (*Gender Trouble* and *Bodies*). Exemplary instances of the second are Watney; Halperin; Dolimore; Abelove, Barale, and Halperin; and Evans. For contestations between proponents of the two modes, see Ryan, and for another critique of (post)modernity in relation to sexuality, see Edwards. In spite of local differences, the two strands converge on the political plane. For expansions of the arguments I make here, see my 'Queerity' and introduction to *Queer Theory*.
6. In a recent interview conducted by Jeffrey Williams, Sedgwick hints that she is moving on a 'hunch' in the direction of the cyberqueer ('Sedgwick Unplugged' 63–64). I believe that such a text as 'White Glasses' shows her to be already there.

WORKS CITED

Abelove, Henry, Michèle Barale, and David M. Halperin, eds. *The Lesbian and Gay Studies Reader*. New York: Routledge, 1993.

Adam, Barry D. *The Rise of a Gay and Lesbian Movement*. Boston: Twayne, 1987.

Barthes, Roland. *The Pleasure of the Text*. Trans. Richard Miller. New York: Hill, 1965.

Basic Instinct. Dir. Paul Verhoeven. TriStar, 1992.

Baudrillard, Jean. *For a Critique of the Political Economy of the Sign*. Trans. Charles Levin. Saint Louis: Telos, 1981.

Baudrillard, Jean, *Simulations*. Trans. Paul Foss and Paul Patton. New York: Semiotext(e), 1983.

Berlant, Lauren, and Elizabeth Freeman. 'Queer Nationality.' *Boundary* 2 19.1 (1992): 149–80.

Boone, Joseph, and Michael Cadden, eds. *Engendering Men: The Question of Male Feminist Criticism*. New York: Routledge, 1990.

Bredbeck, Gregory W. 'The Postmodernist and the Homosexual.' Readings and Schaber 254–59.

Bright, Susie. *Susie Bright's Sexual Reality: A Virtual Sex World Reader*. Pittsburgh: Cleis, 1992.

Bristow, Joseph, ed. *Sexual Sameness: Textual Differences in Lesbian and Gay Writing*. London: Routledge, 1992.

Browning, Frank. *The Culture of Desire: Paradox and Perversity in Gay Lives Today*. New York: Crown, 1993.

Butler, Judith. *Bodies That Matter: On the Discursive Limits of 'Sex.'* New York: Routledge, 1993.

Butler, Judith. *Gender Trouble: Feminism and the Subversion of Identity*. New York: Routledge, 1990.

Butler, Judith. 'Imitation and Gender Insubordination.' Fuss, *Inside/Out* 13–31.

Caniglia, Julie. 'Cyberpunks Hate You.' *Utne Reader* July–Aug. 1993: 88–96.

Clark, Danae. 'Commodity Lesbianism.' *Camera Obscura* 25–26 (1991): 181–201.

Cohen, Ed. 'Are We (Not) What We Are Becoming? Gay "Identity", "Gay Studies," and the Disciplining of Knowledge.' Boone and Cadden 161–75.

Cooper, Dennis. 'Johnny Noxzema to the Gay Community: "You Are the Enemy"' *Village Voice* 30 June 1992: 31–32.

Davis, Erik. 'A Computer, a Universe: Mapping an Online Cosmology.' *Voice Literary Supplement* Mar. 1993: 10–11.

de Lauretis, Teresa, ed. *Queer Theory*. Spec. issue of *Differences* 3.2 (1991): i–xviii, 1–59.

Deleuze, Gilles, and Félix Guattari. *Anti-Oedipus: Capitalism and Schizophrenia*. Trans. Robert Hurley, Mark Seem, and Helen R. Lane. New York: Viking, 1977.

Deleuze, Gilles. *A Thousand Plateaus: Capitalism and Schizophrenia*. Trans. Brian Massumi. Minneapolis: U of Minnesota P, 1987.

Derrida, Jacques. *Of Grammatology*. Trans. Gayatri C. Spivak. Baltimore: Johns Hopkins UP, 1976.

Derrida, Jacques. 'Spectres of Marx.' *New Left Review* 205 (1994): 131–60.

Dibbell, Julian. 'Let's Get Digital: The Writer à la Modem.' *Voice Literary Supplement* Mar. 1993: 13–14.

Doan, Laura, ed. *The Lesbian Postmodern*. New York: Columbia UP, 1994.

Dollimore, Jonathan. *Sexual Dissidence: Augustine to Wilde, Freud to Foucault*. Oxford: Clarendon, 1991.

Ebert, Teresa L. 'Ludic Feminism, the Body, Performance, and Labor: Bringing *Materialism* Back into Feminist Cultural Studies.' *Cultural Critique* 23 (1992–93): 5–50.

Edelman, Lee. *Homographesis: Essays in Gay Literary and Cultural Theory*. New York: Routledge, 1993.

Edwards, Tim. *Erotic Politics: Gay Male Sexuality, Masculinity, and Feminism*. London: Routledge, 1994.

Evans, David T. *Sexual Citizenship: The Material Construction of Sexualities*. London: Routledge, 1993.

Fish, Stanley. *There is No Such Thing as Free Speech ... and It's a Good Thing, Too*. New York: Oxford UP, 1993.

Foucault, Michel. *The Care of the Self*. Trans. Robert Hurley. New York: Vintage-Random, 1988. Vol. 3 of *The History of Sexuality*.

Foucault, Michel. *An Introduction*. Trans. Robert Hurley. New York: Vintage-Random, 1980. Vol. 1 of *The History of Sexuality*.

Foucault, Michel. 'Nietzsche, Genealogy, History.' *Language, Counter-memory, Practice: Selected Essays and Interviews*. Trans. Donald F. Bouchard and Sherry Simon. Ithaca: Cornell UP, 1977, 139–64.

Freud, Sigmund. *Civilization and Its Discontents*. Trans. James Strachey. New York: Norton, 1961.

Fuss, Diana. *Essentially Speaking*. New York: Routledge, 1989.

Fuss, Diana, ed. *Inside/Out: Lesbian Theory, Gay Theory*. New York: Routledge, 1991.

Gallop, Jane. *Thinking through the Body*. New York: Columbia UP, 1988.

Halperin, David. *'One Hundred Years of Homosexuality' and Other Essays on Greek Love*. New York: Routledge, 1990.

Haraway, Donna. *Simians, Cyborgs, and Women: The Reinvention of Nature*. New York: Routledge, 1991.

Hocquenghem, Guy. *Homosexual Desire*. Trans. Daniella Dangoor, London: Allison, 1978.

Jameson, Fredric. *Postmodernism; or, The Cultural Logic of Late Capitalism*. Durham: Duke UP, 1991.

Johnson, Barbara. *A World of Difference*. Baltimore: Johns Hopkins UP. 1987.

Keenan, Thomas. 'The Point Is to (Ex)Change It: Reading "Capital" Rhetorically.' *Fetishism as Cultural Discourse*. Ed. Emily Apter. Ithaca: Cornell UP, 1993. 152–85.

Kernan, Alvin. *The Death of Literature*. New Haven: Yale UP, 1990.

Kipnis, Laura. *Ecstasy Unlimited: On Sex, Capital, Gender, and Aesthetics*. Minneapolis: U of Minnesota P, 1993.

Lacan, Jacques. *Ecrits: A Selection*. Trans. Alan Sheridan. New York: Norton, 1977.

Lyotard, Jean-François. *Libidinal Economy*. Trans. Iaian H. Grant. Bloomington: Indiana UP, 1993.

Lyotard, Jean-François, and Jean-Loup Thébaud. *Just Gaming*. Trans. Wlad Godzich. Minneapolis: U of Minnesota P, 1985.

Marx, Karl. *The Communist Manifesto*. Ed. Frederic L. Bender. New York: Norton, 1988.

Meese, Elizabeth A. *(Sem)Erotics: Theorizing Lesbian: Writing*. New York: New York UP, 1992.

Miller, James. *The Passion of Michel Foucault*. New York: Simon, 1993.

Mitchell, Hugh, and Kayode Olafimihan. 'A Queer View.' *Living Marxism* Nov. 1992: 38–39.

Morton, Donald. Introduction. *Queer Theory: A Lesbian and Gay Cultural Studies Reader*. Ed. Morton. Boulder: Westview, 1995.

Morton, Donald. 'The Politics of Queer Theory in the (Post)Modern Moment.' *Genders* 17 (1993): 121–50.

Morton, Donald. 'Queerity and Ludic Sado-masochism.' *Transformation 1* (1995): 180–230.

Mulvey, Laura. 'Some Thoughts on Theories of Fetishism in the Context of Contemporary Culture.' *October* 65 (1993): 3–20.

Munt, Sally, ed. *New Lesbian Criticism: Literary and Cultural Readings*. New York: Columbia UP, 1992.

Parker, Andrew. 'Unthinking Sex: Marx, Engels, and the Scene of Writing.' Warner 19–41.

Penley, Constance, and Andrew Ross. eds. *Technoculture*. Minneapolis: U of Minnesota P, 1991.

Readings, Bill, and Bennett Schaber, eds. *Postmodernism across the Ages: Essays for a Postmodernity That Wasn't Born Yesterday*. Syracuse: Syracuse UP, 1993.

Rheingold, Howard. *Virtual Reality*. New York: Summit, 1991.

Ronell, Avital. *The Telephone Book: Technology, Schizophrenia, Electric Speech*. Lincoln: U of Nebraska P, 1989.

Ryan, Michael. 'Foucault's Fallacy.' *Strategies* 7 (1992–93): 132–54.

Sedgwick, Eve Kosofsky. *Epistemology of the Closet*. Berkeley: U of California P, 1990.

Sedgwick, Eve Kosofsky. 'Sedgwick Unplugged (An Interview with Eve Kosofsky Sedgwick).' By Jeffrey Williams. *Minnesota Review* ns 40 (1993): 53–64.

Sedgwick, Eve Kosofsky. 'White Glasses.' *Yale Journal of Criticism* 5.3 (1992): 193–208.

Smith, Barbara. 'Where's the Revolution?' *Nation* 5 July 1993: 12–16.

Smyth, Cheryl, *Lesbians Talk Queer Notions*. London: Scarlet, 1992.

Toepfer, Karl. *Theatre, Aristocracy, and Pornocracy: The Orgy Calculus*. New York: PAJ, 1991.

Warner, Michael, ed. *Fear of a Queer Planet: Queer Politics and Social Theory*. Minneapolis: U of Minnesota P, 1993.

Watney, Simon. *Policing Desire: Pornography, AIDS, and the Media*. Minneapolis: U of Minnesota P, 1987.

Wiegman, Robin. 'Mapping the Lesbian Postmodern.' Introduction. Doan 1–20.

Wolfe, Susan J., and Julia Penelope, eds. *Sexual Practices/Textual Theory: Lesbian Cultural Criticism*. Cambridge: Blackwell, 1993.

Zavarzadeh, Mas'ud. 'Post-ality.' *Transformation* 1 (1995): 1–50.

Zavarzadeh, Mas'ud, and Donald Morton. *Theory as Resistance: Politics and Culture after (Post)Structuralism*. New York: Guilford, 1994.

17

THE PROMISES OF MONSTERS: A REGENERATIVE POLITICS FOR INAPPROPRIATE/D OTHERS

Donna Haraway

If primates have a sense of humor, there is no reason why intellectuals may not share in it. (Plank, 1989)

A BIOPOLITICS OF ARTIFACTUAL REPRODUCTION

'The Promises of Monsters' will be a mapping exercise and travelogue through mind-scapes and landscapes of what may count as nature in certain local/ global struggles. These contests are situated in a strange, allochronic time – the time of myself and my readers in the last decade of the second Christian millenium – and in a foreign, allotopic place – the womb of a pregnant monster, here, where we are reading and writing. The purpose of this excursion is to write theory, i.e., to produce a patterned vision of how to move and what to fear in the topography of an impossible but all-too-real present, in order to find an absent, but perhaps possible, other present. I do not seek the address of some full presence; reluctantly, I know better. Like Christian in *Pilgrim's Progress*, however, I am committed to skirting the slough of despond and the parasite-infested swamps of nowhere to reach more salubrious environs.[1] The theory is meant to orient, to provide the roughest sketch for travel, by means of moving within and through a relentless artifactualism, which forbids any direct si(gh)tings of nature, to a science fictional, speculative factual, SF place called, simply, elsewhere. At least for those whom this essay addresses, 'nature' outside artifactualism is not so much elsewhere as nowhere, a different matter altogether. Indeed, a reflexive artifactualism offers serious political and analytical hope. This essay's theory is modest. Not a systematic overview, it is a little siting device in a long line of such craft tools. Such sighting devices have been known to reposition worlds for their devotees – and for their opponents. Optical instruments are subject-shifters. Goddess knows, the subject is being changed relentlessly in the late twentieth century.

My diminutive theory's optical features are set to produce not effects of distance, but effects of connection, of embodiment, and of responsibility for an imagined elsewhere that we may yet learn to see and build here. I have high stakes in reclaiming vision from the technopornographers, those theorists of minds, bodies, and planets who insist effectively – i.e., in practice – that sight is the sense made to realize the fantasies of the phallocrats.[2] I think sight can be remade for the activists and advocates engaged in fitting political filters to see the world in the hues of red, green, and ultraviolet, i.e., from the perspectives of a still possible socialism, feminist and anti-racist environmentalism, and science for the people. I take as a self-evident premise that 'science is culture.'[3] Rooted in that premise, this essay is a contribution to the heterogeneous and very lively contemporary discourse of science studies *as* cultural studies. Of course, what science, culture, or nature – and their 'studies' – might mean is far less self-evident.

Nature is for me, and I venture for many of us who are planetary fetuses gestating in the amniotic effluvia of terminal industrialism,[4] one of those impossible things characterized by Gayatri Spivak as that which we cannot not desire. Excruciatingly conscious of nature's discursive constitution as 'other' in the histories of colonialism, racism, sexism, and class domination of many kinds, we nonetheless find in this problematic, ethno-specific, long-lived, and mobile concept something we cannot do without, but can never 'have.' We must find another relationship to nature besides reification and possession. Perhaps to give confidence in its essential reality, immense resources have been expended to stabilize and materialize nature, to police its/her boundaries. Such expenditures have had disappointing results. Efforts to travel into 'nature' become tourist excursions that remind the voyager of the price of such displacements – one pays to see fun-house reflections of oneself. Efforts to preserve 'nature' in parks remain fatally troubled by the ineradicable mark of the founding explusion of those who used to live there, not as innocents in a garden, but as people for whom the categories of nature and culture were not the salient ones. Expensive projects to collect 'nature's' diversity and bank it seem to produce debased coin, impoverished seed, and dusty relics. As the banks hypertrophy, the nature that feeds the storehouses 'disappears.' The World Bank's record on environmental destruction is exemplary in this regard. Finally, the projects for representing and enforcing human 'nature' are famous for their imperializing essences, most recently reincarnated in the Human Genome Project.

So, nature is not a physical place to which one can go, nor a treasure to fence in or bank, nor as essence to be saved or violated. Nature is not hidden and so does not need to be unveiled. Nature is not a text to be read in the codes of mathematics and biomedicine. It is not the 'other' who offers origin, replenishment, and service. Neither mother, nurse, nor slave, nature is not matrix, resource, or tool for the reproduction of man.

Nature is, however, a *topos*, a place, in the sense of a rhetorician's place or topic for consideration of common themes; nature is, strictly, a commonplace. We turn to this topic to order our discourse, to compose our memory. As a topic in this sense, nature also reminds us that in seventeenth-century English the 'topick gods' were the local gods, the gods specific to places and peoples. We need these spirits, rhetorically if we can't have them any other way. We need them in order to reinhabit, precisely, *common* places – locations that are widely shared, inescapably local, worldly, enspirited; i.e., topical. In this sense, nature is the place to rebuild public culture.[5] Nature is also a *trópos*, a trope. It is figure, construction, artifact, movement, displacement. Nature cannot pre-exist its construction. This construction is based on a particular kind of move – a *trópos* or 'turn.' Faithful to the Greek, as *trópos* nature is about turning. Troping, we turn to nature as if to the earth, to the primal stuff – geotropic, physiotropic. Topically, we travel toward the earth, a commonplace. In discoursing on nature, we turn from Plato and his heliotropic son's blinding star to see something else, another kind of figure. I do not turn from vision, but I do seek something other than enlightenment in these sightings of science studies as cultural studies. Nature is a topic of public discourse on which much turns, even the earth.

In this essay's journey toward elsewhere, I have promised to trope nature through a relentless artifactualism, but what does artifactualism mean here? First, it means that nature for us is *made*, as both fiction and fact. If organisms are natural objects, it is crucial to remember that organisms are not born; they are made in world-changing techno-scientific practices by particular collective actors in particular times and places. In the belly of the local/global monster in which I am gestating, often called the postmodern world,[6] global technology appears to *denature* everything, to make everything a malleable matter of strategic decisions and mobile production and reproduction processes (Hayles, 1990). Technological decontextualization is ordinary experience for hundreds of millions if not billions of human beings, as well as other organisms. I suggest that this is not a *denaturing* so much as a *particular production* of nature. The preoccupation with productionism that has characterized so much parochial Western discourse and practice seems to have hypertrophied into something quite marvelous: the whole world is remade in the image of commodity production.[7]

How, in the face of this marvel, can I seriously insist that to see nature as artifactual is an *oppositional*, or better, a *differential* siting?[8] Is the insistence that nature *is* artifactual not more evidence of the extremity of the violation of a nature outside and other to the arrogant ravages of our technophilic civilization, which, after all, we were taught began with the heliotropisms of enlightment projects to dominate nature with blinding light focused by optical technology?[9] Haven't eco-feminists and other multicultural and intercultural radicals begun to convince us that nature is precisely *not* to be seen in the guise of the Eurocentric productionism and anthropocentrism that have threatened to reproduce, literally, all the world in the deadly image of the Same?

I think the answer to this serious political and analytical question lies in two related turns: 1) unblinding ourselves from the sun-worshiping stories about the history of science and technology as paradigms of rationalism; and 2) refiguring the actors in the construction of the ethno-specific categories of nature *and* culture. The actors are not all 'us.' If the world exists for us as 'nature,' this designates a kind of relationship, an achievement among many actors, not all of them human, not all of them organic, not all of them techno-logical.[10] In its scientific embodiments as well as in other forms, nature is made, but not entirely by humans; it is a co-construction among humans and non-humans. This is a very different vision from the postmodernist observation that all the world is denatured and reproduced in images or replicated in copies. That specific kind of violent and reductive artifactualism, in the form of a hyper-productionism actually practiced widely throughout the planet, be-comes *contestable* in theory and other kinds of praxis, without recourse to a resurgent transcendental naturalism. Hyper-productionism refuses the witty agency of all the actors but One; that is a dangerous strategy – for everybody. But transcendental naturalism also refuses a world full of cacophonous agencies and settles for a mirror image sameness that only pretends to dif-ference. The commonplace nature I seek, a public culture, has many houses with many inhabitants which/who can refigure the earth. Perhaps those other actors/actants, the ones who are not human, are our topick gods, organic and inorganic.[11]

It is this barely admissible recognition of the odd sorts of agents and actors which/whom we must admit to the narrative of collective life, including nature, that simultaneously, first, turns us decisively away from enlightenment-derived modern and postmodern premises about nature and culture, the social and technical, science and society and, second, saves us from the deadly point of view of productionism. Productionism and its corollary, humanism, come down to the story line that 'man makes everything, including himself, out of the world that can only be resource and potency to his project and active agency.'[12] This productionism is about man the tool-maker and -user, whose highest technical production is himself; i.e., the story line of phallogocentrism. He gains access to this wondrous technology with a subject-constituting, self-deferring, and self-splitting entry into language, light, and law. Blinded by the sun, in thrall to the father, reproduced in the sacred image of the same, his reward is that he is self-born, an autotelic copy. That is the mythos of enlight-enment transcendence.

Let us return briefly to my remark above that organisms are not born, but they are made. Besides troping on Simone de Beauvoir's observation that one is not born a woman, what work is this statement doing in this essay's effort to articulate a relentless differential/oppositional artifactualism? I wrote that organisms are made as objects of knowledge in world-changing practices of scientific discourse by particular and always collective actors in specific times and places. Let us look more closely at this claim with the aid of the concept of the apparatus of bodily production.[13] Organisms are *biological* embodiments; as natural-technical entities, they are not pre-existing plants, animals, protistes, etc., with boundaries already established and awaiting the right kind of instrument to note them correctly. Organisms emerge from a discursive process. Biology is a discourse, not the living world itself. But humans are not the only actors in the construction of the entities of any scientific discourse; machines (delegates that can produce surprises) and other partners (not 'pre- or extra-discursive objects,' but partners) are active constructors of natural scientific objects. Like other scientific bodies, organisms are not *ideological* constructions. The whole point about discursive construction has been that it is *not* about ideology. Always radically historically specific, always lively, bodies have a different kind of specificity and effectivity; and so they invite a different kind of engagement and intervention.

Elsewhere, I have used the term 'material-semiotic actor' to highlight the object of knowledge as an active part of the apparatus of bodily production, without *ever* implying immediate presence of such objects or, what is the same thing, their final or unique determination of what can count as objective knowledge of a biological body at a particular historical juncture. Like Katie King's objects called 'poems,' sites of literary production where language also is an actor, bodies as objects of knowledge are material-semiotic generative nodes. Their boundaries materialize in social interaction among humans and non-humans, including the machines and other instruments that mediate exchanges at crucial interfaces and that function as delegates for other actors' functions and purposes. 'Objects' like bodies do not pre-exist as such. Similarly, 'nature' cannot pre-exist as such, but neither is its existence ideological. Nature is a commonplace and a powerful discursive construction, effected in the interactions among material-semiotic actors, human and not. The siting/sighting of such entities is not about disengaged discovery, but about mutual and usually unequal structuring, about taking risks, about delegating competences.[14]

The various contending biological bodies emerge at the intersection of biological research, writing, and publishing; medical and other business practices; cultural productions of all kinds, including available metaphors and narratives; and technology, such as the visualization technologies that bring color-enhanced killer T cells and intimate photographs of the developing fetus into high-gloss art books, as well as scientific reports. But also invited into that node of intersection is the analogue to the lively languages that actively intertwine in the production of literary value: the coyote and protean embodiments of a world as witty agent and actor. Perhaps our hopes for accountability for techno-biopolitics in the belly of the monster turn on revisioning the world as coding trickster with whom we must learn to converse. So while the late twentieth-century immune system, for example, is a construct of an elaborate apparatus of bodily production, neither the immune system nor any other of biology's world-changing bodies – like a virus or an ecosystem – is a ghostly fantasy. Coyote is not a ghost, merely a protean trickster.

This sketch of the artifactuality of nature and the apparatus of bodily production helps us toward another important point: the corporeality of theory. Overwhelmingly, theory is bodily, and theory is literal. Theory is not about matters distant from the lived body; quite the opposite. Theory is *anything* but disembodied. The fanciest statements about radical decontextualization as the historical form of nature in late capitalism are tropes for the embodiment, the production, the literalization of experience in that specific mode. This is not a question of reflection or correspondences, but of technology, where the social and the technical implode into each other. Experience is a semiotic process – a semiosis (de Lauretis, 1984). Lives are built; so we had best become good craftspeople with the other worldly actants in the story. There is a great deal of rebuilding to do, beginning with a little more surveying with the aid of optical devices fitted with red, green, and ultraviolet filters.

Repeatedly, this essay turns on figures of pregnancy and gestation. Zoe Sofia (1984) taught me that every technology is a reproductive technology. She and I have meant that literally; ways of life are at stake in the culture of science. I would, however, like to displace the terminology of reproduction with that of generation. Very rarely does anything really get *reproduced*; what's going on is much more polymorphous than that. Certainly people don't reproduce, unless they get themselves cloned, which will always be very expensive and risky, not to mention boring. Even technoscience must be made into the paradigmatic model not of closure, but of that which is contestable and contested. That involves knowing how the world's agents and actants work; how they/we/it come into the world, and how they/we/it are reformed. Science becomes the myth not of what escapes agency and responsibility in a realm above the fray, but rather of accountability and responsibility for translations and solidarities linking the cacophonous visions and visionary voices that characterize the knowledges of the marked bodies of history. Actors, as well as actants, come in many and wonderful forms. And best of all, 'reproduction' – or less inaccurately, the generation of novel forms – need not be imagined in the stodgy bipolar terms of hominids.[15]

If the stories of hyper-productionism and enlightenment have been about the reproduction of the sacred image of the same, of the one true copy, mediated by the luminous technologies of compulsory heterosexuality and masculinist self-birthing, then the differential artifactualism I am trying to envision might issue in something else. Artifactualism is askew of productionism; the rays from my optical device diffract rather than reflect. These diffracting rays compose *interference* patterns, not reflecting images. The 'issue' from this generative technology, the result of a monstrous[16] pregnancy, might be kin to Vietnamese-American filmmaker and feminist theorist Trinh Minh-ha's (1986/7; 1989) 'inappropriate/d others.'[17] Designating the networks of multicultural, ethnic, racial, national, and sexual actors emerging since World War II, Trinh's phrase referred to the historical positioning of those who cannot adopt the mask of either 'self' or 'other' offered by previously dominant, modern Western narratives of identity and politics. To be 'inappropriate/d' does not mean 'not to be in relation with' – i.e., to be in a special reservation, with the status of the authentic, the untouched, in the allochronic and allotopic condition of innocence. Rather to be an 'inappropriate/d other' means to be in critical, deconstructive relationality, in a diffracting rather than reflecting (ratio)nality – as the means of making potent connection that exceeds domination. To be inappropriate/d is not to fit in the *taxon*, to be dislocated from the available maps specifying kinds of actors and kinds of narratives, not to be originally fixed by difference. To be inappropriate/d is to be neither modern nor postmodern, but to insist on the *a*modern. Trinh was looking for a way to figure 'difference' as a 'critical difference within,' and not as special taxonomic marks grounding difference as apartheid. She was writing about people; I wonder if the same observations might apply to humans and to both organic and technological non-humans.

The term 'inappropriate/d others' can provoke rethinking social relationality within artifactual nature – which is, arguably, global nature in the 1990s. Trinh Minh-ha's metaphors suggest another geometry and optics for considering the relations of difference among people and among humans, other organims, and machines than hierarchical domination, incorporation of parts into wholes, paternalistic and colonialist protection, symbiotic fusion, antagonistic opposition, or instrumental production from resource. Her metaphors also suggest the hard intellectual, cultural, and political work these new geometries will require. If Western patriarchal narratives have told that the physical body issued from the first birth, while man was the product of the heliotropic second birth, perhaps a differential, diffracted feminist allegory might have the 'inappropriate/d others' emerge from a third birth into an SF world called elsewhere – a place composed from interference patterns. Diffraction does not produce 'the same' displaced, as reflection and refraction do. Diffraction is a mapping of interference, not of replication, reflection, or reproduction. A diffraction pattern does not map where differences appear, but rather maps where the *effects* of difference appear. Tropically, for the promises of monsters, the first invites the illusion of essential, fixed position, while the second trains us to more subtle vision. Science fiction is generically concerned with the interpenetration of boundaries between problematic selves and unexpected others and with the exploration of possible worlds in a context structured by transnational technoscience. The emerging social subjects called 'inappropriate/d others' inhabit such worlds. SF – science fiction, speculative futures, science fantasy, speculative fiction – is an especially apt sign under which to conduct an inquiry into the artifactual as a reproductive technology that might issue in something other than the sacred image of the same, something inappropriate, unfitting, and so, maybe, inappropriated.

Within the belly of the monster, even inappropriate/d others seem to be interpellated – called through interruption – into a particular location that I have learned to call a cyborg subject position.[18] Let me continue this travelogue and inquiry into artifactualism with an illustrated lecture on the nature of cyborgs as they appear in recent advertisements in *Science*, the journal of the American Association for the Advancement of Science. These ad figures remind us of the corporeality, the mundane materiality, and literality of theory. These commercial cyborg figures tell us what may count as nature in technoscience worlds. Above all, they show us the implosion of the technical, textual, organic, mythic, and political in the gravity wells of science in action. These figures are our companion monsters in the *Pilgrim's Progress* of this essay's travelogue.

Consider Figure 17.1, 'A Few Words about Reproduction from a Leader in the Field,' the advertising slogan for Logic General Corporation's software duplication system. The immediate visual and verbal impact insists on the absurdity of separating the technical, organic, mythic, textual, and political threads in the semiotic fabric of the ad and of the world in which this ad makes sense. Under the unliving, orange-to-yellow rainbow colors of the earth-sun logo of Logic General, the biological white rabbit has its (her? yet, sex and gender are not so settled in this reproductive system) back to us. It has its paws on a keyboard, that inertial, old-fashioned residue of the typewriter that lets our computers feel natural to us, user-friendly, as it were.[19] But the keyboard is

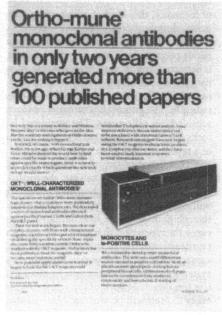

Figure 17.1   Figure 17.2

misleading; no letters are transferred by a mechnical key to a waiting solid surface. The computer–user interface works differently. Even if she doesn't understand the implications of her lying keyboard, the white rabbit is in her natural home; she is fully artifactual in the most literal sense. Like fruit flies, yeast, transgenic mice, and the humble nematode worm, *Caenorhabditis elegans*,[20] this rabbit's evolutionary story transpires in the lab; the lab is its proper niche, its true habitat. Both material system and symbol for the measure of fecundity, this kind of rabbit occurs in no other nature than the lab, that preeminent scene of replication practices.

With Logic General, plainly, we are not in a biological laboratory. The organic rabbit peers at its image, but the image is not her reflection, indeed, *especially* not her reflection. This is not Lacan's world of mirrors; primary identification and maturing metaphoric substitution will be produced with other techniques, other writing technologies.[21] The white rabbit will be translated, her potencies and competences relocated radically. The guts of the computer produce another kind of visual product than distorted, self-birthing reflections. The simulated bunny peers out at us face first. It is she who locks her/its gaze with us. She, also, has her paws on a grid, one just barely reminiscent of a typewriter, but even more reminiscent of an older icon of technoscience – the Cartesian coordinate system that locates the world in the imaginary spaces of rational modernity. In her natural habitat, the virtual rabbit is on a grid that insists on the world as a game played on a chess-like board. This rabbit insists that the truly rational actors will replicate themselves in a virtual world where the best players will not be Man, though he may linger like the horse-drawn carriage that gave its form to the railroad car or the typewriter that gave its illusory shape to the computer interface. The functional privileged signifier in this system will not be so easily mistaken for any primate male's urinary and copulative organ. Metaphoric substitution and other circulations in the very material symbolic domain will be more likely to be effected by a competent mouse. The if-y femaleness of both of the rabbits, of course, gives no confidence that the new players other to Man will be women. More likely, the rabbit that is interpellated into the world in this non-mirror stage, this diffracting moment of subject constitution, will be literate in a quite different grammar of gender. *Both* the rabbits here are cyborgs – compounds – of the organic, technical, mythic, textual, and political – and they call us into a world in which we may not wish to take shape, but through whose 'Miry Slough' we might have to travel to get elsewhere. Logic General is into a very particular kind of *écriture*. The reproductive stakes in this text are future life forms and ways of life for humans and unhumans. 'Call toll free' for 'a few words about reproduction from an acknowledged leader in the field.'

Ortho-mune*{Trade Mark}'s monoclonal antibodies expand our understanding of a cyborg subject's relation to the inscription technology that is the laboratory (Figure 17.2). In only two years, these fine monoclonals generated more than 100 published papers – higher than any rate of literary production by myself or any of my human colleagues in the human sciences. But this alarming rate of publication was achieved in 1982, and has surely been wholly surpassed by new generations of biotech mediators of literary replication. Never has theory been more literal, more bodily, more technically adept. Never has the collapse of the 'modern' distinctions between the mythic, organic, technical, political, and textual into the gravity well, where the unlamented enlightenment transcendentals of Nature and Society also disappeared, been more evident.

LKB Electrophoresis Division has an evolutionary story to tell, a better, more complete one than has yet been told by physical anthropologists, paleontologists, or naturalists about the entities/actors/actants that structure niche space in an extra-laboratory world: 'There are no missing links in MacroGene Workstation' (Figure 17.3). Full of promises, breaching the first of the ever-multiplying final frontiers, the prehistoric monster *Ichthyostega* crawls from the amniotic ocean into the future, onto the dangerous but enticing dry land. Our no-longer-fish, not-yet-salamander will end up fully identified and separated, as man-in-space, finally disembodied, as did the hero of J. D. Bernal's fantasy in *The World, the Flesh, and the Devil*. But for now, occupying the zone between fishes and amphibians, *Ichthyostega* is firmly on the margins, those potent places where theory is best cultured. It behooves us, then, to join this heroic reconstructed beast with LKB, in order to trace out the transferences of competences – the metaphoric-material chain of substitutions – in this quite literal apparatus of bodily production. We are presented with a travel story, a *Pilgrim's Progress*, where there are no gaps, no 'missing links.'

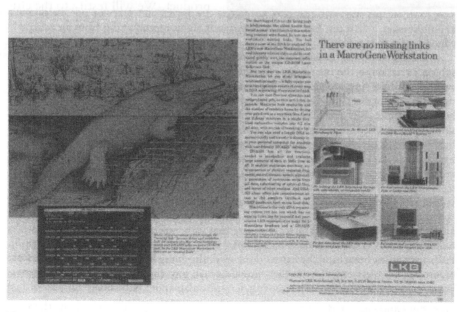

Figure 17.3

From the first non-original actor – the reconstructed *Ichthyostega* – to the final printout of the DNA homology search mediated by LKB's software and the many separating and writing machines pictured on the right side of the advertisement, the text promises to meet the fundamental desire of phallologocentrism for fullness and presence. From the crawling body in the Miry Sloughs of the narrative to the printed code, we are assured of full success – the compression of time into instantaneous and full access 'to the complete GenBank ... on one laser disk.' Like Christian, we have conquered time and space, moving from entrapment in body to fulfillment in spirit, all in the everyday workspaces of the Electrophoresis Division, whose Hong Kong, Moscow, Antwerp, and Washington phone numbers are all provided. Electrophoresis: *pherein* – to bear or carry us relentlessly on.

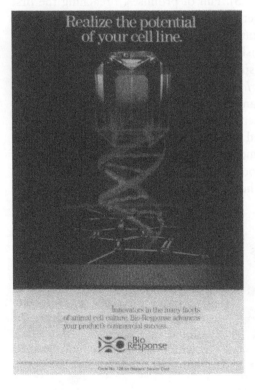

Figure 17.4

Bio-Response, innovators in many facets of life's culture, interpellates the cyborg subject into the barely secularized, evangelical, Protestant Christianity that pervades American techno-culture: 'Realize the potential of your cell line' (Figure 17.4). This ad addresses us directly. We are called into a salvation narrative, into history, into biotechnology, into our true natures: our cell line, ourselves, our successful product. We will testify to the efficacy of this culture system. Colored in the blues, purples, and ultraviolets of the sterilizing commercial rainbow – in which art, science, and business arch in lucrative grace – the virus-like crystalline shape mirrors the luminous crystals of New Age promises. Religion, science, and mysticism join easily in the facets of modern and postmodern commercial bio-response. The simultaneously promising and threatening crystal/virus unwinds its tail to reveal the language-like icon of the Central Dogma, the code structures of DNA that underlie all possible bodily response, all semiosis, all culture. Gem-like, the frozen, spiraling crystals of Bio-Response promise life itself. This is a jewel of great price – available from the Production Services office in Hayward, California. The imbrications of layered signifiers and signifieds forming cascading hierarchies of signs guide us through this mythic, organic, textual, technical, political icon.[22]

Finally, the advertisement from Vega Biotechnologies graphically shows us the final promise, 'the link between science and tomorrow: Guaranteed. Pure' (Figure 17.5). The graph reiterates the ubiquitous grid system that is the signature and matrix, father and mother, of the modern world. The sharp peak is the climax of the search for certainty and utter clarity. But the diffracting apparatus of a monstrous artifactualism can perhaps interfere in this little family drama, reminding us that the modern world never existed and its fantastic guarantees are void. Both the organic and computer rabbits of Logic General might re-enter at this point to challenge all the passive voices of productionism. The oddly duplicated bunnies might resist their logical interpellation and instead hint at a neo-natalogy of inappropriate/d

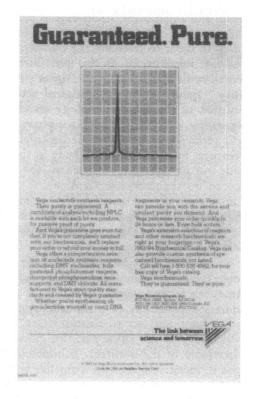

Figure 17.5

others, where the child will not be in the sacred image of the same. Shapeshifting, these interfering cyborgs might craft a diffracted logic of sameness and difference and utter a different word about reproduction, about the link between science and tomorrow, from collective actors in the field.

II. THE FOUR-SQUARE CYBORG: THROUGH ARTIFACTUALISM TO ELSEWHERE

It is time to travel, therefore, with a particular subset of shifted subjects, Cyborgs for Earthly Survival,[23] into the mindscapes and landscapes indicated at the beginning of this essay. To get through the artifactual to elsewhere, it would help to have a little travel machine that also functions as a map. Consequently, the rest of the 'Promises of Monsters' will rely on an artificial device that generates meanings very noisily: A. J. Greimas's infamous semiotic square. The regions mapped by this clackety, structuralist meaning-making machine could never be mistaken for the transcendental realms of Nature or Society. Allied with Bruno Latour, I will put my structuralist engine to amodern purposes: this will not be a tale of the rational progress of science, in potential league with progressive politics, patiently unveiling a grounding nature, nor will it be a demonstration of the social construction of science and nature that locates all agency firmly on the side of humanity. Nor will the

modern be superceded or infiltrated by the postmodern, because belief in
something called the modern has itself been a mistake. Instead, the amodern
refers to a view of the history of science as culture that insists on the absence of
beginnings, enlightenments, and endings: the world has always been in the
middle of things, in unruly and practical conversation, full of action and
structured by a startling array of actants and of networking and unequal
collectives. The much-criticized inability of structuralist devices to provide the
narrative of diachronic history, of progress through time, will be my semiotic
square's greatest virtue. The shape of my amodern history will have a different
geometry, not of progress, but of permanent and multi-patterned interaction
through which lives and worlds get built, human and unhuman. This Pilgrim's
Progress is taking a monstrous turn.

I like my analytical technologies, which are unruly partners in discursive
construction, delegates who have gotten into doing things on their own, to
make a lot of noise, so that I don't forget all the circuits of competences,
inherited conversations, and coalitions of human and unhuman actors that go
into any semiotic excursions. The semiotic square, so subtle in the hands of a
Fredric Jameson, will be rather more rigid and literal here (Greimas, 1966;
Jameson, 1972). I only want it to keep four spaces in differential, relational
separation, while I explore how certain local/global struggles for meanings and
embodiments of nature are occurring within them. Almost a joke on 'elemen-
tary structures of signification' ('Guaranteed. Pure.'), the semiotic square in
this essay nonetheless allows a contestable collective world to take shape for us
out of structures of difference. The four regions through which we will move
are A, Real Space or Earth; B, Outer Space or the Extraterrestrial; not-B, Inner
Space or the Body; and finally, not-A, Virtual Space or the SF world oblique to
the domains of the imaginary, the symbolic, and the real (Figure 17.6).

Somewhat unconventionally, we will move through the square clockwise to
see what kinds of figures inhabit this exercise in science studies as cultural
studies. In each of the first three quadrants of the square, I will begin with a
popular image of nature and science that initially appears both compelling and
friendly, but quickly becomes a sign of deep structures of domination. Then I
will switch to a differential/oppositional image and practice that might promise
something else. In the final quadrant, in virtual space at the end of the journey,
we will meet a disturbing guide figure who promises information about
psychic, historical, and bodily formations that issue, perhaps, from some other
semiotic processes than the psychoanalytic in modern and postmodern guise.
Directed by John Varley's (1986) story of that name, all we will have to do to
follow this disquieting, amodern Beatrice will be to 'Press Enter.' Her job will
be to instruct us in the neo-natology of inappropriate/d others. The goal of this
journey is to show in each quadrant, and in the passage through the machine
that generates them, metamorphoses and boundary shifts that give grounds for
a scholarship and politics of hope in truly monstrous times. The pleasures
promised here are not those libertarian masculinist fantasmics of the infinitely

The Promises of Monsters
Through Artifactualism to Elsewhere ...
A regenerative politics for inappropriate/d others

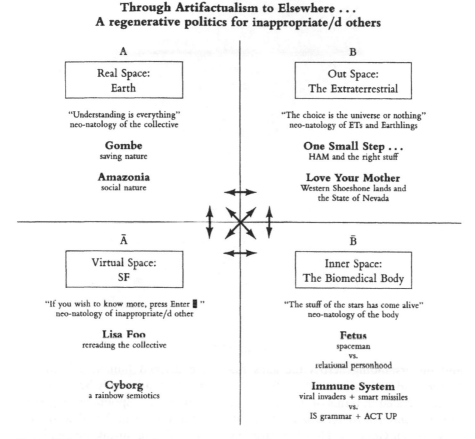

A

| Real Space: |
| Earth |

"Understanding is everything"
neo-natology of the collective

Gombe
saving nature

Amazonia
social nature

B

| Out Space: |
| The Extraterrestrial |

"The choice is the universe or nothing"
neo-natology of ETs and Earthlings

One Small Step ...
HAM and the right stuff

Love Your Mother
Western Shoshone lands and
the State of Nevada

Ā

| Virtual Space: |
| SF |

"If you wish to know more, press Enter ▮"
neo-natology of inappropriate/d other

Lisa Foo
rereading the collective

Cyborg
a rainbow semiotics

B̄

| Inner Space: |
| The Biomedical Body |

"The stuff of the stars has come alive"
neo-natology of the body

Fetus
spaceman
vs.
relational personhood

Immune System
viral invaders + smart missiles
vs.
IS grammar + ACT UP

Figure 17.6

regressive practice of boundary violation and the accompanying *frisson* of brotherhood, but just maybe the pleasure of regeneration in less deadly, chiasmatic borderlands.[24] Without grounding origins and without history's illuminating and progressive tropisms, how might we map some semiotic possibilities for other topick gods and common places?

A. REAL SPACE: EARTH

In 1984, to mark nine years of underwriting the National Geographic Society's television specials, the Gulf Oil Corporation ran an advertisement entitled 'Understanding Is Everything' (Figure 17.7). The ad referred to some of the most watched programs in the history of public television – the nature specials about Jane Goodall and the wild chimpanzees in Tanzania's Gombe National Park. Initially, the gently clasped hands of the ape and the young white woman seem to auger what the text proclaims – communication, trust, responsibility,

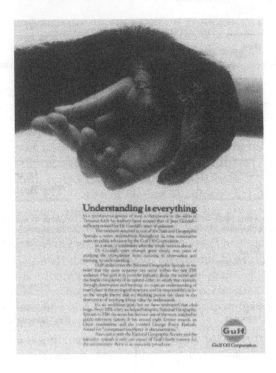

Figure 17.7

and understanding across the gaps that have defined human existence in Nature and Society in 'modern' Western narratives. Made ready by a scientific practice coded in terms of 'years of patience,' through a 'spontaneous gesture of trust' *initiated by the animal*, Goodall metamorphoses in the ad copy from 'Jane' to 'Dr Goodall.' Here is a natural science, coded unmistakenly feminine, to counter the instrumentalist excesses of a military-industrial-technoscience complex, where the code of science is stereotypically anthropocentric and masculine. The ad invites the viewer to forget Gulf's status as one of the Seven Sisters of big oil, ranking eighth among the Forbes 500 in 1980 (but acquired by Chevron by the end of the decade's transnational capitalist restructuring). In response to the financial and political challenges mounted in the early 1970s by the Organization of Oil Exporting Countries (OPEC) and by ecological activism around the globe, by the late 1970s the scandal-ridden giant oil corporations had developed advertising strategies that presented themselves as the world's leading environmentalists – indeed, practically as the mothers of eco-feminism. There could be no better story than that of Jane Goodall and the chimpanzees for narrating the healing touch between nature and society, mediated by a science that produces full communication in a chain that leads innocently 'from curiosity, to observation, to learning, to understanding.'[25] Here is a story of blissful incorporation.

There is another repressed set of codes in the ad as well, that of race and imperialism, mediated by the dramas of gender and species, science and nature. In the National Geographic narrative, 'Jane' entered the garden 'alone' in 1960 to seek out 'man's' closest relatives, to establish a knowing touch across gulfs of time. A natural family is at stake; the PBS specials document a kind of inter-species family therapy. Closing the distance between species through a patient discipline, where first the animals could only be known by their spoor and their calls, then by fleeting sightings, then finally by the animal's direct inviting touch, after which she could name them, 'Jane' was admitted as 'humanity's' delegate back into Eden. Society and nature had made peace; 'modern science' and 'nature' could co-exist. Jane/Dr Goodall was represented almost as a new Adam, authorized to name not by God's creative hand, but by the animal's transformative touch. The people of Tanzania disappear in a story in which the actors are the anthropoid apes and a young British white woman engaged in a thoroughly modern sacred secular drama. The chimpanzees and Goodall are both enmeshed in stories of endangerment and salvation. In the post-World War II era the apes face biological extinction; the planet faces nuclear and ecological annihilation; and the West faces expulsion from its former colonial possessions. If only communication can be established, destruction can be averted. As Gulf Oil insists, 'Our goal is to provoke curiosity about the world and the fragile complexity of its natural order; to satisfy that curiosity through observation and learning; to create an understanding of man's place in the ecological structure, and his responsibility to it – on the simple theory that no thinking person can share in the destruction of anything whose value he understands.' Progress, rationality, and nature join in the great myth of modernity, which is so thoroughly threatened by a dozen looming apocalypses. A cross-species family romance promises to avert the threatened destruction.

Inaudible in the Gulf and National Geographic version, communication and understanding are to emerge in the communion between Jane/Dr Goodall and the spontaneously trusting chimpanzee at just the historical moment when dozens of African nations are achieving their national independence, 15 in 1960 alone, the year Goodall set out for Gombe. Missing from the family romance are such beings as Tanzanians. African peoples seek to establish hegemony over the lands in which they live; to do that the stories of the natural presence of white colonists must be displaced, usually by extremely complex and dangerous nationalist stories. But in 'Understanding Is Everything,' the metonymic 'spontaneous gesture of trust' from the animal hand to the white hand obliterates once again the invisible bodies of people of color who have never counted as able to represent humanity in Western iconography. The white hand will be the instrument for saving nature – and in the process be saved from a rupture with nature. Closing great gaps, the transcendentals of nature and society meet here in the metonymic figure of softly embracing hands from two worlds, whose innocent touch depends on the absence of the 'other world,' the 'third world,' where the drama actually transpires.

In the history of the life sciences, the great chain of being leading from 'lower' to 'higher' life forms has played a crucial part in the discursive construction of race as an object of knowledge and of racism as a living force. After World War II and the partial removal of explicit racism from evolutionary biology and physical anthropology, a good deal of racist and colonialist discourse remained projected onto the screen of 'man's closest relatives,' the anthropoid apes.[26] It is impossible to picture the entwined hands of a white woman and an African ape without evoking the history of racist inconography in biology and in European and American popular culture. The animal hand is metonymically the individual chimpanzee, all threatened species, the third world, peoples of color, Africa, the ecologically endangered earth – all firmly in the realm of Nature, all represented in the leathery hand folding around that of the white girl-woman under the Gulf sun logo shining on the Seven Sisters' commitment to science and nature. The spontaneous gesture of touch in the wilds of Tanzania authorizes a whole doctrine of representation. Jane, as Dr Goodall, is empowered to speak for the chimpanzees. Science speaks for nature. Authorized by unforced touch, the dynamics of representation take over, ushering in the reign of freedom and communication. This is the structure of depoliticizing expert discourse, so critical to the mythic political structures of the 'modern' world and to the mythic political despair of much 'postmodernism,' so undermined by fears about the breakdown of representation.[27] Unfortunately, representation, fraudulent or not, is a very resilient practice.

The clasping hands of the Gulf ad are semiotically similar to the elution peak in the Vega ad of Figure 17.5: 'Guaranteed. Pure.'; 'Understanding Is Everything.' There is no interruption in these stories of communication, progress, and salvation through science and technology. The story of Jane Goodall in Gombe, however, can be made to show its conditions of possibility; even in the footage of the National Geographic specials we see the young woman on a mountain top at night eating from a can of pork and beans, that sign of industrial civilization so crucial to the history of colonialism in Africa, as Orson Welles's voice-over speaks of the lonely quest for contact with nature! In one of Goodall's published accounts of the early days at Gombe, we learn that she and her mother, enroute to the chimpanzee preserve, were stopped on the shores of Lake Tanganyika in the town of Kigoma, across from the no-longer-Belgian Congo, as *uhuru*, freedom, sounded across Africa. Goodall and her mother made 2000 spam sandwiches for fleeing Belgians before embarking for the 'wilds of Tanzania' (Goodall, 1971, p. 27). It is also possible to reconstruct a history of Gombe as a research site in the 1970s. One of the points that stands out in this reconstruction is that people – research staff and their families, African, European, and North American – considerably outnumbered chimpanzees during the years of most intense scientific work. Nature and Society met in one story; in another story, the structure of action and the actants take a different shape.

It is hard, however, to make the story of Jane Goodall and the wild chimpanzees shed its 'modern' message about 'saving nature,' in both senses of nature as salvific and of the scientist speaking for and preserving nature in a drama of representation. Let us, therefore, leave this narrative for another colonized tropical spot in the Real/Earth quadrant in the semiotic square – Amazonia. Remembering that all colonized spots have, euphemistically stated, a special relation to nature, let us structure this story to tell something amodern about nature and society – and perhaps something more compatible with the survival of all the networked actants, human and unhuman. To tell this story we must disbelieve in both nature and society and resist their associated imperatives to represent, to reflect, to echo, to act as a ventriloquist for 'the other.' The main point is there will be no Adam – and no Jane – who gets to name all the beings in the garden. The reason is simple: there is no garden and never has been. No name and no touch is original. The question animating this diffracted narrative, this story based on little differences, is also simple: is there a consequential difference between a political semiotics of articulation and a political semiotics of representation?

The August 1990 issue of *Discover* magazine has a story entitled 'Tech in the Jungle.' A one and one-half page color photo of a Kayapó Indian, in indigenous dress and using a videocamera, dramatically accompanies the opening paragraphs. The caption tells us the man is 'tap[ing] his tribesmen, who had gathered in the central Brazilian town of Altamira to protest plans for a hydroelectric dam on their territory' (Zimmer, 1990, 42–5). All the cues in the *Discover* article invite us to read this photo as the drama of the meeting of the 'traditional' and the 'modern,' staged in this popular North American scientific publication for audiences who have a stake in maintaining belief in those categories. We have, however, as disbelieving members of those audiences, a different political, semiotic responsibility, one made easier by another publication, Susanna Hecht and Alexander Cockburn's *The Fate of the Forest* (1989; see also T. Turner, 1990) through which I propose to suggest articulations and solidarities with the *filming practice* of the Kayapó man, rather than to read the *photograph of him*, which will not be reproduced in this essay.[28]

In their book, which was deliberately packaged, published, and marketed in a format and in time for the 1989 December gift-giving season, a modest act of cultural politics not to be despised, Hecht and Cockburn have a central agenda. They insist on deconstructing the image of the tropical rain forest, especially Amazonia, as 'Eden under glass.' They do this in order to insist on locations of responsibility and empowerment in current conservation struggles, on the outcome of which the lives and ways of life for people and many other species depend. In particular, they support a politics not of 'saving nature' but of 'social nature,' not of national parks and walled-off reserves, responding with a technical fix to whatever particular danger to survival seems most inescapable, but of a different organization of land and people, where the practice of justice restructures the concept of nature.

The authors tell a relentless story of a 'social nature' over many hundreds of years, at every turn co-inhabited and co-constituted by humans, land, and other organisms. For example, the diversity and patterns of tree species in the forest cannot be explained without the deliberate, long-term practices of the Kayapó and other groups, whom Hecht and Cockburn describe, miraculously avoiding romanticizing, as 'accomplished environmental scientists.' Hecht and Cockburn avoid romanticizing because they do not invoke the category of the modern as the special zone of science. Thus, they do not have to navigate the shoals threatening comparisons of, according to taste, mere or wonderful 'ethnoscience' with real or disgusting 'modern science.' The authors insist on visualizing the forest as the dynamic outcome of human as well as biological history. Only after the dense indigenous populations – numbering from six to twelve million in 1492 – had been sickened, enslaved, killed, and otherwise displaced from along the rivers could Europeans represent Amazonia as 'empty' of culture, as 'nature,' or, in later terms, as a purely 'biological' entity.

But, of course, the Amazon was not and did not become 'empty,' although 'nature' (like 'man') is one of those discursive constructions that operates as a technology for making the world over into its image. First, there are indigenous people in the forest, many of whom have organized themselves in recent years into a regionally grounded, world-historical subject prepared for local/global interactions, or, in other terms, for building new and powerful collectives out of humans and unhumans, technological and organic. With all of the power to reconstitute the real implied in discursive construction, they have become a new discursive subject/object, the Indigenous Peoples of the Amazon, made up of national and tribal groups from Colombia, Ecuador, Brazil, and Peru, numbering about one million persons, who in turn articulate themselves with other organized groups of the indigenous peoples of the Americas. Also, in the forest are about 200,000 people of mixed ancestry, partly overlapping with the indigenous people. Making their living as petty extractors – of gold, nuts, rubber, and other forest products – they have a history of many generations in the Amazon. It is a complex history of dire exploitation. These people are also threatened by the latest schemes of world banks or national capitals from Brasília to Washington.[29] They have for many decades been in conflict with indigenous peoples over resources and ways of life. Their presence in the forest might be the fruit of the colonial fantasies of the *bandeirantes*, romantics, curators, politicians, or speculators; but their fate is entwined intimately with that of the other always historical inhabitants of this sharply contested world. It is from these desperately poor people, specifically the rubber tappers union, that Chico Mendes, the world-changing activist murdered on December 22, 1988, came.[30]

A crucial part of Mendes's vision for which he was killed was the union of the extractors and the indigenous peoples of the forest into, as Hecht and Cockburn argue, the 'true defenders of the forest.' Their position as defenders derives not from a concept of 'nature under threat,' but rather from a *relationship* with 'the forest as the integument in their own elemental struggle to survive' (p. 196).[31] In other words, their authority derives *not* from the power to represent from a distance, *nor* from an ontological natural status, but from a constitutive social relationality in which the forest is an integral partner, part of natural/social embodiment. In their claims for authority over the fate of the forest, the resident peoples are articulating a social collective entity among humans, other organisms, and other kinds of non-human actors.

Indigenous people are resisting a long history of forced 'tutelage,' in order to confront the powerful representations of the national and international environmentalists, bankers, developers, and technocrats. The extractors, for example, the rubber tappers, are also independently articulating their collective viewpoint. Neither group is willing to see the Amazon 'saved' by their exclusion and permanent subjection to historically dominating political and economic forces. As Hecht and Cockburn put it, 'The rubber tappers have not risked their lives for extractive reserves so they could live on them as debt peons' (p. 202). 'Any program for the Amazon begins with basic human rights: an end to debt bondage, violence, enslavement, and killings practiced by those who would seize the lands these forest people have occupied for generations. Forest people seek legal recognition of native lands and extractive reserves held under the principle of collective property, worked as individual holdings with individual returns' (p. 207).

At the second Brazilian national meeting of the Forest People's Alliance at Rio Branco in 1989, shortly after Mendes's murder raised the stakes and catapaulted the issues into the international media, a program was formulated in tension with the latest Brazilian state policy called *Nossa Natureza*. Articulating quite a different notion of the first person plural relation to nature or natural surroundings, the basis of the program of the Forest People's Alliance is control by and for the peoples of the forest. The core matters are direct control of indigenous lands by native peoples; agrarian reform joined to an environmental program; economic and technical development; health posts; raised incomes; locally controlled marketing systems; an end to fiscal incentives for cattle ranchers, agribusiness, and unsustainable logging; an end to debt peonage; and police and legal protection. Hecht and Cockburn call this an 'ecology of justice' that rejects a technicist solution, in whatever benign or malignant form, to environmental destruction. The Forest People's Alliance does not reject scientific or technical know-how, their own and others'; instead, they reject the 'modern' political epistemology that bestows jurisdiction on the basis of technoscientific discourse. The fundamental point is that the Amazonian Biosphere is an irreducibly human/non-human collective entity.[32] There *will be* no nature without justice. Nature and justice, contested discursive objects embodied in the material world, will become extinct or survive together.

Theory here is exceedingly corporeal, and the body is a collective; it is an historical artifact constituted by human as well as organic and technological unhuman actors. Actors are entities which do things, have effects, build worlds in concatenation with other *unlike* actors.[33] Some actors, for example specific human ones, can try to reduce other actors to resources – to mere ground and matrix for their action; but such a move is contestable, not the necessary relation of 'human nature' to the rest of the world. Other actors, human and unhuman, regularly resist reductionisms. The powers of domination do fail sometimes in their projects to pin other actors down; people can work to enhance the relevant failure rates. Social nature is the nexus I have called artifactual nature. The human 'defenders of the forest' do not and have not lived in a garden; it is from a knot in the always historical and heterogeneous nexus of social nature that they articulate their claims. Or perhaps, it is within such a nexus that I and people like me narrate a possible politics of articulation rather than representation. It is our responsibility to learn whether such a fiction is one with which the Amazonians might wish to connect in the interests of an alliance to defend the rain forest and its human and non-human ways of life – because assuredly North Americans, Europeans, and the Japanese, among others, cannot watch from afar as if we were not actors, willing or not, in the life and death struggles in the Amazon.

In a review of *Fate of the Forest*, Joe Kane, author of another book on the tropical rain forest marketed in time for Christmas in 1989, the adventure trek *Running the Amazon* (1989),[34] raised this last issue in a way that will sharpen and clarify my stakes in arguing against a politics of representation generally, and in relation to questions of environmentalism and conservation specifically. In the context of worrying about ways that social nature or socialist ecology sounded too much like the multi-use policies in national forests in the United States, which have resulted in rapacious exploitation of the land and of other organisms, Kane asked a simple question: '[W]ho speaks for the jaguar?' Now, I care about the survival of the jaguar – and the chimpanzee, and the Hawaiian land snails, and the spotted owl, and a lot of other earthlings. I care a great deal; in fact, I think I and my social groups are particularly, but not uniquely, *responsible* if jaguars, and many other non-human, as well as human, ways of life should perish. But Kane's question seemed wrong on a fundamental level. Then I understood why. His question was precisely like that asked by some pro-life groups in the abortion debates: Who speaks for the fetus? What is wrong with both questions? And how does this matter relate to science studies as cultural studies?

Who speaks for the jaguar? Who speaks for the fetus? Both questions rely on a political semiotics of representation.[35] Permanently speechless, forever requiring the services of a ventriloquist, never forcing a recall vote, in each case the object or ground of representation is the realization of the representative's fondest dream. As Marx said in a somewhat different context, 'They cannot represent themselves; they must be represented.'[36] But for a political semiology of representation, nature and the unborn fetus are even better, epistemologically, than subjugated human adults. The effectiveness of such representation depends on distancing operations. The represented must be disengaged from surrounding and constituting discursive and non-discursive nexuses and relocated in the authorial domain of the representative. Indeed, the effect of this magical operation is to disempower precisely those – in our case, the pregnant woman and the peoples of the forest – who are 'close' to the now-represented 'natural' object. Both the jaguar and the fetus are carved out of one collective entity and relocated in another, where they are reconstituted as objects of a particular kind – as the ground of a representational practice that *forever* authorizes the ventriloquist. Tutelage will be eternal. The represented is reduced to the permanent status of the recipient of action, never to be a co-actor in an articulated practice among unlike, but joined, social partners.

Everything that used to surround and sustain the represented object, such as pregnant women and local people, simply disappears or re-enters the drama as an agonist. For example, the pregnant woman becomes *juridically* and *medically*, two very powerful discursive realms, the 'maternal environment' (Hubbard, 1990). Pregnant women and local people are the *least* able to 'speak for' objects like jaguars or fetuses because they get discursively reconstituted as beings with opposing 'interests.' Neither woman nor fetus, jaguar nor Kayapó Indian is an actor in the drama of representation. One set of entities becomes the represented, the other becomes the environment, often threatening, of the represented object. The *only* actor left is the spokesperson, the one who represents. The forest is no longer the integument in a co-constituted social nature; the woman is in no way a partner in an intricate and intimate dialectic of social relationality crucial to her own personhood, as well as to the possible personhood of her social – *but unlike* – internal co-actor.[37] In the liberal logic of representation, the fetus and the jaguar must be protected precisely from those closest to them, from their 'surround.' The power of life and death must be delegated to the epistemologically most disinterested ventriloquist, and it is crucial to remember that all of this *is* about the power of life and death.

Who, within the myth of modernity, is less biased by competing interests or polluted by excessive closeness than the expert, especially the scientist? Indeed, even better than the lawyer, judge, or national legislator, the scientist is the perfect representative of nature, that is, of the permanently and constitutively speechless objective world. Whether he be a male or a female, his passionless distance is his greatest virtue; this discursively constituted, structurally gendered distance legitimates his professional privilege, which in these cases, again, is the power to testify about the right to life and death. After Edward Said quoted Marx on representation in his epigraph to *Orientalism*, he quoted Benjamin Disraeli's *Tancred*, 'The East is a career.' The separate, objective world – non-social nature – is a career. Nature legitimates the scientist's career, as the Orient justifies the representational practices of the Orientalist, even as precisely 'Nature' and the 'Orient' are the *products* of the constitutive practice of scientists and orientalists.

These are the inversions that have been the object of so much attention in science studies. Bruno Latour sketches the double structure of representation through which scientists establish the objective status of their knowledge. First, operations shape and enroll new objects or allies through visual displays or other means called inscription devices. Second, scientists speak as if they were the mouthpiece for the speechless objects that they have just shaped and enrolled as allies in an agonistic field called science. Latour defines the actant as that which is represented; the objective world *appears* to be the actant solely by virtue of the operations of representation (Latour, 1987, pp. 70–74, 90). The authorship rests with the representor, even as he claims independent object status for the represented. In this doubled structure, the simultaneously semiotic and political ambiguity of representation is glaring. First, a chain of substitutions, operating through inscription devices, relocates power and action in 'objects' divorced from polluting contextualizations and named by formal abstractions ('the fetus'). Then, the reader of inscriptions speaks for his docile constituencies, the objects. This is not a very lively world, and it does not finally offer much to jaguars, in whose interests the whole apparatus supposedly operates.

In this essay I have been arguing for another way of seeing actors and actants – and consequently another way of working to position scientists and science in important struggles in the world. I have stressed actants as collective entities doing things in a structured and structuring field of action; I have framed the issue in terms of articulation rather than representation. Human beings use names to point to themselves and other actors and easily mistake the names for the things. These same humans also think the traces of inscription devices are like names – pointers to things, such that the inscriptions and the things can be enrolled in dramas of substitution and inversion. But the things, in my view, do not pre-exist as ever-elusive, but fully pre-packaged, referents for the names. Other actors are more like tricksters than that. Boundaries take provisional, never-finished shape in articulatory practices. The potential for the unexpected from unstripped human and unhuman actants enrolled in articulations – i.e., the potential for generation – remains both to trouble and to empower technoscience. Western philosophers sometimes take account of the inadequacy of names by stressing the 'negativity' inherent in all representations. This takes us back to Spivak's remark cited early in this paper about the important things that we cannot not desire, but can never posses – or represent, because representation depends on possession of a passive resource, namely, the silent object, the *stripped* actant. Perhaps we can, however, 'articulate' with humans and unhumans in a social relationship, which for us is always language-mediated (among other semiotic, i.e., 'meaningful,' mediations). But, for our unlike partners, well, the action is 'different,' perhaps 'negative' from our linguistic point of view, but crucial to the generativity of the collective. It is the empty space, the undecidability, the wiliness of other actors, the 'negativity,' that give me confidence in the *reality* and therefore ultimate *unrepresentability* of social nature and that make me suspect doctrines of representation and objectivity.

My crude characterization does not end up with an 'objective world' or 'nature,' but it certainly does insist on the *world*. This world must always be articulated, from people's points of view, through 'situated knowledges' (Haraway, 1988; 1991). These knowledges are friendly to science, but do not provide any grounds for history-escaping inversions and amnesia about how articulations get made, about their political semiotics, if you will. I think the world is precisely what gets lost in doctrines of representation and scientific objectivity. It is *because* I care about jaguars, among other actors, including the overlapping but non-identical groups called forest peoples and ecologists, that I reject Joe Kane's question. Some science studies scholars have been terrified to criticize their constructivist formulations because the only alternative seems to be some retrograde kind of 'going back' to nature and to philosophical realism.[38] But above all people, these scholars should know that 'nature' and 'realism' are precisely the consequences of representational practices. Where we need to move is not 'back' to nature, but *elsewhere*, through and within an artifactual social nature, which these very scholars have helped to make expressable in current Western scholarly practice. That knowledge-building practice might be articulated to other practices in 'pro-life' ways that aren't about the fetus or the jaguar as nature fetishes and the expert as their ventriloquist.

Prepared by this long detour, we can return to the Kayapó man videotaping his tribesmen as they protest a new hydroelectric dam on their territory. The National Geographic Society, *Discover* magazine, and Gulf Oil – and much philosophy and social science – would have us see his practice as a double boundary crossing between the primitive and the modern. His representational practice, signified by his use of the latest technology, places him in the realm of the modern. He is, then, engaged in an entertaining contradiction – the preservation of an unmodern way of life with the aid of incongruous modern technology. But, from the perspective of a political semiotics of articulation, the man might well be forging a recent collective of humans and unhumans, in this case made up of the Kayapó, videocams, land, plants, animals, near and distant audiences, and other constituents; but no boundary violation is involved. The way of life is not unmodern (closer to nature); the camera is not modern or postmodern (in society). Those categories should no longer make sense. Where there is no nature and no society, there is no pleasure, no entertainment to be had in representing the violation of the boundary between them. Too bad for nature magazines, but a gain for inappropriate/d others.

The videotaping practice does not thereby become innocent or uninteresting; but its meanings have to be approached differently, in terms of the kinds of collective action taking place and the claims they make on others – such as ourselves, people who do not live in the Amazon. We *are all* in chiasmatic borderlands, liminal areas where new shapes, new kinds of action and responsibility, are gestating in the world. The man using that camera is forging a practical claim on us, morally and epistemologically, as well as on the other forest people to whom he will show the tape to consolidate defense of the forest. His practice invites further articulation – on terms shaped by the forest people. They will no longer be represented as Objects, not because they cross a line to represent themselves in 'modern' terms as Subjects, but because they powerfully form articulated collectives.

In May of 1990, a week-long meeting took place in Iquitos, a formerly prosperous rubber boom-town in the Peruvian Amazon. COICA, the Coordinating Body for the Indigenous Peoples of the Amazon, had assembled forest people (from all the nations constituting Amazonia), environmental groups from around the world (Greenpeace, Friends of the Earth, the Rain Forest Action Network, etc.), and media organizations (*Time* magazine, CNN, NBC, etc.) in order 'to find a common path on which we can work to preserve the Amazon forest' (Arena-De Rosa, 1990, pp. 1–2). Rain forest protection was formulated as a necessarily joint human rights-ecological issue. The fundamental demand by indigenous people was that they must be part of *all* international negotiations involving their territories. 'Debt for nature' swaps were particular foci of controversy, especially where indigenous groups end up worse off than in previous agreements with their governments as a result of bargaining between banks, external conservation groups, and national states. The controversy generated a proposal: instead of a swap of debt-for-nature, forest people would support swaps of debt-for-indigenous-controlled territory, in which non-indigenous environmentalists would have a 'redefined role in helping to develop the plan for conservation management of the particular region of the rain forest' (Arena-De Rosa, 1990). Indigenous environmentalists would also be recognized not for their quaint 'ethnoscience,' but for their *knowledge*.

Nothing in this structure of action rules out articulations by scientists or other North Americans who care about jaguars and other actors; but the patterns, flows, and intensities of power are most certainly changed. That is what articulation does; it is always a non-innocent, contestable practice; the partners are never set once and for all. There is no ventriloquism here. Articulation is work, and it may fail. All the people who care, cognitively, emotionally, and politically, must articulate their position in a field constrained by a new collective entity, made up of indigenous people and other human and unhuman actors. Commitment and engagement, not their invalidation, in an emerging collective are the conditions of joining knowledge-producing and world-building practices. This is situated knowledge in the New World; it builds on common places, and it takes unexpected turns. So far, such knowledge has not been sponsored by the major oil corporations, banks, and logging interests. That is precisely one of the reasons why there is so much work for North Americans, Europeans, and Japanese, among others, to do in articulation with those humans and non-humans who live in rain forests and in many other places in the semiotic space called earth.

B. OUTER SPACE: THE EXTRATERRESTRIAL

Since we have spent so much time on earth, a prophylactic exercise for residents of the alien 'First World,' we will rush through the remaining three quadrants of the semiotic square. We move from one topical commonplace to another, from earth to space, to see what turns our journeys to elsewhere might take.

An ecosystem is always of a particular type, for example, a temperate grassland or a tropical rain forest. In the iconography of late capitalism, Jane Goodall did not go to that kind of ecosystem. She went to the 'wilds of Tanzania,' a mythic 'ecosystem' reminiscent of the original garden from which her kind had been expelled and to which she returned to commune with the wilderness's present inhabitants to learn how to survive. This wilderness was close in its dream quality to 'space,' but the wilderness of Africa was coded as dense, damp, bodily, full of sensuous creatures who touch intimately and intensely. In contrast, the extraterrestrial is coded to be fully general; it is about escape from the bounded globe into an anti-ecosystem called, simply, space. Space is not about 'man's' origins on earth but about 'his' future, the two key allochronic times of salvation history. Space and the tropics are both utopian topical figures in Western imaginations, and their opposed properties dialectically signify origins and ends for the creature whose mundane life is supposedly outside both: modern or postmodern man.

The first primates to approach that abstract place called 'space' were monkeys and apes. A rhesus monkey survived an 83 mile-high flight in 1949. Jane Goodall arrived in 'the wilds of Tanzania' in 1960 to encounter and name the famous Gombe Stream chimpanzees introduced to the National Geographic television audience in 1965. However, other chimpanzees were vying for the spotlight in the early 1960s. On January 31, 1961, as part of the United States man-in-space program, the chimpanzee HAM, trained for his task at Holloman Air Force Base, 20 minutes by car from Alamogordo, New Mexico, near the site of the first atom bomb explosion in July 1945, was shot into suborbital flight (Figure 17.8). HAM's name inevitably recalls Noah's youngest and only black son. But this chimpanzee's name was from a different kind of text. His name was an acronym for the scientific-military institution that launched him, Holloman Aero-Medical; and he rode an arc that traced the birth path of modern science – the parabola, the conic section. HAM's parabolic path is rich with evocations of the history of Western science. The path of a projectile that does not escape gravity, the parabola is the shape considered so deeply by Galileo, at the first mythic moment of origins of modernity, when the unquantifiable sensuous and countable mathematical properties of bodies were separated from each other in scientific knowledge. It describes the path of ballistic weapons, and it is the trope for 'man's' doomed projects in the writings of the existentialists in the 1950s. The parabola traces the path of Rocket Man at the end of World War II in Thomas Pynchon's *Gravity's Rainbow* (1973). An understudy for man, HAM went only to the

Figure 17.8 Ham awaits release in his couch aboard the recovery
vessel LSD *Donner* after his successful Mercury Project launch
(photograph by Henry Borroughs)

boundary of space, in suborbital flight. On his return to earth, he was named.
He had been known only as #65 before his successful flight. If, in the official
birth-mocking language of the Cold War, the mission had to be 'aborted,' the
authorities did not want the public worrying about the death of a famous and
named, even if not quite human, astronaut. In fact, #65 did have a name among
his handlers, Chop Chop Chang, recalling the stunning racism in which the
other primates have been made to participate.[39] The space race's surrogate
child was an 'understudy for man in the conquest of space' (Eimerl and De
Vore, 1965, p. 173). His hominid cousins would transcend that closed parabolic figure, first in the ellipse of orbital flight, then in the open trajectories of
escape from earth's gravity.

HAM, his human cousins and simian colleagues, and their englobing and
interfacing technology were implicated in a reconstitution of masculinity in
Cold War and space race idioms. The movie *The Right Stuff* (1985) shows the
first crop of human astronau(gh)ts struggling with their affronted pride when
they realize their tasks were competently performed by their simian cousins.
They and the chimps were caught in the same theater of the Cold War, where

the masculinist, death-defying, and skill-requiring heroics of the old jet aircraft test pilots became obsolete, to be replaced by the media-hype routines of projects Mercury, Apollo, and their sequelae. After chimpanzee Enos completed a fully automated orbital flight on November 29, 1961, John Glenn, who would be the first human American astronaut to orbit earth, defensively 'looked toward the future by affirming his belief in the superiority of astronauts over chimponauts.' *Newsweek* announced Glenn's orbital flight of February 20, 1962, with the headline, 'John Glenn: One Machine That Worked Without Flaw.'[40] Soviet primates on both sides of the line of hominization raced their U.S. siblings into extraterrestrial orbit. The space ships, the recording and tracking technologies, animals, and human beings were joined as cyborgs in a theater of war, science, and popular culture.

Henry Burroughs's famous photograph of an interested and intelligent, actively participating HAM, watching the hands of a white, laboratory-coated, human man release him from his contour couch, illuminated the system of meanings that binds humans and apes together in the late twentieth century (Weaver, 1961). HAM is the perfect child, reborn in the cold matrix of space. *Time* described chimponaut Enos in his 'fitted contour couch that looked like a cradle trimmed with electronics'.[41] Enos and HAM were cyborg neonates, born of the interface of the dreams about a technicist automaton and masculinist autonomy. There could be no more iconic cyborg than a telemetrically implanted chimpanzee, understudy for man, launched from earth in the space program, while his conspecific in the jungle, 'in a spontaneous gesture of trust,' embraced the hand of a woman scientist named Jane in a Gulf Oil ad showing 'man's place in the ecological structure.' On one end of time and space, the chimpanzee in the wilderness modeled communication for the stressed, ecologically threatened and threatening, modern human. On the other end, the ET chimpanzee modeled social and technical cybernetic communication systems, which permit postmodern man to escape both the jungle and the city, in a thrust into the future made possible by the social-technical systems of the 'information age' in a global context of threatened nuclear war. The closing image of a human fetus hurtling through space in Stanley Kubrick's *2001: A Space Odyssey* (1968) completed the voyage of discovery begun by the weapon-wielding apes at the film's gripping opening. It was the project(ile) of self-made, reborn man, in the process of being raptured out of history. The Cold War was simulated ultimate war; the media and advertising industries of nuclear culture produced in the bodies of animals – paradigmatic natives and aliens – the reassuring images appropriate to this state of pure war (Virilio and Lotringer, 1983).[42]

In the aftermath of the Cold War, we face not the end of nuclearism, but its dissemination. Even without our knowing his ultimate fate as an adult caged chimpanzee, the photograph of HAM rapidly ceases to entertain, much less to edify. Therefore, let us look to another cyborg image to figure possible emergencies of inappropriate/d others to challenge our rapturous mythic brothers, the postmodern spacemen.

At first sight, the T-shirt worn by anti-nuclear demonstrators at the Mother's and Others' Day Action in 1987 at the United States's Nevada nuclear test site seems in simple opposition to HAM in his electronic cradle (Figure 17.9). But a little unpacking shows the promising semiotic and political complexity of the image and of the action. When the T-shirt was sent to the printer, the name of the event was still the 'Mother's Day Action,' but not long after some planning participants objected. For many, Mother's Day was, at best, an ambivalent time for a women's action. The overdetermined gender coding of patriarchal nuclear culture all too easily makes women responsible for peace while men fiddle with their dangerous war toys without semiotic dissonance. With its commercialism and multi-leveled reinforcement of compulsory heterosexual reproduction, Mother's Day is also not everybody's favorite feminist holiday. For others, intent on reclaiming the holiday for other meanings, mothers, and by extension women in general, do have a special obligation to preserve children, and so the earth, from military destruction. For them, the earth is metaphorically mother and child, and in both figurations, a subject of nurturing and birthing. However, this was not an all-women's (much less all-mothers') action, although women organized and shaped it. From discussion,

Figure 17.9   Mother's Day 1987 Nevada Test Site Action T-shirt

the designation 'Mother's and Others' Day Action' emerged. But then, some thought that meant mothers and men. It took memory exercises in feminist analysis to rekindle shared consciousness that mother does not equal woman and vice versa. Part of the day's purpose was to recode Mother's Day to signify men's obligations to nurture the earth and all its children. In the spirit of this set of issues, at a time when Baby M and her many debatable – and unequally positioned – parents were in the news and the courts, the all-female affinity group which I joined took as its name the Surrogate Others. These surrogates were not understudies for man, but were gestating for another kind of emergence.

From the start, the event was conceived as an action that linked social justice and human rights, environmentalism, anti-militarism, and anti-nuclearism. On the T-shirt, there is, indeed, the perfect icon of the union of all issues under environmentalism's rubric: the 'whole earth,' the lovely, cloud-wrapped, blue, planet earth is simultaneously a kind of fetus floating in the amniotic cosmos and a mother to all its own inhabitants, germ of the future, matrix of the past and present. It is a perfect globe, joining the changeling matter of mortal bodies and the ideal eternal sphere of the philosophers. This snapshot resolves the dilemma of modernity, the separation of Subject and Object, Mind and Body. There is, however, a jarring note in all this, even for the most devout. That particular image of the earth, of Nature, could only exist if a camera on a satellite had taken the picture, which is, of course, precisely the case. Who speaks for the earth? Firmly in the object world called nature, this bourgeois, family-affirming snapshot of mother earth is about as uplifting as a loving commercial Mother's Day card. And yet, it *is* beautiful, and it is ours; it must be brought into a different focus. The T-shirt is part of a complex collective entity, involving many circuits, delegations, and displacements of competencies. Only in the context of the space race in the first place, and the militarization and commodification of the whole earth, does it make sense to relocate that image as the special sign of an anti-nuclear, anti-militaristic, earth-focused politics. The relocation does not cancel its other resonances; it contests for their out-come.

I read Environmental Action's 'whole earth' as a sign of an irreducible artifactual social nature, like the Gaia of SF writer John Varley and biologist Lynn Margulis. Relocated on this particular T-shirt, the satellite's eye view of planet earth provokes an ironic version of the question, who speaks for the earth (for the fetus, the mother, the jaguar, the object world of nature, all those who must be represented)? For many of us, the irony made it possible to participate – indeed, to participate as fully committed, if semiotically unruly, eco-feminists. Not everybody in the Mother's and Others' Day Action would agree; for many, the T-shirt image meant what it said, love your mother who is the earth. Nuclearism is misogyny. The field of readings in tension with each other is also part of the point. Eco-feminism and the non-violent direct action movement have been based on struggles over differences, not on identity. There is hardly a need for affinity groups and their endless process if sameness prevailed. Affinity is precisely *not* identity; the sacred image of the same is not gestating on this Mother's and Others' Day. Literally, enrolling the satellite's camera and the peace action in Nevada into a new collective, this Love Your Mother image is based on diffraction, on the processing of small but consequential differences. The processing of differences, semiotic action, is about ways of life.

The Surrogate Others planned a birthing ceremony in Nevada, and so they made a birth canal – a sixteen-foot long, three-foot diameter, floral polyester-covered worm with lovely dragon eyes. It was a pleasingly artifactual beast, ready for connection. The worm-dragon was laid under the barbed-wire boundary between the land on which the demonstrators could stand legally and the land on which they would be arrested as they emerged. Some of the Surrogate Others conceived of crawling through the worm to the forbidden side as an act of solidarity with the tunneling creatures of the desert, who had to share their subsurface niches with the test site's chambers. This surrogate birthing was definitely not about the obligatory heterosexual nuclear family compulsively reproducing itself in the womb of the state, with or without the underpaid services of the wombs of 'surrogate mothers.' Mother's and Others' Day was looking up.

It wasn't only the desert's non-human organisms with whom the activists were in solidarity as they emerged onto the proscribed territory. From the point of view of the demonstrators, they were quite legally on the test-site land. This was so not out of some 'abstract' sense that the land was the people's and had been usurped by the war state, but for more 'concrete' reasons: all the demonstrators had written permits to be on the land signed by the Western Shoshone National Council. The 1863 Treaty of Ruby Valley recognized the Western Shoshone title to ancestral territory, including the land illegally invaded by the U.S. government to build its nuclear facility. The treaty has never been modified or abrogated, and U.S. efforts to buy the land (at 15 cents per acre) in 1979 was refused by the only body authorized to decide, the Western Shoshone National Council. The county sheriff and his deputies, surrogates for the federal government, were, in 'discursive' and 'embodied' fact, trespassing. In 1986 the Western Shoshone began to issue permits to the anti-nuclear demonstrators as part of a coalition that joined anti-nuclearism and indigenous land rights. It is, of course, hard to make citizens' arrests of the police when they have you handcuffed and when the courts are on their side. But it is quite possible to join this ongoing struggle, which is very much 'at home,' and to articulate it with the defense of the Amazon. That articulation requires collectives of human and unhuman actors of many kinds.

There were many other kinds of 'symbolic action' at the test site that day in 1987. The costumes of the sheriff's deputies and their nasty plastic handcuffs were also symbolic action – highly embodied symbolic action. The 'symbolic action' of brief, safe arrest is also quite a different matter from the 'semiotic' conditions under which most people in the U.S., especially people of color and the poor, are jailed. The difference is not the presence or absence of 'symbolism,' but the force of the respective collectives made up of humans and unhumans, of people, other organisms, technologies, institutions. I am not unduly impressed with the power of the drama of the Surrogate Others and the other affinity groups, nor, unfortunately, of the whole action. But I do take seriously the work to relocate, to diffract, embodied meanings as crucial work to be done in gestating a new world.[43] It is cultural politics, and it is techno-science politics. The task is to build more powerful collectives in dangerously unpromising times.

NOT-B. INNER SPACE: THE BIOMEDICAL BODY

The limitless reaches of outer space, joined to Cold War and post-Cold War nuclear technoscience, seem vastly distant from their negation, the enclosed and dark regions of the inside of the human body, domain of the apparatuses of biomedical visualization. But these two quadrants of our semiotic square are multiply tied together in technoscience's heterogeneous apparatuses of bodily production. As Sarah Franklin noted, 'The two new investment frontiers, outer space and inner space, vie for the futures market.' In this 'futures market,' two entities are especially interesting for this essay: the fetus and the immune system, both of which are embroiled in determinations of what may count as nature and as human, as separate natural object and as juridical subject. We have already looked briefly at some of the matrices of discourse about the fetus in the discussion of earth (who speaks for the fetus?) and outer space (the planet floating free as cosmic germ). Here, I will concentrate on contestations for what counts as a self and an actor in contemporary immune system discourse.

The equation of Outer Space and Inner Space, and of their conjoined discourses of extraterrestrialism, ultimate frontiers, and high technology war, is literal in the official history celebrating 100 years of the National Geographic Society (Bryan, 1987). The chapter that recounts the magazine's coverage of the Mercury, Gemini, Apollo, and Mariner voyages is called 'Space' and introduced with the epigraph, 'The Choice Is the Universe – or Nothing.' The final chapter, full of stunning biomedical images, is titled 'Inner Space' and introduced with the epigraph, 'The Stuff of the Stars Has Come Alive.'[44] The photography convinces the viewer of the fraternal relation of inner and outer space. But, curiously, in outer space, we see spacemen fitted into explorer craft or floating about as individuated cosmic fetuses, while in the supposed earthy space of our own interiors, we see non-humanoid strangers who are the means by which our bodies sustain our integrity and individuality, indeed our humanity in the face of a world of others. We seem invaded not just by the threatening 'non-selves' that the immune system guards against, but more fundamentally by our own strange parts.

Lennart Nilsson's photographs, in the coffee table art book *The Body Victorious* (1987), as well as in many medical texts, are landmarks in the

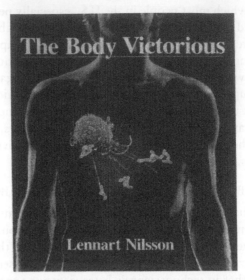

Figure 17.10   Design for Lennart Nilsson book

photography of the alien inhabitants of inner space[45] (Figure 17.10). The blasted scenes, sumptuous textures, evocative colors, and ET monsters of the immune landscape are simply *there*, inside *us*. A white extruding tendril of a pseudopodinous macrophage ensnares bacteria; the hillocks of chromosomes lie flattened on a blue-hued moonscape of some other planet; an infected cell buds myriads of deadly virus particles into the reaches of inner space where more cells will be victimized; the auto-immune disease-ravaged head of a femur glows against a sunset on a dead world; cancer cells are surrounded by the lethal mobil squads of killer T-cells that throw chemical poisons into the self's malignant traitor cells.

A diagram of the 'Evolution of Recognition Systems' in a recent immunology textbook makes clear the intersection of the themes of literally 'wonderful' diversity, escalating complexity, the self as a defended stronghold, and extra-terrestrialism in inner space (Figure 17.11). Under a diagram culminating in the evolution of the mammals, represented without comment by a mouse and a *fully-suited spaceman*, is this explanation: 'From the humble amoeba searching for food (top left) to the mammal with its sophisticated humoral and cellular immune mechanisms (bottom right), the process of "self versus non-self recognition" shows a steady development, keeping pace with the increasing need of animals to maintain their integrity in a hostile environment. The decision at which point "immunity" appeared is thus a purely semantic one' (Playfair, 1984, emphasis in the original). These are the 'semantics' of defense and invasion. The perfection of the fully defended, 'victorious' self is a chilling fantasy, linking phagocytotic amoeba and space-voyaging man cannibalizing the earth in an evolutionary teleology of post-apocalypse extraterrestrialism.

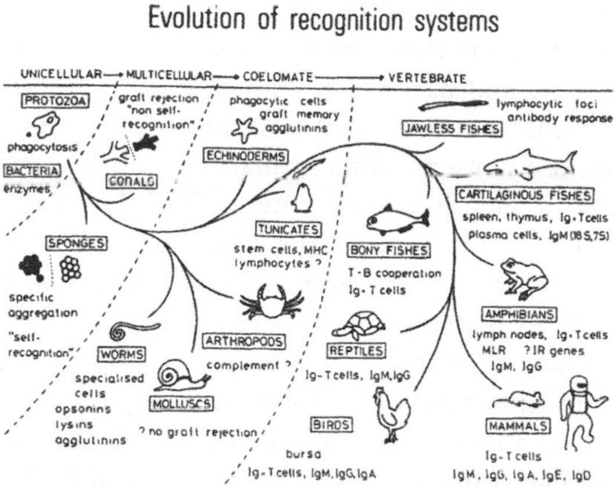

Evolution of recognition systems

Figure 17.11   From a recent immunology textbook

When is a self enough of a self that its boundaries become central to institutionalized discourses in biomedicine, war, and business?

Images of the immune system as a battlefield abound in science sections of daily newspapers and in popular magazines, e.g., *Time* magazine's 1984 graphic for the AIDS virus's 'invasion' of the cell-as-factory. The virus is a tank, and the viruses ready for export from the expropriated cells are lined up ready to continue their advance on the body as a productive force. The *National Geographic* explicitly punned on Star Wars in its graphic called 'Cell Wars' (Jaret, 1986). The militarized, automated factory is a favorite convention among immune system technical illustrators and photographic processors. The specific historical markings of a Star Wars-maintained individuality are enabled by high-technology visualization technologies, which are also basic to conducting war and commerce, such as computer-aided graphics, artificial intelligence software, and specialized scanning systems.

It is not just imagers of the immune system who learn from military cultures; military cultures draw symbiotically on immune system discourse, just as strategic planners draw directly from and contribute to video game practices and science fiction. For example, arguing for an elite special force within the parameters of 'low-intensity conflict' doctrine, a U.S. army officer wrote: 'The most appropriate example to describe how this system would work is the most complex biological model we know – the body's immune system. Within the body there exists a remarkably complex corps of internal bodyguards. In absolute numbers they are small – only about one percent of the body's cells. Yet they consist of reconnaissance specialists, killers, reconstitution specialists, and communicators that can seek out invaders, sound the alarm, reproduce rapidly, and swarm to the attack to repel the enemy. ... In this regard, the June

1986 issue of *National Geographic* contains a detailed account of how the body's immune system functions' (Timmerman, 1987).

The circuits of competencies sustaining the body as a defended self – personally, culturally, and nationally – spiral through the fantasy entertainment industry, a branch of the apparatus of bodily production fundamental to crafting the important consensual hallucinations about 'possible' worlds that go into building 'real' ones. In Epcot Center of Walt Disney World, we may be interpellated as subjects in the new Met Life Pavilion, which is 'devoted to dramatizing the intricacies of the human body.' A special thrill ride, called 'Body Wars,' promises that we will 'experience the wonders of life,' such as encountering 'the attack of the platelets.'[46] This lively battle simulator is promoted as 'family entertainment.' The technology for this journey through the human body uses a motion-based simulator to produce three-dimensional images for a stationary observer. As in other forms of high-tech tourism, we can go everywhere, see everything and leave no trace. The apparatus has been adopted to teach medical anatomy at the University of Colorado Health Sciences Center. Finally, we should not forget that more Americans travel to the combined Disney worlds than voyage in most other myth-realizing machines, like Washington, D.C.[47] Met Life cautions those who journey on 'Body Wars' that they may experience extreme vertigo from the simulated motion. Is that merely 'symbolic action' too?

In the embodied semiotic zones of earth and outer space, we saw the diffraction patterns made possible by recomposed visualizing technologies, relocated circuits of competencies that promise to be more user-friendly for inappropriate/d others. So also, the inner spaces of the biomedical body are central zones of technoscientific contestation, i.e., of science as culture in the amodern frame of social nature. Extremely interesting new collectives of human and unhuman allies and actors are emerging from these processes. I will briefly sketch two zones where promising monsters are undergoing symbiogenesis in the nutrient media of technoscientific work: 1) theories of immune function based on laboratory research, and 2) new apparatuses of knowledge production being crafted by Persons with AIDS (PWAs) and their heterogeneous allies. Both sets of monsters generate distinctly diffracted views of the self, evident in beliefs and practices in relation to vulnerability and mortality.

Like non-violent direct action and environmentalism, immune system discourse is about the unequally distributed chances of life and death. Since sickness and mortality are at the heart of immunology, it is hardly surprising that conditions of battle prevail. Dying is not an easy matter crying out for 'friendly' visualization. But battle is not the only way to figure the process of mortal living. Persons coping with the life-threatening consequences of infection with the HIV virus have insisted that they are *living* with AIDS, rather than accepting the status of *victims* (or prisoners of war?). Similarly, laboratory scientists also have built research programs based on non-militaristic, relational embodiments, rather than on the capabilities of the defended self of atomic individuals. They do this in order to construct IS articulations more effectively, not in order to be nice folks with pacifist metaphors.

Let me attempt to convey the flavor of the artifactual bodily object called the human immune system, culled from major textbooks and research reports published in the 1980s. These characterizations are part of working systems for interacting with the immune system in many areas of practice, including business decisions, clinical medicine, and lab experiments. With about 10 to the 12th cells, the IS has two orders of magnitude more cells than the nervous system. IS cells are regenerated throughout life from pluripotent stem cells. From embryonic life through adulthood, the immune system is sited in several morphologically dispersed tissues and organs, including the thymus, bone marrow, spleen, and lymph nodes; but a large fraction of its cells are in the blood and lymph circulatory systems and in body fluids and spaces. If ever there were a 'distributed system,' this is one! It is also a highly adaptable communication system with many interfaces.

There are two major cell lineages to the system: (1) The first is the *lymphocytes*, which include the several types of T cells (helper, suppressor, killer, and variations of all these) and the B cells (each type of which can produce only one sort of the vast array of potential circulating antibodies). T and B cells have particular specificities capable of recognizing almost any molecular array of the right size that can ever exist, no matter how clever industrial chemistry gets. This specificity is enabled by a baroque somatic mutation mechanism, clonal selection, and a polygenic receptor or marker system. (2) The second immune cell lineage is the *mononuclear phagocyte system*, including the multi-talented macrophages, which, in addition to their other recognition skills and connections, also appear to share receptors and some hormonal peptide products with neural cells. Besides the cellular compartment, the immune system comprises a vast array of circulating acellular products, such as antibodies, lymphokines, and complement components. These molecules mediate communication among components of the immune system, but also between the immune system and the nervous and endocrine systems, thus linking the body's multiple control and coordination sites and functions. The genetics of the immune system cells, with their high rates of somatic mutation and gene product splicings and rearrangings to make finished surface receptors and antibodies, makes a mockery of the notion of a constant genome even within 'one' body. The hierarchical body of old has given way to a network-body of amazing complexity and specificity. The immune system is everywhere and nowhere. Its specificities are indefinite if not infinite, and they arise randomly; yet these extraordinary variations are the critical means of maintaining bodily coherence.

In the early 1970s, winning a Nobel Prize for the work, Niels Jerne proposed a theory of immune system self-regulation, called the network theory, which deviates radically from notions of the body victorious and the defended self. 'The network theory differs from other immunological thinking because it endows the immune system with the ability to regulate itself using only itself' (Golub, 1987; Jerne, 1985).[48] Jerne proposed that any antibody molecule must be able to act functionally as both antibody to some antigen *and* as antigen for the production of an antibody to itself, at another region of 'itself.' These sites have acquired a nomenclature sufficiently daunting to thwart popular understanding of the theory, but the basic conception is simple. The concatenation of internal recognitions and responses would go on indefinitely, in a series of interior mirrorings of sites on immunoglobulin molecules, such that the immune system would always be in a state of dynamic internal responding. It would never be passive, 'at rest,' awaiting an activating stimulus from a hostile outside. In a sense, there could be no *exterior* antigenic structure, no 'invader,' that the immune system had not already 'seen' and mirrored internally. Replaced by subtle plays of partially mirrored readings and responses, self and other lose their rationalistic oppositional quality. A radical conception of *connection* emerges unexpectedly at the core of the defended self. Nothing in the model prevents therapeutic action, but the entities in the drama have different kinds of interfaces with the world. The therapeutic logics are unlikely to be etched into living flesh in patterns of DARPA's latest high-tech tanks and smart missiles.

Some of those logics are being worked out in and by the bodies of persons with AIDS and ARC. In their work to sustain life and alleviate pain in the context of mortal illness, PWAs engage in many processes of knowledge-building. These processes demand intricate code switching, language bridging, and alliances among worlds previously held apart. These 'generative grammars' are matters of life and death. As one activist put it, 'ACT UP's humor is no joke' (Crimp and Rolston, 1990, p. 20; see also Crimp, 1983). The AIDS Coalition to Unleash Power (ACT UP) is a collective built from many articulations among unlike kinds of actors – for example, activists, biomedical machines, government bureaucracies, gay and lesbian worlds, communities of color, scientific conferences, experimental organisms, mayors, international information and action networks, condoms and dental dams, computers, doctors, IV drug-users, pharmaceutical companies, publishers, virus components, counselors, innovative sexual practices, dancers, media technologies, buying clubs, graphic artists, scientists, lovers, lawyers, and more. The actors, however, are not all equal. ACT UP has an animating center – PWAs, who are to the damage wrought by AIDS and the work for restored health around the world as the indigenous peoples of the Amazon are to forest destruction and environmentalism. These are the actors with whom others must articulate. That structure of action is a fundamental consequence of learning to visualize the heterogeneous, artifactual body that is our 'social nature,' instead of narrowing our vision that 'saving nature' and repelling alien invaders from an unspoiled organic eden called the autonomous self. Saving nature is, finally, a deadly project. It relies on perpetuating the structure of boundary violation and the falsely liberating *frisson* of transgression. What happened in the first Eden should have made that clear.

So, if the tree of knowledge cannot be forbidden, we had all better learn how to eat and feed each other with a little more savvy. That is the difficult process being engaged by PWAs, Project Inform, ACT UP, NIH, clinical practitioners, and many more actors trying to build responsible mechanisms for producing effective knowledge in the AIDS epidemic.[49] Unable to police the same boundaries separating insiders and outsiders, the world of biomedical research will never be the same again. The changes range across the epistemological, the commercial, the juridical, and the spiritual domains. For example, what is the status of knowledge produced through the new combinations of decision-making in experimental design that are challenging previous research conventions? What are the consequences of the *simultaneous* challenges to expert monopoly of knowledge *and* insistence on both the rapid improvement of the biomedical knowledge base and the equitable mass distribution of its fruits? How will the patently amodern hybrids of healing practices cohabit in the emerging social body? And, who will live and die as a result of these very non-innocent practices?

NOT-A. VIRTUAL SPACE: SF [50]

Articulation is not a simple matter. Language is the effect of articulation, and so are bodies. The articulata are jointed animals; they are not smooth like the perfect spherical animals of Plato's origin fantasy in the *Timaeus*. The articulata are cobbled together. It is the condition of being articulate. I rely on the articulata to breathe life into the artifactual cosmos of monsters that this essay inhabits. Nature may be speechless, without language, in the human sense, but nature is highly articulate. Discourse is only one process of articulation. An articulated world has an undecidable number of modes and sites where connections can be made. The surfaces of this kind of world are not frictionless curved planes. Unlike things can be joined – and like things can be broken apart – and vice versa. Full of sensory hairs, evaginations, invaginations, and indentations, the surfaces which interest me are dissected by joints. Segmented invertebrates, the articulata are insectoid and worm-like, and they inform the inflamed imaginations of SF filmmakers and biologists. In obsolete English, to articulate meant to make terms of agreement. Perhaps we should live in such an 'obsolete,' amodern world again. To articulate is to signify. It is to put things together, scary things, risky things, contingent things. I want to live in an articulate world. We articulate; therefore, we are. Who 'I' am is a very limited, in the endless perfection of (clear and distinct) Self-contemplation. Unfair as always, I think of it as the paradigmatic psychoanalytic question. 'Who am I?' is about (always unrealizable) identity; always wobbling, it still pivots on the law of the father, the sacred image of the same. Since I am a moralist, the real question must have more virtue: who are 'we'? That is an inherently more open question, one always ready for contingent, friction-generating articulations. It is a remonstrative question.

In optics, the virtual image is formed by the apparent, but not actual, convergence of rays. The virtual seems to be the counterfeit of the real; the virtual has effects by seeming, not being. Perhaps that is why 'virtue' is still given in dictionaries to refer to women's chastity, which must always remain doubtful in patriarchal optical law. But then, 'virtue' used to mean manly spirit and valor too, and God even named an order of angels the Virtues, though they were of only middling rank. Still, no matter how big the effects of the virtual are, they seem somehow to lack a proper ontology. Angels, manly valor, and women's chastity certainly constitute, at best, a virtual image from the point of view of late twentieth-century 'postmoderns.' For them, the virtual is precisely *not* the real; that's why 'postmoderns' *like* 'virtual reality.' It seems transgressive. Yet, I can't forget that an obsolete meaning of 'virtual' was having virtue, i.e., the inherent power to produce effects. 'Virtu,' after all, is excellence or merit, and it is still a common meaning of virtue to refer to having efficacy. The 'virtue' of something is its 'capacity.' The virtue of (some) food is that it nourishes the body. Virtual space *seems* to be the negation of real space; the domains of SF *seem* the negation of earthly regions. But perhaps this negation is the real illusion.

'Cyberspace, absent its high-tech glitz, is the idea of virtual consensual community. ... A virtual community is first and foremost a community of belief.'[51] For William Gibson (1986), cyberspace is 'consensual hallucination experienced daily by billions. ... Unthinkable complexity.' Cyberspace seems to be the consensual hallucination of too much complexity, too much articulation. It is the virtual reality of paranoia, a well-populated region in the last quarter of the Second Christian Millenium. Paranoia is the belief in the unrelieved density of connection, requiring, if one is to survive, withdrawal and defense unto death. The defended self re-emerges at the heart of relationality. Paradoxically, paranoia is the condition of the impossibility of remaining articulate. In virtual space, the virtue of articulation – i.e., the power to produce connection – threatens to overwhelm and finally to engulf all possibility of effective action to change the world.

So, in our travels into virtual space, if we are to emerge from our encounter with the artifactual articulata into a livable elsewhere, we need a guide figure to navigate around the slough of despond. Lisa Foo, the principal character in a Hugo and Nebula award-winning short story by John Varley (1986), will be our unlikely Beatrice through the System.

'If you wish to know more, press enter' (p. 286).[52]

With that fatal invitation, Varley's profoundly paranoid story begins and ends. The Tree of Knowledge is a Web, a vast system of computer connections generating, as an emergent property, a new and terrifyingly unhuman collective entity. The forbidden fruit is knowledge of the workings of this powerful Entity, whose deadly essence is extravagant connection. All of the human characters are named after computers, programs, practices, or concepts – Victor Apfel, Detective Osborne, and the hackers Lisa Foo and Charles Kluge.

The story is a murder mystery. With a dubious suicide note, called up by responding to the command 'press enter' on the screen of one of the dozens of personal computers in his house, which is also full of barrels of illicit drugs, Kluge has been found dead by his neighbor, Apfel. Apfel is a reclusive middle-aged epileptic, who had been a badly treated prisoner-of-war in Korea, leaving him with layers of psychological terror, including a fear and hatred of 'orientals.' When Los Angeles homicide Detective Osborne's men prove totally inept at deciphering the elaborate software running Kluge's machines, Lisa Foo, a young Vietnamese immigrant, now a U.S. citizen, is called in from Cal Tech; and she proceeds to play Sherlock Holmes to Osborne's Lestrade. The story is narrated from Apfel's point of view, but Foo is the tale's center and, I insist, its pivotal actor.

Insisting, I wish to exercise the license that is built into the anti-elitist reading conventions of SF popular cultures. SF conventions invite – or at least permit more readily than do the academically propagated, respectful consumption protocols for literature – rewriting as one reads. The books are cheap; they don't stay in print long; why not rewrite them as one goes? Most of the SF I like motivates me to engage actively with images, plots, figures, devices, linguistic moves, in short, with worlds, not so much to make them come out 'right,' as to make them move 'differently.' These worlds motivate me to test their virtue, to see if their articulations work – and what they work for. Because SF makes identification with a principal character, comfort within the patently con-structed world, or a relaxed attitude toward language, especially risky reading strategies, the reader is likely to be more generous and more suspicious – *both generous and suspicious*, exactly the receptive posture I seek in political semiosis generally. It is a strategy closely aligned with the oppositional and differential consciousness theorized by Chela Sandoval and by other feminists insistent on navigating mined discursive waters.

Our first view of Lisa Foo is through Apfel's eyes; and for him, '[l]eaving out only the moustache, she was a dead ringer for a cartoon Tojo. She had the glasses, ears, and the teeth. But her teeth had braces, like piano keys wrapped in barbed wire. And she was five-eight or five-nine and couldn't have weighed more than a hundred and ten. I'd have said a hundred, but added five pounds for each of her breasts, so improbably large on her scrawny frame that all I could read of the message on her T-shirt was "POCK LIVE." It was only when she turned sideways that I saw the esses before and after' (pp. 241–42). Using such messages among the many other languages accessed by this intensely literate figure, Foo communicated constantly through her endless supply of T-shirts. Her breasts turned out to be silicone implants, and as Foo said, 'I don't think I've ever been so happy with anything I ever bought. Not even the car [her Ferrari]' (p. 263). From Foo's childhood perspective, 'the West ... [is] the place where you buy tits' (p. 263).

When Foo and Apfel became lovers, in one of the most sensitively structured heterosexual, cross-racial relationships in print anywhere, we also learn that

Foo's body was multiply composed by the history of Southeast Asia. Varley gave her a name that is an 'orientalized' version of the computer term 'fu bar' – 'fucked up beyond all recognition.' Her Chinese grandmother had been raped in Hanoi by an occupying Japanese soldier in 1942. In Foo's mother's Vietnam, 'Being Chinese was bad enough, but being half Chinese and half Japanese was worse … My father was half French and half Annamese. Another bad combination' (p. 275). Her mother was killed in the Tet offensive when Foo was ten. The girl became a street hustler and child prostitute in Saigon, where she was 'protected' by a pedophilic white U.S. major. Refusing to leave Saigon with him, after Saigon 'fell,' Foo ended up in Pol Pot's Cambodia, where she barely survived the Khmer Rouge work camps. She escaped to Thailand, and 'when I finally got the Americans to notice me, my Major was still looking for me' (p. 276). Dying of a cancer that might have been the result of his witnessing the atom bomb tests in Nevada early in his career, he sponsored her to the U.S. Her intelligence and hustling got her 'tits by Goodyear' (p. 275), a Ferrari, and a Cal Tech education. Foo and Apfel struggle together within their respective legacies of multiple abuse, sexual and otherwise, and criss-crossing racisms. They are both multi-talented, but scarred, survivors. This story, its core figure and its narrator, will not let us dodge the scary issues of race/racism, gender/ sexism, historical tragedy, and technoscience within the region of time we politely call 'the late twentieth century.' There is no safe place here; there are, however, many maps of possibility.

But, there is entirely *too much* connection in 'Press Enter,' and it is only the beginning. Foo is deeply in love with the power-knowledge systems to which her skills give her access. 'This is money, Yank, she said, and her eyes glittered' (p. 267). As she traces the fascinating webs and security locks, which began in military computer projects but which have taken on a vastly unhuman life of their own, her love and her skills bring her too deep into the infinitely dense connections of the System, where she, like Kluge before her, is noticed. Too late, she tries to withdraw. Soon after, a clearly fake suicide note appears on her T-shirt on her ruined body. Investigation showed that she had rewired the microwave oven in Kluge's house to circumvent its security checks. She put her head in the oven, and she died shortly after in the hospital, her eyes and brain congealed and her breasts horribly melted. The promise of her name, 'fu bar,' was all-too-literally fulfilled – fucked up beyond all recognition. Apfel, who had been brought back into articulation with life in his love with Lisa Foo, retreated totally, stripping his house of all its wiring and any other means of connecting with the techno-webs of a world he now saw totally within the paranoid terms of infinite and alien connection. At the end, the defended self, alone, permanently hides from the alien Other.

It is possible to read 'Press Enter' as a conventional heterosexual romance, bourgeois detective fiction, technophobic-technophilic fantasy, dragon-lady story, and, finally, white masculinist narrative whose condition of possibility is access to the body and mind of a woman, especially a 'Third World' woman, who, here as elsewhere in misogynist and racist culture, is violently destroyed. Not just violently – superabundantly, without limit. I think such a reading does serious violence to the subtle tissues of the story's writing. Nonetheless, 'Press Enter' induces in me, and in other women and men who have read the story with me, an irreconcilable pain and anger: Lisa Foo should not have been killed that way. It really is not alright. The text and the body lose all distinction. I fall out of the semiotic square and into the viciously circular thing-in-itself. More than anything else, that pornographic, gendered and colored death, that excessive destruction of her body, that total undoing of her being – that extravagant final connection – surpasses the limits of pleasure in the conventions of paranoid fiction and provokes the necessity of active rewriting as reading. I cannot read this story without rewriting it; that is one of the lessons of transnational, intercultural, feminist literacy. And the conclusion forces rewriting not just of itself, but of the whole human and unhuman collective that is Lisa Foo. The point of the differential/oppositional rewriting is not to make the story come out 'right,' whatever that would be. The point is to rearticulate the figure of Lisa Foo to unsettle the closed logics of a deadly racist misogyny. Articulation must remain open, its densities accessible to action and intervention. When the system of connections closes in on itself, when symbolic action becomes perfect, the world is frozen in a dance of death. The cosmos is finished, and it is One. Paranoia is the only possible posture; generous suspicion is foreclosed. To 'press enter' is, in that world, a terrible mistake.

The whole argument of 'The Promises of Monsters' has been that to 'press enter' is not a fatal error, but an inescapable possibility for changing maps of the world, for building new collectives out of what is not quite a plethora of human and unhuman actors. My stakes in the textual figure of Lisa Foo, and of many of the actors in Varley's SF, are high. Built from multiple interfaces, Foo can be a guide through the terrains of virtual space, but only if the fine lines of tension in the articulated webs that constitute her being remain in play, open to the unexpected realization of an unlikely hope. It's not a 'happy ending' we need, but a non-ending. That's why none of the narratives of masculinist, patriarchal apocalypses will do. The System is not closed; the sacred image of the same is not coming. The world is not full.

The final image of this excessive essay is *Cyborg*, a 1989 painting by Lynn Randolph, in which the boundaries of a fatally transgressive world, ruled by the Subject and the Object, give way to the borderlands, inhabited by human and unhuman collectives (Figure 17.12).[53] These borderlands suggest a rich topography of combinatorial possibility. That possibility is called the Earth, here, now, this elsewhere, where real, outer, inner, and virtual space implode. The painting maps the articulations among cosmos, animal, human, machine, and landscape in their recursive sidereal, bony, electronic, and geological skeletons. Their combinatorial logic is embodied; theory is corporeal; social nature is articulate. The stylized DIP switches of the integrated circuit board on the human figure's chest are devices that set the defaults in a form intermediate between hardwiring and software control – not unlike the mediating structural-functional anatomy of the feline and hominid forelimbs, especially the flexible, homologous hands and paws. The painting is replete with organs of touch and mediation, as well as with organs of vision. Direct in their gaze at the viewer, the eyes of the woman and the cat center the whole composition. The spiraling skeleton of the Milky Way, our galaxy, appears behind the cyborg figure in three different graphic displays made possible by high-technology visualizing apparatuses. In the place of virtual space in my semiotic square, the fourth square is an imaging of the gravity well of a black hole. Notice the tic-tac-toe game, played with the European male and female astrological signs (Venus won this game); just to their right are some calculations that might appear in the mathematics of chaos. Both sets of symbols are just below a calculation found in the Einstein papers. The mathematics and games are like logical skeletons. The keyboard is jointed to the skeleton of the planet Earth, on which a pyramid rises in the left mid-foreground. The whole painting has the quality of a meditation device. The large cat is like a spirit animal, a white tiger perhaps. The woman, a young Chinese student in the United States, figures that which is human, the universal, the generic. The 'woman of color,' a very particular, problematic, recent collective identity, resonates with local and global conversations.[54] In this painting, she embodies the still oxymoronic simultaneous statuses of woman, 'Third World' person, human, organism, communications technology, mathematician, writer, worker, engineer, scientist, spiritual guide, lover of the Earth. This is the kind of 'symbolic action' transnational feminisms have made legible. S/he is not finished.

Figure 17.12   Lynn Randolph, *Cyborg*, 1989

We have come full circle in the noisy mechanism of the semiotic square, back to the beginning, where we met the commercial cyborg figures inhabiting technoscience worlds. Logic General's oddly recursive rabbits, forepaws on the keyboards that promise to mediate replication and communication, have given way to different circuits of competencies. If the cyborg has changed, so might the world. Randolph's cyborg is in conversation with Trinh Minh-ha's inappropriate/d other, the personal and collective being to whom history has forbidden the strategic illusion of self-identity. This cyborg does not have an Aristotelian structure; and there is no master-slave dialectic resolving the struggles of resource and product, passion and action. S/he is not utopian nor imaginary; s/he is virtual. Generated, along with other cyborgs, by the collapse into each other of the technical, organic, mythic, textual, and political, s/he is constituted by articulations of critical differences within and without each figure. The painting might be headed, 'A few words about articulation from the actors in the field.' Privileging the hues of red, green, and ultraviolet, I want to read Randolph's *Cyborg* within a rainbow political semiology, for wily transnational technoscience studies as cultural studies.

NOTES

1. 'They drew near to a very Miry Slough ... The name of this Slow was Dispond' (John Bunyan, *Pilgrim's Progress*, 1678; quoted in the Oxford English Dictionary). The non-standardization of spelling here should also mark, at the beginning of the 'Promises of Monsters,' the suggestiveness of words at the edge of the regulatory technologies of writing.

2. Sally Hacker, in a paper written just before her death ('The Eye of the Beholder: An Essay on Technology and Eroticism,' manuscript, 1989), suggested the term 'pornotechnics' to refer to the embodiment of perverse power relations in the artifactual body. Hacker insisted that the heart of pornotechnics is the military as an institution, with its deep roots and wide reach into science, technology, and erotics. 'Technical exhilaration' is profoundly erotic; joining sex and power is the designer's touch. Technics and erotics are the cross hairs in the focusing device for scanning fields of skill and desire. See also Hacker (1989). Drawing from Hacker's arguments, I believe that control over technics is the enabling practice for class, gender, and race supremacy. Realigning the join of technics and erotics must be at the heart of anti-racist feminist practice. (cf. Haraway, 1989b; Cohn, 1987).

3. See the provocative publication that replaced *Radical Science Journal, Science as Culture*, Free Association Books, 26 Freegrove Rd., London N7 9RQ, England.

4. This incubation of ourselves as planetary fetuses is not quite the same thing as pregnancy and reproductive politics in post-industrial, post-modern, or other posted locations, but the similarities will become more evident as this essay proceeds. The struggles over the outcomes are linked.

5. Here I borrow from the wonderful project of the Journal, *Public Culture*, Bulletin of the Center for Transnational Cultural Studies, The University Museum, University of Pennsylvania, Philadelphia, PA 19104. In my opinion, this journal embodies the best impulses of cultural studies.

6. I demure on the label 'postmodern' because I am persuaded by Bruno Latour that within the historical domains where science has been constructed, the 'modern' never existed, if by modern we mean the rational, enlightened mentality (the subject, mind, etc.) actually proceeding with an objective method toward adequate representations, in mathematical equations if possible, of the object – i.e., 'natural' – world. Latour

argues that Kant's *Critique*, which set off at extreme poles Things-in-Themselves from the Transcendental Ego, is what made us believe ourselves to be 'modern,' with escalating and dire consequences for the repertoire of explanatory possibilities of 'nature' and 'society' for Western scholars. The separation of the two transcendances, the object pole and subject pole, structures '"the political Constitution of Truth." I call it "modern," defining modernity as the complete separation of the representation of things – science and technology – from the representation of humans – politics and justice.' (Latour, forthcoming, a).

Debilitating though such a picture of scientific activity should seem, it has guided research in the disciplines (history, philosophy, sociology, anthropology), studying science with a pedagogical and prophylactic vengeance, making culture seem other to science; science alone could get the goods on nature by unveiling and policing her unruly embodiments. Thus, science studies, focused on the edifying object of 'modern' scientific practice, has seemed immune from the polluting infections of cultural studies – but surely no more. To rebel against or to lose faith in rationalism and enlightenment, the infidel state of respectively modernists and postmodernists, is not the same thing as to show that rationalism was the emperor that had no clothes, that never was, and so there never was its other either. (There is a nearly inevitable terminological confusion here among modernity, the modern, and modernism. I use modernism to refer to a cultural movement that rebelled against the premises of modernity, while postmodernism refers less to rebellion than loss of faith, leaving nothing to rebel against.) Latour calls his position *a*modern and argues that scientific practice is and has been amodern, a sighting that makes the line between real scientific (West's) and ethnoscience and other cultural expressions (everything else) disappear. The difference reappears, but with a significantly different geometry – that of scales and volumes, i.e., the size differences among 'collective' entities made of humans and non-humans – rather than in terms of a line between rational science and ethnoscience.

This modest turn or tropic change does not remove the study of scientific practice from the agenda of cultural studies and political intervention, but places it decisively on the list. Best of all, the focus gets fixed clearly on inequality, right where it belongs in science studies. Further, the addition of science to cultural studies does not leave the notions of culture, society, and politics untouched, far from it. In particular, we cannot make a critique of science and its constructions of nature based on an ongoing belief in culture or society. In the form of social constructionism, that belief has grounded the major strategy of left, feminist, and anti-racist science radicals. To remain with that strategy, however, is to remain bedazzled by the ideology of enlightenment. It will not do to approach science as cultural or social construction, as if culture and society were transcendent categories, any more than nature or the object is. Outside the premises of enlightenment – i.e., of the modern – the binary pairs of culture and nature, science and society, the technical and the social all lose their co-constitutive, oppositional quality. Neither can explain the other. 'But instead of providing the explanation, Nature and Society are now accounted for as the historical consequences of the movement of collective things. All the interesting realities are no longer captured by the two extremes but are to be found in the substitution, cross over, translations, through which actants shift their competences' (Latour, 1990, p. 170). When the pieties of belief in the modern are dismissed, both members of the binary pairs collapse into each other as into a black hole. But what happens to them in the black hole is, by definition, not visible from the shared terrain of modernity, modernism, or postmodernism. It will take a superluminal SF journey into elsewhere to find the interesting new vantage points. Where Latour and I fundamentally agree is that in that gravity well, into which Nature and Society as transcedentals disappeared, are to be found actors/actants of many and wonderful kinds. Their relationships constitute the artifactualism I am trying to sketch.

7. For quite another view of 'production' and 'reproduction' than that enshrined in so much Western political and economic (and feminist) theory, see Marilyn Strathern (1988, pp. 290–308).

8. Chela Sandoval develops the distinctions between oppositional and differential consciousness in her forthcoming doctoral dissertation, University of California at Santa Cruz. See also Sandoval (1990).

9. My debt is extensive in these paragraphs to Luce Irigaray's wonderful critique of the allegory of the cave in *Spœculum de l'autre femme* (1974). Unfortunately, Irigaray, like almost all white Europeans and Americans after the mid-nineteenth-century consolidation of the myth that the 'West' originated in a classical Greece unsullied by Semitic and African roots, transplants, colonizations, and loans, never questioned the 'original' status of Plato's fathership of philosophy, enlightenment, and rationality. If Europe was colonized first by Africans, that historical narrative element would change the story of the birth of Western philosophy and science. Martin Bernal's extraordinarily important book, *Black Athena*, Vol. 1, *The Fabrication of Ancient Greece, 1785–1985* (1987), initiates a groundbreaking re-evaluation of the founding premises of the myth of the uniquences and self-generation of Western culture, most certainly including those pinnacles of Man's self-birthing, science and philosophy. Bernal's is an account of the determinative role of racism and Romanticism in the fabrication of the story of Western rationality. Perhaps ironically, Martin Bernal is the son of J. D. Bernal, the major pre-World War II British biochemist and Marxist whose four-volume *Science in History* movingly argued the superior rationality of a science freed from the chains of capitalism. Science, freedom, and socialism were to be, finally, the legacy of the West. For all its warts, that surely would have been better than Reagan's and Thatcher's version! See Gary Wersky, *The Invisible College: The Collective Biography of British Socialist Scientists in the 1930s* (1978).

   Famous in his own generation for his passionate heterosexual affairs, J. D. Bernal, in the image of enlightenment second birthing so wryly exposed by Irigaray, wrote his own vision of the future in *The Word, the Flesh, and the Devil* as a science-based speculation that had human beings evolving into disembodied intelligences. In her manuscript (May, 1990) 'Talking about Science in Three Colors: Bernal and Gender Politics in the Social Studies of Science,' Hilary Rose discusses this fantasy and its importance for 'science, politics, and silences.' J. D. Bernal was also actively supportive of independent women scientists. Rosalind Franklin moved to his laboratory after her nucleic acid crystallographic work was stolen by the flamboyantly sexist and heroic James Watson on his way to the immortalizing, luminous fame of the *Double Helix* of the 1950s and 60s and its replicant of the 1980s and 90s, the Human Genome Project. The *story* of DNA has been an archetypical tale of blinding modern enlightenment and untrammeled, disembodied, autochthonous origins. See Ann Sayre (1975); Mary Jacobus (1982); Evelyn Fox Keller (1990).

10. For an argument that nature is a *social* actor, see Elizabeth Bird (1987).

11. Actants are not the same as actors. As Terence Hawkes (1977, p. 89) put it in his introduction to Greimas, actants operate at the level of function, not of character. Several characters in a narrative may make up a single actant. The structure of the narrative generates its actants. In considering what kind of entity 'nature' might be, I am looking for a coyote and historical grammar of the world, where deep structure can be quite a surprise, indeed, a veritable trickster. Non-humans are not necessarily 'actors' in the human sense, but they are part of the functional collective that makes up an actant. Action is not so much an ontological as a semiotic problem. This is perhaps as true for humans as non-humans, a way of looking at things that may provide exits from the methodological individualism inherent in concentrating constantly on who the agents and actors are in the sense of liberal theories of agency.

12. In this productionist story, women make babies, but this is a poor if necessary substitute for the real action in reproduction – the second birth through self-birthing, which requires the obstetrical technology of optics. One's relation to the phallus determines whether one gives birth to oneself, at quite a price, or serves, at an even greater price, as the conduit or passage for those who will enter the light of self-birthing. For a refreshing demonstration that women do not make babies everywhere, see Marilyn Strathern (1988), pp. 314–18.

13. I borrow here from Katie King's notion of the apparatus of literary production, in which the poem congeals at the intersection of business, art, and technology. See King (1990). See also Donna Haraway (1991), chaps. 8–10.

14. Latour has developed the concept of delegation to refer to the translations and exchanges between and among people doing science and their machines, which act as 'delegates' in a wide array of ways. Marx considered machines to be 'dead labor,' but that notion, while still necessary for some crucial aspects of forced and reified delegation, is too unlively to get at the many ways that machines are part of *social* relations 'through which actants shift competences' (Latour, 1990, p. 170). See also Bruno Latour (forthcoming). Latour, however, as well as most of the established scholars in the social studies of science, ends up with too narrow a concept of the 'collective,' one built up out of only machines and scientists, who are considered in a very narrow time and space frame. But circulations of skills turn out to take some stranger turns. First, with the important exception of his writing and teaching in collaboration with the primatologist Shirley Strum, who has fought hard in her profession for recognition of primates as savvy social actors, Latour pays too little attention to the non-machine, *other* non-humans in the interactions. See Strum (1987).

   The 'collective,' of which 'nature' in any form is one example from my point of view, is always an artifact, always social, not because of some transcendental Social that explains science or vice versa, but because of its heterogeneous actants/actors. Not only are not all of those actors/actants people; I agree there *is* a sociology of machines. But that is not enough; not all of the other actors/actants were *built* by people. The artifactual 'collective' includes a witty actor that I have sometimes called coyote. The interfaces that constitute the 'collective' must include those between humans and artifacts in the form of instruments and machines, a genuinely *social* landscape. But the interface between machines and *other* non-humans, as well as the interface between humans and *non-machine* non-humans must also be counted in. Animals are fairly obvious actors, and their interfaces with people and machines are easier to admit and theorize. See Donna Haraway (1989a); Barbara Noske (1989); Paradoxically, from the perspective of the kind of artifactualism I am trying to sketch, animals lose their *object* status that has reduced them to things in so much Western philosophy and practice. They inhabit neither nature (as object) nor culture (as surrogate human), but instead inhabit a place called elsewhere. In Noske's terms (p. xi), they are other 'worlds, whose otherworldliness must not be disenchanted and cut to our size but respected for what it is.' Animals, however, do not exhaust the coyote world of non-machine non-humans. The domain of machine and non-machine non-humans (the unhuman, in my terminology) joins people in the building of the artifactual collective called nature. None of these actants can be considered as simply resource, ground, matrix, object, material, instrument, frozen labor; they are all more unsettling than that. Perhaps my suggestions here come down to re-inventing an old option within a non-Eurocentric Western tradition indebted to Egyptian Hermeticism that insists on the active quality of the world and on 'animate' matter. See Martin Bernal (1987, pp. 121–60); Frances Yates (1964). Worldly and enspirited, coyote nature is a collective, cosmopolitan artifact crafted in stories with heterogeneous actants.

   But there is a second way in which Latour and other major figures in science studies work with an impoverished 'collective.' Correctly working to resist a 'social' explanation of 'technical' practice by exploding the binary, these scholars have a tendency covertly to reintroduce the binary by worshipping only one term – the 'technical.' Especially, *any* consideration of matters like masculine supremacy or racism or imperialism or class structures are inadmissible because they are the old 'social' ghosts that blocked real explanation of science in action. See Latour (1987). As Latour noted, Michael Lynch is the most radical proponent of the premise that there is no social explanation of a science but the technical content itself, which assuredly includes the interactions of people with each other in the lab and with their machines, but excludes a great deal that I would include in the 'technical' content of science if one really doesn't want to evade a binary by worshipping one of its old poles. Lynch (1985); Latour (1990, p. 169n). I agree with Latour and Lynch that practice creates its own context, but they draw a suspicious line around what gets to count as 'practice.' They *never* ask how the *practices* of masculine supremacy, or many other systems of structured inequality, get *built* into and out of working machines. How and in what directions these transferences of 'competences' work should be a focus of rapt attention. Systems of exploitation might be crucial parts of the 'technical content' of science. But the SSS scholars tend to dismiss such questions with the assertion that they lead to the bad old days when science was asserted by radicals simply to 'reflect' social relations. But in my view, such transferences of competences, or delegations, have nothing to do with reflections or harmonies of social organization and cosmologies, like 'modern science.' Their unexamined, consistent, and defensive prejudice seems part of Latour's (1990, pp. 164–69) stunning misreading of several moves in Sharon Traweek's *Beam Times and Life Times: The World of High Energy Physicists* (1988). See also Hilary Rose, 'Science in Three Colours: Bernal and Gender Politics in the Social Studies of Science,' unpublished manuscript, May 2, 1990.

The same blind spot, a retinal lesion from the old phallogocentric heliotropism that Latour *did* know how to avoid in other contexts, for example in his trenchant critique of the modern and postmodern, seems responsible for the abject failure of the social studies of science as an organized discourse to take account of the last twenty years of feminist inquiry. What counts as 'technical' and what counts as 'practice' should remain far from self-evident in science in action. For all of their extraordinary creativity, so far the mappings from most SSS scholars have stopped dead at the fearful seas where the worldly practices of inequality lap at the shores, infiltrate the estuaries, and set the parameters of reproduction of scientific practice, artifacts, and knowledge. If only it were a question of reflections between social relations and scientific constructions, how easy it would be to conduct 'political' inquiry into science! Perhaps the tenacious prejudice of the SSS professionals is the punishment for the enlightenment transcendental, the social, that did inform the rationalism of earlier generations of radical science critique and is still all too common. May the topick gods save us from both the reified technical and the transcendental social!

15. See Lynn Margulis and Dorion Sagan (1986). This wonderful book does the cell biology and evolution for a host of inappropriate/d others. In its dedication, the text affirms 'the combinations, sexual and parasexual, that bring us out of ourselves and make us more than we are alone' (p. v). That should be what science studies as cultural studies do, by showing how to visualize the curious collectives of humans and unhumans that make up naturalsocial (one word) life. To stress the point that all the actors in these generative, dispersed, and layered collectives do not have human form and function – and should not be anthropomorphized – recall that the Gaia hypothesis with which Margulis is associated is about the tissue of the planet as a living entity, whose metabolism and genetic exchange are effected through webs of prokaryotes. Gaia is a society; Gaia is nature; Gaia did not read the *Critique*. Neither, probably, did John Varley. See his Gaea hypothesis in the SF book, *Titan* (1979). Titan is an alien that is a world.

16. Remember that *monsters* have the same root as *to demonstrate*; monsters signify.
17. Trinh T. Minh-ha, ed., 1986/7, *She, the Inappropriate/d Other*. See also her *Woman, Native, Other: Writing Postcoloniality and Feminism* (1989).
18. Interpellate: I play on Althusser's account of the call which constitutes the production of the subject in ideology. Althusser is, of course, playing on Lacan, not to mention on God's interruption that calls Man, his servant, into being. Do we have a vocation to be cyborgs? Interpellate: *Interpellatus*, past participle for 'interrupted in speaking' – effecting transformations like Saul into Paul. Interpellation is a special kind of interruption, to say the least. Its key meaning concerns a procedure in a parliament for asking a speaker who is a member of the government to provide an explanation of an act or policy, usually leading to a vote of confidence. The following ads interrupt us. They insist on an explanation in a confidence game; they force recognition of how transfers of competences are made. A cyborg subject position results from and leads to interruption, diffraction, reinvention. It is dangerous and replete with the promises of monsters.
19. In *King Solomon's Ring*, Konrad Lorenz pointed out how the railroad car kept the appearance of the horse drawn carriage, despite the different functional requirements and possibilities of the new technology. He meant to illustrate that biological evolution is similarily conservative, almost nostalgic for the old, familiar forms, which are reworked to new purposes. Gaia was the first serious bricoleuse.
20. For a view of the manufacture of particular organisms as flexible model systems for a universe of research practice, see Barbara R. Jasny and Daniel Koshland, Jr, eds, *Biological Systems* (1990). As the advertising for the book states, 'The information presented will be especially useful to graduate students and to all researchers interested in learning the limitations and assets of biological systems currently in use,' *Science* 248 (1990), p. 1024. Like all forms of protoplasm collected in the extra-laboratory world and brought into a technoscientific niche, the organic rabbit (not to mention the simulated one) and its tissues have a probable future of a particular sort – as a commodity. Who should 'own' such evolutionary products? If seed protoplasm is collected in peasants' fields in Peru and then used to breed valuable commercial seed in a 'first world' lab, does a peasant cooperative or the Peruvian state have a claim on the profits? A related problem about proprietary interest in 'nature' besets the biotechnology industry's development of cell lines and other products derived from removed human tissue, e.g., as a result of cancer surgery. The California Supreme Court recently reassured the biotechnology industry that a patient, whose cancerous spleen was the source of a product, Colony Stimulating Factor, that led to a patent that brought its scientist-developer stock in a company worth about $3 million, did not have a right to a share of the bonanza. Property in the self, that lynchpin of liberal existence, does not seem to be the same thing as proprietary rights in one's body or its products – like fetuses or other cell lines in which the courts take a regulatory interest. See Marcia Barinaga (1990, p. 239).
21. Here and throughout this essay, I play on Katie King's play on Jacques Derrida's *Of Grammatology*, (1976). See King (1990), and King (in progress), where she develops her description, which is also a persuasive enabling construction, of a discursive field called 'feminism and writing technologies.'
22. Roland Barthes, *Mythologies* (1972) is my guide here and elsewhere.
23. Peace-activist and scholar in science studies, Elizabeth Bird came up with the slogan and put it on a political button in 1986 in Santa Cruz, California.
24. I am indebted to another guide figure throughout this essay, Gloria Anzaldúa, *Borderlands, La Frontera: The New Mestiza* (1987) and to at least two other travelers in embodied virtual spaces, Ramona Fernandez, 'Trickster Literacy: Multiculturalism and the (Re) Invention of Learning,' Qualifying Essay, History of Consciousness, University of California at Santa Cruz, 1990, and Allucquére R. Stone, 'Following Virtual Communities,' unpublished essay, History of Conscious-

ness, University of California at Santa Cruz. The ramifying 'virtual consensual community' (Sandy Stone's term in another context) of feminist theory that incubates at UCSC densely infiltrates my writing.

25. For an extended reading of National Geographic's Jane Goodall stories, *always to be held in tension with other versions of Goodall and the chimpanzees at Gombe*, see Haraway, 'Apes in Eden, Apes in Space,' in *Primate Visions* (1989a, pp. 133–95). Nothing in my analysis should be taken as grounds to oppose primate conservation or to make claims about the other Jane Goodalls; those are complex matters that deserve their own careful, materially specific consideration. My point is about the semiotic and political frames within which survival work might be approached by geopolitically differentiated actors.

26. My files are replete with recent images of cross-species ape-human family romance that fail to paper over the underlying racist iconography. The most viciously racist image was shown to me by Paula Treichler: an ad directed to physicians by the HMO, Premed, in Minneapolis, from the *American Medical News*, August 7, 1987. A white-coated white man, stethoscope around his neck, is putting a wedding ring on the hand of an ugly, very black, gorilla-suited female dressed in a white wedding gown. White clothing does not mean the same thing for the different races, species, and genders! The ad proclaims, 'If you've made an unholy HMO alliance, perhaps we can help.' The white male physician (man) tied to the black female patient (animal) in the inner cities by HMO marketing practices in relation to medicaid policies must be freed. There is no woman in this ad; there is a hidden threat disguised as an ape female, dressed as the vampirish bride of scientific medicine (a single white tooth gleams menacingly against the black lips of the ugly bride) – another illustration, if we needed one, that black women do not have the *discursive* status of woman/human in white culture. 'All across the country, physicians who once had visions of a beautiful marriage to an HMO have discovered the honeymoon is over. Instead of quality care and a fiscally sound patient-base, they end up accepting reduced fees and increased risks.' The codes are transparent. Scientific medicine has been tricked into a union with vampirish poor black female patients. Which risks are borne by whom goes unexamined. The clasped hands in this ad carry a different surface message from the Gulf ad's, but their enabling semiotic structures share too much.

27. At the oral presentation of this paper at the conference on 'Cultural Studies Now and in the Future,' Gloria Watkins/bell hooks pointed out the painful current U.S. discourse on African-American men as 'an endangered species.' Built into that awful metaphor is a relentless history of animalization and political infantilization. Like other 'endangered species,' such people cannot speak for themselves, but must be spoken for. They must be represented. Who speaks for the African-American man as 'an endangered species'? Note also how the metaphor applied to black *men* justifies anti-feminist and misogynist rhetoric about and policy toward black women. They actually become one of the forces, if not the chief threat, endangering African-American men.

28. Committing only a neo-imperialist venial sin in a footnote, I yield to voyeuristic temptation just a little: in *Discover* the videocam and the 'native' have a relation symmetrical to that of Goodall's and the chimpanzee's hands. Each photo represents a touch across time and space, and across politics and history, to tell a story of salvation, of saving man and nature. In this version of cyborg narrative, the touch that joins portable high technology and 'primitive' human parallels the touch that joins animal and 'civilized' human.

29. It is, however, important to note that the present man in charge of environmental affairs in the Amazon in the Brazilian government has taken strong, progressive stands on conservation, human rights, destruction of indigenous peoples, and the links of ecology and justice. Further, current proposals and policies, like the government's plan called *Nossa Natureza* and some international aid and

conservations organizations' activities and ecologists' understandings, have much to recommend them. In addition, unless arrogance exceeds all bounds, *I* can hardly claim to adjudicate these complex matters. The point of my argument is not that whatever comes from Brasília or Washington is bad and whatever from the forest residents is good – a patently untrue position. Nor is it my point that nobody who doesn't come from a family that has lived in the forest for generations has any place in the 'collectives, human and unhuman,' crucial to the survival of lives and ways of life in Amazonia and elsewhere. Rather, the point is about the self-constitution of the indigenous peoples as *principal* actors and agents, with whom others must interact – in coalition and in conflict – not the reverse.

30. For the story of Mendes's life work and his murder by opponents of an extractive reserve off limits to logging, see Andrew Revkin (1990).

31. Further references are parenthetical in the text.

32. Similar issues confront Amazonians in countries other than Brazil. For example, there are national parks in Colombia from which native peoples are banned from their historical territory, but to which loggers and oil companies have access under park multi-use policy. This should sound very familiar to North Americans, as well.

33. Revising and displacing his statements, I am again in conversation with Bruno Latour here, who has insisted on the social status of both human and non-human actors. 'We use actor to mean anything that is made by some other actor the source of an action. It is in no way limited to humans. It does not imply will, voice, self-consciousness or desire.' Latour makes the crucial point that 'figuring' (in words or in other matter) non-human actors as if they were like people is a semiotic operation; non-figural characterizations are quite possible. The likeness or unlikeness of actors is an interesting problem opened up by placing them firmly in the shared domain of social interaction. Bruno Latour (forthcoming).

34. Kane's review appeared in the *Voice Literary Supplement*, February 1990, and Hecht and Cockburn replied under the title 'Getting Historical,' *Voice Literary Supplement*, March 1990, p. 26.

35. My discussion of the politics of representation of the fetus depends on twenty years of feminist discourse about the location of responsibility in pregnancy and about reproductive freedom and constraint generally. For particularly crucial arguments for this essay, see Jennifer Terry (1989); Valerie Hartouni (1991); and Rosalind Pollock Petchesky (1987).

36. *The Eighteenth Brumaire of Louis Bonaparte*. Quoted in Edward Said (1978, p. xiii), as his opening epigraph.

37. Marilyn Strathern describes Melanesian notions of a child as the 'finished repository of the actions of multiple others,' and not, as among Westerners, a resource to be constructed into a fully human being through socialization by others. Marilyn Strathern, 'Between Things: A Melanesianist's Comment on Deconstructive Feminism,' unpublished manuscript. Western feminists have been struggling to articulate a phenomenology of pregnancy that rejects the dominant cultural framework of productionism/reproductionism, with its logic of passive resource and active technologist. In these efforts the woman-fetus nexus is refigured as a knot of relationality within a wider web, where liberal individuals are not the actors, but where complex collectives, including non-liberal social persons (singular and plural), are. Similar refigurings appear in eco-feminist discourse.

38. See the fall 1990 newsletter of the Society for the Social Study of Science, *Technoscience* 3, no. 3, pp. 20, 22, for language about 'going back to nature.' A session of the 4S October meetings is titled 'Back to Nature.' Malcolm Ashmore's abstract, 'With a Reflexive Sociology of Actants, There Is No Going Back,' offers 'fully comprehensive insurance against going back,' instead of other competitors' less good 'ways of not going back to Nature (or Society or Self).' All of this occurs in the context of a crisis of confidence among many 4S scholars that their very fruitful research programs of the last 10 years are running into dead ends. They are. I will

refrain from commenting on the blatant misogyny in the Western scholar's textualized terror of 'going back' to a phantastic nature (figured by science critics as 'objective' nature. Literary academicians figure the same terrible dangers slightly differently; for both groups such a nature is definitively pre-social, monstrously not-human, and a threat to their careers). Mother nature always waits, in these adolescent boys' narratives, to smother the newly individuated hero. He forgets this weird mother is his creation; the forgetting, or the inversion, is basic to ideologies of scientific objectivity and of nature as 'eden under glass.' It also plays a yet-to-be-examined role in some of the best (most reflexive) science studies. A theoretical gender analysis is indispensible to the reflexive task.

39. *Time*, February 10, 1961, p. 58. The caption under HAM's photograph read 'from Chop Chop Chang to No. 65 to a pioneering role.' For HAM's flight and the Holloman chimps' training see Weaver (1961) and *Life Magazine*, February 10, 1961. *Life* headlined, 'From Jungles to the Lab: The Astrochimps.' All were captured from Africa; that means many other chimps died in the 'harvest' of babies. The astrochimps were chosen over other chimps for, among other things, 'high IQ.' Good scientists all.

40. *Time*, December 8, 1961, p. 50; *Newsweek*, March 5, 1962, p. 19.

41. *Time*, December 8, 1961, p. 50.

42. See also Chris Gray, 'Postmodern War,' Qualifying Exam, History of Consciousness, UCSC, 1988.

43. For indispensable theoretical and participant-observation writings on eco-feminism, social movements, and non-violent direct action, see Barbara Epstein (1991).

44. For a fuller discussion of the immune system, see Haraway, 'The Biopolitics of Postmodern Bodies,' in *Simians, Cyborgs, and Women* (1991).

45. Recall that Nilsson shot the famous and discourse-changing photographs of fetuses (really abortuses) as glowing back-lit universes floating free of the 'maternal environment.' Nilsson (1977).

46. Advertising copy for the Met Life Pavilion. The exhibit is sponsored by the Metropolitan Life and Affiliated Companies. In the campground resort at Florida's Walt Disney World, we may also view the 'endangered species island,' in order to learn the conventions for 'speaking for the jaguar' in an eden under glass.

47. Ramona Fernandez, 'Trickster Literacy,' Qualifying Exam, History of Consciousness, UCSC, 1990, wrote extensively on Walt Disney World and the multiple cultural literacies required and taught on-site for successfully traveling there. Her essay described the visualizing technology and medical school collaboration in its development and use. See the *Journal of the American Medical Association* 260, no. 18 (November 18, 1988), pp. 2776–83.

48. Building an unexpected collective, Jerne (1985) drew directly from Noam Chomsky's theories of structural linguistics. The 'textualized' semiotic body is not news by the late twentieth century, but what kind of textuality is put into play still matters!

49. See, for example, the recent merger of Project Inform with the Community Research Alliance to speed the community-based testing of promising drugs – and the NIH's efforts to deal with these developments: *PI Perspective*, May 1990. Note also the differences between President Bush's Secretary of Health and Human Services, Lewis Sullivan, and Director of the National Institute of Allergy and Infectious Diseases, Anthony Fauci, on dealing with activists and PWAs. After ACT UP demonstrations against his and Bush's policies during the secretary's speech at the AIDS conference in San Francisco in June 1990, Sullivan said he would have no more to do with ACT UP and instructed government officials to limit their contacts. (Bush had been invited to address the international San Francisco conference, but his schedule did not permit it. He was in North Carolina raising money for the ultra-reactionary senator Jesse Helms at the time of the conference.) In July 1990, at the ninth meeting of the AIDS Clinical Trials Group (ACTG), at which patient activists participated for the first time, Fauci said that he would work to include the

AIDS constituency at every level of the NIAID process of clinical trials. He urged scientists to develop the skills to discuss freely in those contexts ('Fauci,' 1990). Why is constructing this kind of scientific articulation 'softer'? I leave the answer to readers' imaginations informed by decades of feminist theory.

50. This quadrant of the semiotic square is dedicated to A. E. Van Vogt's Players of Null-A (1974), for their non-Aristotelian adventures. An earlier version of 'The Promises of Monsters' had the imagination, not SF, in virtual space. I am indebted to a questioner who insisted that the imagination was a nineteenth-century faculty that is in political and epistemological opposition to the arguments I am trying to formulate. As I am trying vainly to skirt psychoanalysis, I must also skirt the slough of the romantic imagination.

51. Allucquére R. Stone, 'Following Virtual Communities,' unpublished manuscript, History of Consciousness, UCSC, 1990.

52. Thanks to Barbara Ige, graduate student in the Literature Board, UCSC, for conversations about our stakes in the figure of Lisa Foo.

53. Oil on canvas, 36" by 28", photo by D. Caras. In conversation with the 1985 essay 'A Manifesto for Cyborgs' (in Haraway, 1991), Randolph painted her *Cyborg* while at the Bunting Institute and exhibited it there in a spring 1990 solo exhibition, titled 'A Return to Alien Roots.' The show incorporated, from many sources, 'traditional religious imagery with a postmodern secularized context.' Randolph paints 'images that empower women, magnify dreams, and cross racial, class, gender, and age barriers' (exhibition brochure). Living and painting in Texas, Randolph was an organizer of the Houston Area Artists' Call Against U.S. Intervention in Central America. The human model for *Cyborg* was Grace Li, from Beijing, who was at the Bunting Institute in the fateful year of 1989.

54. I borrow this use of 'conversation' and the notion of transnational feminist literacy from Katie King's (in progress) concept of women and writing technologies.

BIBLIOGRAPHY

Anzaldúa, Gloria (1987) *Borderlands, La Frontera: The New Mestiza.* San Francisco: Spinsters/Aunt Lute.

Anzaldúa, Gloria (ed.) (1990) *Making Face, Making Soul: Haciendo Caras.* San Francisco: Aunt Lute.

Arena-DeRosa, James (1990) 'Indigenous leaders host U.S. environmentalists in the Amazon', *Oxfam America News*, Summer/Fall, 1–2.

Aristotle (1968–1970) *Physics* (2 Volumes). Trans. P. H. Wicksteed and F. M. Cornford. London: Heinemann.

Balsamo, Anne (1988) 'Reading Cyborgs Writing Feminism', *Communication*, 10, 331–44.

Barthes, Roland (1972) *Mythologies.* Trans. A. Lavers. London: Cape.

Bender, G. and Druckey, T. (eds) (1994) *Culture on the Brink.* Seattle: Bay Press.

Benedikt, M. (ed.) (1991) *Cyberspace: First Steps.* Cambridge, MA: MIT Press.

Benjamin, M. (ed.) (1993) *A Question of Identity.* New Jersey: Rutgers University Press.

Bernal, Martin (1987) *Black Athena, Vol 1., The Fabrication of Ancient Greece, 1785–1985.* London: Free Association Books.

Bijker, W. and Law, J. (eds) (1994) *Shaping Technology/Building Society: Studies in Sociotechnical Change.* Cambridge, MA: MIT Press.

Bird, E. (1987) 'The social construction of nature: Theoretical approaches to the history of environmental problems', *Environmental Review*, 11(4), 255–64.

Bryan, C. (1987) *The National Geographic Society: 100 Years of Adventure and Discovery.* New York: Abrams.

Caillois, Roger (1984) 'Mimicry and Legendary Psychasthenia'. Trans. John Sheply. *October*, 31, 12–32.

Cohn, C. (1987) 'Sex and death 1990 in the rational world of defence intellectuals', *Signs*, 12(4), 687–718.

Collins, Patricia Hill (1990) *Black Feminist Thought: Knowledge, Consciousness, and the Politics of Empowerment.* London: Harper Collins.

Crimp, Douglas (1983) 'On the museum's ruin', in Hal Foster (ed.), *The Anti-Aesthetic: Essays on Postmodern Culture.* Port Townsend, WA: Bay Press, pp. 43–56.

Crimp, D. and Rolston, A. (1990) *AIDS DEMOGRAPHICS.* Seattle: Bay Press.

de Lauretis, Teresa (1984) *Alice Doesn't: Feminism, Semiotics, Cinema.* Bloomington: Indiana University Press.

Derrida, Jaques (1976) *Of Grammatology.* Trans. G. Spivak. Baltimore: Johns Hopkins University Press.

Doane, Mary Ann (1990) 'Technophilia: Technology, Repesentation, and the Feminine', in M. Jacobus, E. Fox Keller and S. Shuttleworth (eds) *Body/Politics: Women and the Discourses of Science.* New York and London: Routledge, pp. 163–76.

Eimerl, Sarel and Irven De Vore (1965) *The Primates.* New York: Time, Inc.

Epstein, Barbara (1991) *Title.* Berkeley: University of California Press.

'Fauci gets softer on no activists' (1990) *Science*, 249, 244.

Featherstone, M. and Burrows, R. (eds) (1995) *Cyberspace/Cyberbodies/Cyberpunk.*

London, Thousand Oaks, New Delhi: Sage.

Hal Foster (ed.) (1983) *The Anti-Aesthetic: Essays on Postmodern Culture*. Port Townsend, WA: Bay Press.

Foster, Thomas (1993) 'Meat Puppets or Robopaths? Cyberpunk and the Question of Embodiment', *Genders*, 18, 11–31.

Fraassen, Bas van (1970) *An Introduction to the Philosophy of Time and Space*. New York: Random House.

Franklin, Sarah (1988) 'Life Story: The gene as fetish object on TV', *Science as Culture*, 3, 92–100.

Freud, Sigmund (1914) 'On Narcissism: an Introduction', *Standard Edition of the Complete Psychological Works of Sigmund Freud*. Oxford: The Hogarth Press, Vol. 14, pp. 73–102.

Freud, Sigmund (1923) 'The Ego and the Id', Vol. 19, pp. 13–66.

Gibson, William (1986) *Neuromancer*. London: Grafton.

Golub, Edward (1987) *Immunology: A Synthesis*. Sunderland, MA: Sinauer.

Goodall, Jane (1971) *In the Shadow of Man*. Boston: Houghton Mifflin.

González, Jennifer (1995) 'Envisioning Cyborg Bodies: Notes From Current Research', in C. Gray (ed.), *The Cyborg Handbook*. London and New York: Routledge, pp. 267–79.

Greimas, A.J. (1966) *Semantique Structurale*. Paris: Larousse.

Grossberg, L., Nelson, C. and Treichler, P. (eds) (1992) *Cultural Studies*. New York and London: Routledge.

Grosz, Elizabeth (1995) 'Space, Time and Bodies', in E. Grosz, *Space, Time and Perversion*. New York and London: Routledge, pp. 83–101.

Grosz, E. (1995) *Space, Time and Perversion*. New York and London: Routledge.

Guillaume, Paul (1971) *Imitation in Children*. Trans. Elaine Halperin. Chicago: University of Chicago Press.

Hacker, Sally (1989) *Pleasure, Power, and Technology: Some Tales of Gender, Engineering, and the Cooperative Workplace*. Boston: Unwin Hyman.

Haraway, Donna (1988) 'Situated Knowledge', *Feminist Studies*, 14(3), 575–99.

Haraway, Donna (1989a) *Primate Visions: Gender, Race, and Nature in the World of Modern Science*. New York: Routledge.

Haraway, Donna (1989b) 'Technics, erotics, vision, touch: Fantasies of the designer body', Talk presented at the meeting of the Society for the History of Technology, 13 October.

Haraway, Donna (1991) *Simians, Cyborgs, and Women: The Reinvention of Nature*. New York: Routledge.

Haraway, Donna (1992) 'The Promises of Monsters: A Regenerative Politics for Inappropriate/d Others', in L. Grossberg, C. Nelson and P. Treichler (eds), *Cultural Studies*. New York and London: Routledge, pp. 295–337.

Hartouni, Valerie (1991) 'Containing women: Reproductive discourse in the 1980s' in C. Penley and A. Ross (eds), *Technoculture*. Minneapolis: University of Minnesota Press.

Hawkes, Terence (1977) *Structuralism and Semiotics*. Berkeley: University of California.

Hayles, N. Katherine (1990) *Chaos Bound: Orderly Disorder in Contemporary Literature and Science*. Ithaca: Cornell University Press.

Hayles, N. Katherine (1993) 'The Life Cycle of Cyborgs: Writing the Posthuman', in M. Benjamin (ed.), *A Question of Identity*. New Jersey: Rutgers University Press, pp. 152–70.

Hecht, Susanna and Alexander Cockburn (1989) *The Fate of the Forest: Developers, Destroyers, and Defenders of the Amazon*. New York: Verso.

Hollinger, Veronica (1990) 'Cybernetic Deconstructions: Cyberpunk and Postmodernism', *Mosaic*, 23, Part 2, 29–44.

Hubbard, Ruth (1990) 'Technology and childbearing', in *The Politics of Women's*

Biology. New Brunswick, NJ: Rutgers University Press.

Irigaray, Luce (1993) *An Ethics of Sexual Difference*. Trans. Carolyn Burke and Gillian C. Gill. Ithaca: Cornell University Press.

Irigaray, Luce (1974) *Speculum de L'autre femme*. Paris: Minuit. (1985) *Speculum of the Other Woman*. Trans. G. Gill. Ithaca: Cornell University Press.

Jacobus, Mary (1982) 'Is there a woman in this text?', *New Literary History*, 14, 117–41.

Jacobus, M., Fox Keeler, E. and Shuttleworth, S. (eds) (1990) *Body/politics: Women and the Discourse of Science*. New York: Routledge.

Jameson, Fredric (1972) *The Prison-House of Language*. Princeton: Princeton University Press.

Jaret, Peter (1986) 'Our immune system: The wars within', *National Geographic*, 169(6), 701–35.

Jasny, Babara and Daniel Koshland, Jr (eds) (1990) *Biological Systems*. Washington, DC: AAAS Books.

Jerne, N. (1985) 'The generative grammar of the immune system', *Science*, 229, 1057–9.

Kane, Joe (1989) *Running the Amazon*. New York: Knopf.

Keller, E. (1990) 'From secrets of life to sectres of death', in M. Jacobus, E. Fox Keeler and S. Shuttleworth (eds), *Body/politics: Women and the Discourse of Science*. New York: Routledge.

King, Katie (1990) 'A feminist apparatus of literary production', *Text*, 5, 91–103.

King, Katie (1986) 'The situation of lesbianism as feminism's magical sign: Contests for meaning and the U.S. women's movement 1968–1972', *Communication*, 9(1), 65–91.

Lacan, Jacques (1953) 'Some Reflections on the Ego', *International Journal of Psychoanalysis*, 34.

Lacan, Jacques (1977) *Écrits*. A Selection. Trans. Alan Sheridan. London: Tavistock.

Latour, Bruno (1987) *Science in Action: How to Follow Scientists and Engineers Through Society*. Cambridge, MA: Harvard University Press.

Latour, Bruno (1992) 'One More Turn after the social turn ... Easing science studies into the non-modern world', in E. McMullen (ed.), *The Social Dimensions of Science*. Notre Dame, Indiana: University of Notre Dame Press.

Latour, Bruno (1990) 'Postmodern? No, simply Amodern! Steps towards an Anthropology of science,' *Studies in the History and Philosophy of Science*, 21(1), 145–71.

Latour, Bruno (1994) 'Where are the missing masses? Sociology of a few mundane artifacts', in W. Bijker and J. Law (eds), *Shaping Technology/Building Society: Studies in Sociotechnical Change*. Cambridge, MA: MIT Press.

Le Doeuff, Michele (1989) *The Philosophical Imaginary*. Trans. Colin Gordon. Stanford: Stanford University Press.

Lorde, Audre (1984) *Sister Outsider*. Freedom, CA: The Crossing Press.

Lynch, Michael (1985) *Art and Artifact in Laboratory Science: A Study of Shop Work and Shop Talk in a Research Laboratory*. London: Routledge.

Margulis, Lynn and Sagan, Dorian (1986) *Origins of Sex: Three Billion Years of Genetic Recombination*. New Haven: Yale University Press.

McCaffery, L. (ed.) (1991) *Storming the Reality Studio: A Casebook of Cyberpunk and Postmodern Fiction*. Durham, NC and London: Duke University Press.

McMullen, E. (ed.) (1992) *The Social Dimensions of Science*. Notre Dame, Indiana: University of Notre Dame Press.

Merleau-Ponty, Maurice (1964) *Signs*. Trans. Richard C. McCleary. Evanston: Northwest University Press.

Minh-ha, Trinh T. (ed.) (1986/7) *She, The Inappropriated Other. Discourse*, 8.

Minh-ha, Trinh T. (1989) *Woman, Native, Other: Writing Postcoloniality and Feminism*. Bloomington: Indiana University Press.

Morton, Donald (1995) 'Birth of the Cyberqueer', *PMLA*, 110, 369–81.

Nerlich, Graham (1976) *The Shape of Space*. Cambridge: Cambridge University Press.

Nilsson, Lennart (1987) *The Body Victorious: The Illustrated Story of our Immune System and Other Defences of the Human Body*. New York: Delacourt.

Nixon, Nicola (1992) 'Cyberpunk: Preparing the Ground For Revolution or Keeping the Boys Satisfied?', *Science Fiction Studies*, 19, 219–35.

Noske, Barbara (1989) *Humans and Other Animals: Beyond the Boundaries of Anthropology*. London: Pluto Press.

Pearce, L. and Stacey, J. (eds) (1995) *Romance Revisited*. London: Lawrence & Wishart.

Penley, C. and Ross, A. (eds) (1991) *Technoculture*. Minneapolis: University of Minnesota Press.

Petchesky, Rosalind (1987) 'Fetal images: The power of visual culture in the politics of reproduction', *Feminist Studies*, 13(2), 263–92.

Plant, Sadie (1995) 'The Future Looms: Weaving Women and Cybernetics', in M. Featherstone and R. Burrows (eds), *Cyberspace/Cyberbodies/Cyberpunk*. London, Thousand Oaks, New Delhi: Sage, pp. 45–64.

Revkin, Andrew (1990) *The Burning Season*. New York: Houghton Mifflin.

Said, E. W. (1978) *Orientalism*. New York: Pantheon Books.

Sandoval, Chela (1990) 'Feminism and Racism', in G. Anzaldúa (ed.), *Making Face, Making Soul: Haciendo Caras*. San Francisco: Aunt Lute.

Sandoval, Chela (1995) 'New Sciences: Cyborg Feminism and the Methodology of the Oppressed', in C. Gray (ed.), *The Cyborg Handbook*. London and New York: Routledge, pp. 407–21.

Sayre, Ann (1975) *Rosalind Franklin and DNA*. New York: Norton.

Slusser, George (1991) 'Literary MTV', in L. McCaffery (ed.), *Storming the Reality Studio: A Casebook of Cyberpunk and Postmodern Fiction*. Durham, NC and London: Duke University Press, pp. 334–42.

Sofia, Zoë (1984) 'Exterminating Fetuses: Abortion, Disarmament, and the Sexo-Semiotics of Extraterrestialism', *Diacritics*, 14(2), 47–59.

Sofia, Zoë (1992) 'Virtual Corporeality: A Feminist View', *Australian Feminist Studies*, 15, 11–24.

Spivak, G. C. (1987) *In other Worlds: Essays in Cultural Politics*. London and New York: Methuen.

Spivak, G. C. (1990) *The Post-Colonial Critic: Interviews, Strategies, Dialogues*. S. Harasym (ed.). New York: Routledge.

Springer, Claudia (1991) 'The Pleasure of the interface', *Screen*, 32(3), 303–23.

Stone, Allucqere Rosanne (1991) 'Will the Real Body Please Stand Up? Boundary Stories about Virtual Cultures', in M. Benedikt (ed.), *Cyberspace: First Steps*. Cambridge, MA: MIT Press, pp. 81–118.

Strathern, Marilyn (1988) *The Gender of the Gift*. Berkeley: University of California Press.

Strum, S. (1987) *Almost Home: A Journey into the World of Baboons*. New York: Random House.

Terry, Jennifer (1989) 'The body invaded: medical surveillance of women as reproducers', *Socialist Review*, 19(3), 13–43.

Timmerman, Col. Frederick, (1987) 'Future Warriors', *Military Review*, Sept., 44–55.

Turner, T. (1990) 'Visual media, cultural politics, and anthropological practice: Some implications of recent uses of film and video among the Kaiapo of Brazil', *Commission on Visual Anthropology Review*, Spring.

Varley, John (1979) *Titan*. New York: Berkeley Books.

Varley, John (1986) 'Press Enter', in *Blue Champagne*. New York: Berkeley Books.

Virilio, Paul and Sylvere Lotringer (1983) *Pure War*. New York: Semiotext(e).

Weaver, Kenneth (1961) 'Countdown for space', *National Geographic*, 119(5), 702–34.

Wersky, Gary (1978) *The Invisible College: The Collective Biography of British*

Socialist Scientists in the 1930s. London: Allen Lane.

Wolmark, Jenny (1995) 'The Postmodern Romances of Feminist Science Fiction', in L. Pearce and J. Stacey (eds), *Romance Revisited*. London: Lawrence & Wishart, pp. 158–68.

Woodward, Kathleen (1994) 'From Virtual Cyborgs to Biological Time Bombs: Technocriticism and the Material Body', in G. Bender and T. Druckey (eds), *Culture on the Brink*, Seattle: Bay Press, pp. 47–64.

Yates, Frances (1964) *Giordano Bruno and the Hermetic Tradition*. London: Routledge.

Zimmer, Carl (1990) 'Tech in the jungle', *Discover*, Aug., 42.

INDEX